Advanced Engineering and Computational Methodologies for Intelligent Mechatronics and Robotics

Shahin Sirouspour
McMaster University, Canada

Managing Director:	Lindsay Johnston
Editorial Director:	Joel Gamon
Book Production Manager:	Jennifer Yoder
Publishing Systems Analyst:	Adrienne Freeland
Assistant Acquisitions Editor:	Kayla Wolfe
Typesetter:	Erin O'Dea
Cover Design:	Jason Mull

Published in the United States of America by
Information Science Reference (an imprint of IGI Global)
701 E. Chocolate Avenue
Hershey PA 17033
Tel: 717-533-8845
Fax: 717-533-8661
E-mail: cust@igi-global.com
Web site: http://www.igi-global.com

Library of Congress Cataloging-in-Publication Data

Advanced engineering and computational methodologies for intelligent mechatronics and robotics / Shahin Sirouspour, editor.
     pages cm
  Includes bibliographical references and index.
  Summary: "This book presents the latest innovations and technologies in the fields of mechatronics and robotics, offering innovations that are applied to a wide range of applications for robotic-assisted manufacturing, complex systems, and more"-- Provided by publisher.
  ISBN 978-1-4666-3634-7 (hardcover) -- ISBN 978-1-4666-3635-4 (ebook) -- ISBN 978-1-4666-3636-1 (print & perpetual access) 1. Mechatronics. 2. Robotics. I. Sirouspour, Shahin, 1973- editor of compilation.
  TJ163.12.A38 2013
  629.8'92--dc23
                                2012046678

British Cataloguing in Publication Data
A Cataloguing in Publication record for this book is available from the British Library.

The views expressed in this book are those of the authors, but not necessarily of the publisher.

Table of Contents

Detailed Table of Contents

Chapter 1

Christoph Edeler, University of Oldenburg, Germany

Sergej Fatikow, University of Oldenburg, Germany

In this paper a new method to generate forces with stick-slip micro drives is described. The forces are generated if the runner of the stick-slip drive operates against an obstacle. It is shown that the generated force can be varied selectively without additional sensors and that virtually any force between zero and a limiting force given by certain parameters can be generated. For the investigated micro actuator this force is typically in the range up to hundreds of mN. For this reason, the method has the potential to expand the application fields of stick-slip positioners. After the presentation of the testbed containing the measured linear axis, measurements showing the principle and important parameters are discussed. Furthermore, it is shown that the force generation can be qualitatively simulated using state-of-the-art friction models. Finally, the results are discussed and an outlook is given.

Chapter 2

Yongmin Zhong, Curtin University of Technology, Australia

Bijan Shirinzadeh, Monash University, Australia

Xiaobu Yuan, University of Windsor, Canada

This paper presents a new methodology based on neural dynamics for optimal robot path planning by drawing an analogy between cellular neural network (CNN) and path planning of mobile robots. The target activity is treated as an energy source injected into the neural system and is propagated through the local connectivity of cells in the state space by neural dynamics. By formulating the local connectivity of cells as the local interaction of harmonic functions, an improved CNN model is established to propagate the target activity within the state space in the manner of physical heat conduction, which guarantees that the target and obstacles remain at the peak and the bottom of the activity landscape of the neural network. The proposed methodology cannot only generate real-time, smooth, optimal, and collision-free paths without any prior knowledge of the dynamic environment, but it can also easily respond to the real-time changes in dynamic environments. Further, the proposed methodology is parameter-independent and has an appropriate physical meaning.

Chapter 3

Luca Carbonari, Polytechnic University of Marche, Italy
Luca Bruzzone, University of Genova, Italy
Massimo Callegari, Polytechnic University of Marche, Italy

This article describes the impedance control of an in-parallel actuated orientation platform. The algorithm is based on a representation of platform orientation which exploits the equivalent axis of rotation: this approach is more intuitive and easier to visualize than conventional methods based on Cardan or Euler angles. Moreover, since for small angular displacements the Mozzi's axis lies very close to angular velocity, impedance control algorithms based on such representation provides better performances and smoother motions. Results of numerical simulations and experimental tests are shown and commented with reference to the spherical parallel machine.

Chapter 4

X. Jia, Tianjin University, China
Y. Tian, Tianjin University, China
D. Zhang, Tianjin University, China
J. Liu, Hebei University of Technology, China

In order to investigate the influence of the stiffness of the compliant prismatic pair, a planar four-bar parallelogram, in a fully compliant parallel mechanism, the stiffness model of the passive compliant prismatic pair in a compliant parallel positioning stage is established using the compliant matrix method and matrix transformation. The influences of the constraints and the compliance of the connecting rods on the flexibility characteristics of the prismatic pair are studied based on the developed model. The relative geometric parameters are changed to show the rules of the stiffness variation and to obtain the demands for simplification in the stiffness modeling of the prismatic pair. Furthermore, the finite element analysis has been conducted to validate the analytical model.

Chapter 5

Yongjie Zhao, Shantou University, China

Inverse dynamic analysis of the 8-PSS redundant parallel manipulator is carried out in the exhaustive decoupled way. The required output of the torque, the power and the work of the driving motor are achieved. The whole actuating torque is divided into four terms which are caused by the acceleration, the velocity, the gravity, and the external force. It is also decoupled into the components contributed by the moving platform, the strut, the slider, the lead screw, the motor rotor-coupler, and the external force. The required powers contributed by the component of torque caused by the acceleration term, the velocity term, the gravity term, the external force term, and the powers contributed by the moving platform, the strut, the slider, the lead screw, and the motor rotor-coupler are computed respectively. For a prescribed trajectory, the required output work generated by the ith driving motor is obtained by the presented numerical integration method. Simulation for the computation of the driving motor's output torque, power and work is illustrated.

This paper presents a new random weighting method for estimation of one-sided confidence intervals in discrete distributions. It establishes random weighting estimations for the Wald and Score intervals. Based on this, a theorem of coverage probability is rigorously proved by using the Edgeworth expansion for random weighting estimation of the Wald interval. Experimental results demonstrate that the proposed random weighting method can effectively estimate one-sided confidence intervals, and the estimation accuracy is much higher than that of the bootstrap method.

Wheeled mobile rovers are being used in various missions for planetary surface exploration. In this paper a six-wheeled rover with rocker-bogie structure has been analyzed for planar case. The detailed kinematic model of the rover was built and the dynamic model was derived based on bond graph. The simulation studies were performed for obstacle climbing capability of the rover. It was observed from the study that rover can pass through plane surface, inclined surface, and inclined ditch without any control on the actuators of the rover. However, it fails to cross a vertical ditch so a velocity controller was designed. It consists of a proportional integral (PI) controller and reduced model of the rover. It is found from simulation and animation studies that with the proposed velocity controller the rover is able to cross the vertical ditch.

One of the most difficult problems in reverse engineering is the processing of unstructured data. NURBS (Non-uniform Rational B-splines) surfaces are a popular tool for surface modeling. However, they cannot be directly created from unstructured data, as they are defined on a four-sided domain with explicit parametric directions. Therefore, in reverse engineering, it is necessary to process unstructured data into structured data which enables the creation of NURBS surfaces. This paper presents a methodology to processing unstructured data into the structured data for creating NURBS surfaces. A projection based method is established for constructing 3D triangulation from unstructured data. An optimization method is also established to optimize the 3D triangulation to ensure that the resulted NURBS surfaces have a better form. A triangular surface interpolation method is established for constructing triangular surfaces from the triangulation. This method creates five-degree triangular surfaces with C1 continuity. A series of segment data are obtained by cutting the triangular surfaces with a series of parallel planes. Finally, the structured data is obtained by deleting repetitive data points in each segment data. Results demonstrate the efficacy of the proposed methodology.

Bond graphs are suitable tools for modeling many types of dynamical systems and can model these systems consisting of mechanical, electrical, fluidic, and pneumatic sub-systems. The advantage of a bond graph is that it can model non-linear systems and combinational systems. In this paper, the authors utilize bond graphs for modeling mechatronics systems. Mechatronics systems consist of mechanics, electronics, and intelligent software. Many of these systems have digital sections that are constructed by logical circuits (hardware by transistors and now mostly by chips). The authors present a methodology to implement these mechatronics systems by bond graphs.

This paper examines the problem of realizing a 6-DOF motion platform by proposing a closed loop kinematic architecture that benefits from an anthropological serial manipulator design. In contrast to standard motion platforms based on linear actuators, a mechanism with actuator design inspired from anthropological kinematic structure offers a relatively larger motion envelope and higher dexterity making it a viable motion platform for micromanipulations. The design consists of a motion plate connected through only revolute hinges for the passive joints, and three legs located at the base as the active elements. In this hybrid kinematic structure, each leg is connected to the top (motion) plate through three revolute hinges and to the bottom (fixed) plate through a single revolute joint forming a closed-loop kinematic chain. The paper describes the mathematical modeling of the proposed design and demonstrates its simulation model using SimMechanics and xPC Target for real-time simulations and visualization of the motion cues.

The combination of a rigid and a flexible link in a space robot is an interesting field of study from modeling and control point of view. This paper presents the bond graph modeling and overwhelming trajectory control of a rigid-flexible space robot in its work space using the Jacobian based controller. The flexible link is modeled as Euler Bernoulli beam. Bond graph modeling is used to model the dynamics of the system and to devise the control strategy, by representing the dynamics of both rigid and flexible links in a unified manner. The scheme has been verified using simulation for a rigid-flexible space manipulator with two links.

Umesh Bhagat, Monash University, Australia

Bijan Shirinzadeh, Monash University, Australia

Yanling Tian, Tianjian University, China

This paper presents an experimental study of laser interferometry-based closed-loop motion tracking for flexure-based four-bar micro/nano manipulator. To enhance the accuracy of micro/nano manipulation, laser interferometry-based motion tracking control is established with experimental facility. The authors present and discuss open-loop control, model-based closed-loop control, and robust motion tracking closed-loop control for flexure-based mechanism. A comparative error analysis for closed-loop control with capacitive position sensor and laser interferometry feedback is discussed and presented. Model-based closed-loop control shows improvement in position and motion tracking over open-loop control. Robust control demonstrates high precise and accurate motion tracking of flexure-based mechanism compared to the model-based control. With this experimental study, this paper offers evidence that the laser interferometry-based closed-loop control can minimize positioning and tracking errors during dynamic motion, hence realizing high precision motion tracking and accurate position control.

Eklas Hossain, University of Wisconsin-Milwaukee, USA

Md Raisuddin Khan, International Islamic University of Malaysia, Malaysia

Riza Muhida, International Islamic University of Malaysia, Malaysia

Ahad Ali, Lawrence Technological University, USA

Visually impaired people are faced with challenges in detecting information about terrain. This paper presents a new walking support system for the blind to navigate without any assistance from others or using a guide cane. In this research, a belt, wearable around the waist, is equipped with four ultrasonic sensors and one sharp infrared sensor. Based on mathematical models, the specifications of the ultrasonic sensors are selected to identify optimum orientation of the sensors for detecting stairs and holes. These sensors are connected to a microcontroller and laptop for analyzing terrain. An algorithm capable of classifying various types of obstacles is developed. After successful tests using laptop, the microcontroller is used for the walking system, named 'Belt for Blind', to navigate their environment. The unit is also equipped with a servo motor and a buzzer to generate outputs that inform the user about the type of obstacle ahead. The device is light, cheap, and consumes less energy. However, this device is limited to standard pace of mobility and cannot differentiate between animate and inanimate obstacles. Further research is recommended to overcome these deficiencies to improve mobility of blind people.

Kazi Mostafa, National Sun Yat-sen University, Taiwan

Innchyn Her, National Sun Yat-sen University, Taiwan

Jonathan M. Her, National Taiwan University, Taiwan

Natural multiped gaits are believed to evolve from countless generations of natural selection. However, do they also prove to be better choices for walking machines? This paper compares two surefooted gaits, one natural and the other artificial, for six-legged animals or robots. In these gaits four legs are used to support the body, enabling greater stability and tolerance for faults. A standardized hexapod model was carefully examined as it moved in arbitrary directions. The study also introduced a new factor in addition to the traditional stability margin criterion to evaluate the equilibrium of such gaits. Contrary to the common belief that natural gaits would always provide better stability during locomotion, these results show that the artificial gait is superior to the natural gait when moving transversely in precarious conditions.

Chapter 15

Giuseppe Carbone, University of Cassino, Italy

Enrique Villegas, University of Cassino, Italy

Marco Ceccarelli, University of Cassino, Italy

This paper addresses problems for design and validation of force control loops for a 3-DOF parallel manipulator in drilling applications. In particular, the control design has been investigated for a built prototype of CaPaMan2bis at LARM (Laboratory of Robotics and Mechatronics of Cassino). Two control loops have been developed, each one with two types of controllers. The first one is a Constrained Control Loop, which limits the force that is applied to an object to stay below a given value. The second one is a Standard Control Loop with external force feedback, which keeps the force at a given value. The control loops have been implemented on CaPaMan2bis by a Virtual Instrument in LABVIEW Software. CaPaMan2bis has been attached to a serial robot to make dynamic tests. The results of the experimental tests show the effectiveness and quick response of both algorithms after a careful calibration process.

Chapter 16

Dereje Shiferaw, Indian Institute of Technology Roorkee, India & Graphic Era University, India

Anamika Jain, Graphic Era University, India

R. Mitra, Indian Institute of Technology Roorkee, India

This paper presents the design and analysis of a high performance robust controller for the Stewart platform manipulator. The controller is a variable structure controller that uses a linear sliding surface which is designed to drive both tracking and synchronization errors to zero. In the controller the model based equivalent control part of the sliding mode controller is computed in task space and the discontinuous switching controller part is computed in joint space and hence it is a hybrid of the two approaches. The hybrid implementation helps to reduce computation time and to achieve high performance in task space without the need to measure or estimate 6DOF task space positions. Effect of actuator friction, backlash and parameter variation due to loading have been studied and simulation results confirmed that the controller is robust and achieves better tracking accuracy than other types of sliding mode controllers and simple PID controller.

Chapter 17

Hamoon Hadian, Isfahan University of Technology, Iran

Yasser Amooshahi, Isfahan University of Technology, Iran

Abbas Fattah, University of Delaware, USA

This paper addresses the kinematics and dynamics modeling of a 4-DOF cable-driven parallel manipulator with new architecture and a typical Computed Torque Method (CTM) controller is developed for dynamic model in SimMechanics. The novelty of kinematic architecture and the closed loop formulation is presented. The workspace model of mechanism's dynamic is obtained in an efficient and compact form by means of natural orthogonal complement (NOC) method which leads to the elimination of the nonworking kinematic-constraint wrenches and also to the derivation of the minimum number of equations. To verify the dynamic model and analyze the dynamical properties of novel 4-DOF cable-driven parallel manipulator, a typical CTM control scheme in joint-space is designed for dynamic model in SimMechanics.

Chapter 18

Hamoon Hadian, Isfahan University of Technology, Iran
Abbas Fattah, University of Delaware, USA

In this paper, the authors study the kinematic isotropic configuration of spatial cable-driven parallel robots by means of four different methods, namely, (i) symbolic method, (ii) geometric workspace, (iii) numerical workspace and global tension index (GTI), and (iv) numerical approach. The authors apply the mentioned techniques to two types of spatial cable-driven parallel manipulators to obtain their isotropic postures. These are a 6-6 cable-suspended parallel robot and a novel restricted three-degree-of-freedom cable-driven parallel robot. Eventually, the results of isotropic conditions of both cable robots are compared to show their applications.

Chapter 19

G. Satheesh Kumar, Indian Institute of Technology Madras, India
T. Nagarajan, Universiti Teknologi Petronas (UTP), Malaysia

Reconfiguration of Stewart platform for varying tasks accentuates the importance for determination of optimum geometry catering to the specified task. The authors in their earlier work (Satheesh et al., 2008) have indicated the non availability of an efficient holistic methodology for determining the optimum geometry. Further, they have proposed a solution using the variable geometry approach through the formulation of dimensionless parameters in combination with generic parameters like configuration and joint vector. The methodology proposed provides an approach to develop a complete set of design tool for any new reconfigurable Stewart platform for two identified applications viz., contour generation and vibration isolation. This paper details the experimental investigations carried out to validate the analytical results obtained on a developed Stewart platform test rig and error analysis is performed for contour generation. The experimental natural frequency of the developed Stewart platform has also been obtained.

Chapter 20

Madusudanan Sathia Narayanan, University at Buffalo, USA
Srikanth Kannan, University at Buffalo, USA
Xiaobo Zhou, University at Buffalo, USA
Frank Mendel, University at Buffalo, USA
Venkat Krovi, University at Buffalo, USA

There is considerable scientific and commercial interest in understanding the mechanics of mastication. In this paper, the authors develop quantitative engineering tools to enable this process by: (i) designing a general purpose mastication simulator test-bed based on parallel architecture manipulator, capable of producing the requisite motions and forces; and (ii) validating this simulator with a range of test-foods, undergoing various mastication cycles under controlled and monitored circumstances. Such an implementation provides a test bed to quantitatively characterize the mastication based on "chewability index".

Due to the inherent advantages of locating actuators at the base (ground) in terms of actuator efforts and structural rigidity as well as benefits of using prismatic sliders compared to revolute actuators, the 6-P-U-S system was chosen. A detailed symbolic kinematic analysis was then conducted. For the practical implementation of the test-bed, the analytical Jacobian was examined for singularities and the design was adapted to ensure singularity free operation. A comprehensive parametric study was undertaken to obtain optimal design parameters for desired workspace and end effector forces. Experiments captured jaw motion trajectories using the high speed motion capture system which served as an input to the hardware-in-the-loop simulator platform.

Preface

The fields of mechatronics and robotics engineering have witnessed enormous progress in recent years. Unprecedented advances in computing hardware and software, sensor and actuator technology, and science of controls, signal processing, artificial intelligence, and computing have spurred the rapid growth of the applications of mechatronic and robotic systems. Application areas of these technologies span from advanced manufacturing systems in the automotive and microelectronic industries to the emerging fields of medical and personal care. Manufacturing processes have benefited from greater integration of new sensing and information processing technologies into more sophisticated and intelligent systems that can operate at scales considered impossible before. Medical and surgical robots have emerged as intelligent mechatronic devices that allow medical practitioners not only to improve existing procedures, but also to carry out operations that would have been unimaginable only a few years ago. Advances in robotics and mechatronics have led to new prosthetic and rehabilitative devices that help the injured and disabled to recover faster and regain some of their lost functionality. Robots also have gained popularity for personal and home care, in part due to the ageing population in the developed world.

Robots have proven effective in many other terrestrial applications. Unmanned remotely operated robotic vehicles are increasing used in disaster recovery and search and rescue operations, mining, underwater exploration, and military combat and intelligence operations. Remotely operated robots played a crucial role in assessing the damage to the nuclear power plants after the 2011 earthquake and tsunami in Japan. In recent years, the United States military has deployed an increasing number of unmanned mobile robots and aerial drones for combat and intelligence operations. In space, robotic rovers are exploring the surface of Mars to help gather critical scientific information about the planet that one day may lead to its colonization. Robotic manipulators carry out many of the critical assembly and maintenance operations on the International Space Station (ISS) in the hazardous space environment.

A cycle of sensing, perception, control, and action characterizes modern intelligent robotic and mechatronic systems. They incorporate core hardware and software processing and control elements to provide a desired functionality within a task environment. The essential hardware components of these systems are sensors, actuators, and computing machinery. Signal and image processing, state estimation and navigation algorithms, controls, and artificial and computational intelligence are examples of software elements. It is precisely because of this complexity that the mechatronics and robotics engineering builds on such a diverse set of disciplines as electrical and computer engineering, mechanical engineering, computer science, and system engineering, among others. Robotics scientists and engineers have often benefited from latest advances in these disciplines, but they have also had to contend with unique application and system dependent challenges that inevitably arise in the development and integration of complex robotic and mechatronics systems. To effectively utilize, maintain, and upgrade these systems, manufacturing professionals at all levels must have a comprehensive understanding of the mechatronic and robotic principles at work across multiple disciplines.

The chapters in this critical reference explore the emerging technologies and techniques used in manufacturing systems as well as additional topics in the fields of neural networks, assistive technologies, robotics, computer architecture, and 3D modeling, among others. In all, this book presents practitioners, researchers, developers, managers, and engineers with insight into the latest developments in modern intelligent mechatronics, robotics, and automation systems, combining theory and practice to develop innovations in, and enhance knowledge of, mechanical and electrical engineering.

Many robots and mechatronic devices, including some used in medical applications, have to operate in small workspaces, which severely constrains their size. Piezo-electric motor drives have become popular for use in applications requiring miniaturization since they can be made in very compact form factor compared to other types of actuators including conventional electric motors. In Chapter 1, "Open Loop Force Control of Piezo-Actuated Stick-Slip Drives," by Christoph Edeler and Sergej Fatikow, a new method for force generation in stick-slip piezo-electric actuators is presented. While stick-slip drives have been commonly used in motion drives, the literature on their force control is rather scant. The authors of this article demonstrate the feasibility of producing a variable output force with such actuators. A particular actuator studied in the article can output a force in the range of hundreds of mN, promising to expand applications of this type of actuators in micro- and nano-handling miniaturized robots. An interesting point is that the actuator force is produced in open-loop manner eliminating the need for any additional sensor. The authors discuss a model and its important parameters that can be helpful in calibrating the actuator.

Next, Yongmin Zhongm et al. present a new methodology based on neural dynamics for optimal robot path planning in static and dynamic environments in their chapter "Optimal Robot Path Planning with Cellular Neural Network." This new method builds on concepts in Cellular Neural Networks (CNN) by proposing a new network in which the local connectivity of neurons is a nonlinear harmonic function as opposed a linear distance function in the exiting neural network models. The dynamics of this modified cellular neural network ensure that the target and obstacle remain at the top and bottom of the activity landscape for the network. The result is an algorithm that not only produces real-time smooth and collision free paths for a mobile robot, but it also can respond to dynamic changes in the task environment and take into account safety considerations. This is all achieved without any prior knowledge of target and obstacle movements, learning procedures, and explicit searching over the global work space or searching collision paths.

"Impedance Control of a Spherical Parallel Platform" by Luca Carbonari et al. presents an impedance control method for a 3DOF parallel manipulator with purely rotational motion at the end-effector. Instead of using a conventional Euler angles representation, the proposed controller employs the axis/angle of representation of the rotation. This new impedance control approach for an orientation only manipulator yields a smoother and more natural motion and decouples the impedance along different rotational degrees of freedom.

In "Stiffness Modeling and Analysis of Passive Four-Bar Parallelogram in Fully Compliant Parallel Positioning Stage," X. Jia et al. explore the role of a prismatic pair on the overall stiffness of a 3DOF compliant positioning stage. Parallel manipulators with compliant mechanisms can be useful in applications requiring high precision capabilities in a compact package. They can provide higher stiffness and response bandwidth than serial mechanisms and are compatible with operation in vacuum and clean room. Replacing conventional joints with flexure hinges allows for miniaturization of the manipulator structure and can eliminates backlash, friction and accumulation of error. The analysis based on the compliant matrix method and matrix transformation in this article sheds light on the impact of various design parameters of a flexure-based parallelogram on the overall compliance/stiffness of the 3DOF manipulator. A comparison between the results of analytical model with those from finite-element simulations reveals that the compliance of the connecting bars and the constraints need be taken into account for accurate modeling.

In "Computation of the Output Torque, Power, and Work of the Driving Motor for a Redundant Parallel Manipulator" by Yongjie Zhao, inverse dynamic analysis of the 8-PSS redundant parallel manipulator is carried out in the exhaustive decoupled way. The required output of the torque, the power and the work of the driving motor are achieved. The whole actuating torque is divided into four terms which are caused by the acceleration, the velocity, the gravity, and the external force. It is also decoupled into the components contributed by the moving platform, the strut, the slider, the lead screw, the motor rotor-coupler, and the external force. The required powers contributed by the component of torque caused by the acceleration term, the velocity term, the gravity term, the external force term, and the powers contributed by the moving platform, the strut, the slider, the lead screw, and the motor rotor-coupler are computed respectively. For a prescribed trajectory, the required output work generated by the ith driving motor is obtained by the presented numerical integration method. Simulation for the computation of the driving motor's output torque, power, and work is illustrated.

Chapter 6, "Random Weighting Estimation of One-Sided Confidence Intervals in Discrete Distributions," by Yalin Jiao et al., presents a new random weighting method for estimation of one-sided confidence intervals in discrete distributions. It establishes random weighting estimations for the Wald and Score intervals. Based on this, a theorem of coverage probability is rigorously proved by using the Edgeworth expansion for random weighting estimation of the Wald interval. Experimental results demonstrate that the proposed random weighting method can effectively estimate one-sided confidence intervals, and the estimation accuracy is much higher than that of the bootstrap method.

Wheeled mobile rovers are being used in various missions for planetary surface exploration. In Pushpendra Kumar and Pushparaj Mani Pathak's chapter, "Dynamic Modeling, Simulation, and Velocity Control of Rocker-Bogie Rover for Space Exploration," a six-wheeled rover with rocker-bogie structure has been analyzed for planar case. The detailed kinematic model of the rover was built and the dynamic model was derived based on bond graph. The simulation studies were performed for obstacle climbing capability of the rover. It was observed from the study that the rover can pass through plane surface, inclined surface, and inclined ditch without any control on the actuators of the rover. However, it fails to cross a vertical ditch so a velocity controller was designed. It consists of a Proportional Integral (PI) controller and reduced model of the rover. It is found from simulation and animation studies that with the proposed velocity controller the rover is able to cross the vertical ditch.

Robotics and mechatronics engineers often have to deal with the problem of processing of unstructured raw sensory data into a structured form that is more suitable for analysis and decision making. In "Processing of 3D Unstructured Measurement Data for Reverse Engineering," Yongmin Zhong looks at a particular interesting problem in which the sensor outputs unstructured data in the form of a cloud of points in the 3-Dimensional (3D) space. Examples of such sensors are 3D dimensional surface scanners or computed tomography machines, which generate a large set of vertices in the 3D coordinate system. While a careful choice of scanning directions for objects with well-structured shapes can help avoid unstructured data, this may not be possible for objects with complex shapes. The article presents a solution for transforming such unstructured data points into a structured form suitable for application of the Non-Uniform Rational B-Splines (NURBS) surface modeling tools. Example applications of the proposed method for surface reconstruction of the aircraft component and a mold demonstrate its effectiveness.

In the next chapter, by Majid Habibi and Alireza B.Novinzadeh, is "Modeling and Simulation of Digital Systems Using Bond Graphs." Bond graphs are suitable tools for modeling many types of dynamical systems and can model these systems consisting of mechanical, electrical, fluidic, and pneumatic sub-systems. The advantage of a bond graph is that it can model non-linear systems and combinational systems. In this chapter, the authors utilize bond graphs for modeling mechatronics systems. Mechatronics systems consist of mechanics, electronics, and intelligent software. Many of these systems have digital sections that are constructed by logical circuits (hardware by transistors and now mostly by chips). The authors present a methodology to implement these mechatronics systems by bond graphs.

"Modeling, Simulation, and Motion Cues Visualization of a Six-DOF Motion Platform for Micro-Manipulations" by Umar Asif and Javaid Iqbal examines the problem of realizing 6-DOF motion platform by proposing a closed loop kinematic architecture that benefits from an anthropological serial manipulator design. In contrast to standard motion platforms based on linear actuators, a mechanism with actuator design inspired from anthropological kinematic structure offers a relatively larger motion envelope and higher dexterity, making it a viable motion platform for micromanipulations. The design consists of a motion plate connected through only revolute hinges for the passive joints, and three legs located at the base as the active elements. In this hybrid kinematic structure, each leg is connected to the top (motion) plate through three revolute hinges and to the bottom (fixed) plate through a single revolute joint forming a closed-loop kinematic chain. This chapter describes the mathematical modeling of the proposed design and demonstrates its simulation model using SimMechanics and xPC Target for real-time simulations and visualization of the motion cues.

The combination of a rigid and a flexible link in a space robot is an interesting field of study from a modeling and control point of view. "Bond Graph Modeling and Computational Control Analysis of a Rigid-Flexible Space Robot in Work Space" by Amit Kumar et al. presents the bond graph modeling and overwhelming trajectory control of a rigid-flexible space robot in its work space using the Jacobian-based controller. The flexible link is modeled as Euler Bernoulli beam. Bond graph modeling is used to model the dynamics of the system and to devise the control strategy by representing the dynamics of both rigid and flexible links in a unified manner. The scheme has been verified using simulation for a rigid-flexible space manipulator with two links.

Umesh Bhagat et al.'s "Experimental Study of Laser Interferometry Based Motion Tracking of a Flexure-Based Mechanism" presents an experimental study of laser interferometry-based closed-loop motion tracking for flexure-based four-bar micro/nano manipulator. To enhance the accuracy of micro/nano manipulation, laser interferometry-based motion tracking control is established with experimental facility. The authors present and discuss open-loop control, model-based closed-loop control, and robust motion tracking closed-loop control for flexure-based mechanism. A comparative error analysis for closed-loop control with capacitive position sensor and laser interferometery feedback is discussed and presented. Model-based closed-loop control shows improvement in position and motion tracking over open-loop control. Robust control demonstrates high precise and accurate motion tracking of flexure-based mechanism compared to the model-based control. With this experimental study, this chapter offers evidence that the laser interferometry-based closed-loop control can minimize positioning and tracking errors during dynamic motion, hence realizing high precision motion tracking and accurate position control.

Chapter 13

Visually impaired people are faced with challenges in detecting information about terrain. "Analysis and Implementation for a Walking Support System for Visually Impaired People" by Eklas Hossain presents a new walking support system for the blind to navigate without any assistance from others or using a guide cane. In this research, a belt, wearable around the waist, is equipped with four ultrasonic sensors and one sharp infrared sensor. Based on mathematical models, the specifications of the ultrasonic sensors are selected to identify optimum orientation of the sensors for detecting stairs and holes. These sensors are connected to a microcontroller and laptop for analyzing terrain. An algorithm capable of classifying various types of obstacles is developed. After successful tests using a laptop, the microcontroller is used for the walking system, named "Belt for Blind," to navigate their environment. The unit is also equipped with a servo motor and a buzzer to generate outputs that inform the user about the type of obstacle ahead. The device is light, cheap, and consumes less energy. However, this device is limited to standard pace of mobility and cannot differentiate between animate and inanimate obstacles. Further research is recommended to overcome these deficiencies to improve mobility of blind people.

In Chapter 14, Kazi Mostafa et al. ask "Which is Better? A Natural or an Artificial Surefooted Gait for Hexapods." Natural multiped gaits are believed to evolve from countless generations of natural selection. However, do they also prove to be better choices for walking machines? This chapter compares two surefooted gaits, one natural and the other artificial, for six-legged animals or robots. In these gaits four legs are used to support the body, enabling greater stability and tolerance for faults. A standardized hexapod model was carefully examined as it moved in arbitrary directions. The study also introduced a new factor in addition to the traditional stability margin criterion to evaluate the equilibrium of such gaits. Contrary to the common belief that natural gaits would always provide better stability during locomotion, these results show that the artificial gait is superior to the natural gait when moving transversely in precarious conditions.

The next chapter, "Design and Validation of Force Control Loops for a Parallel Manipulator" by Giuseppe Carbone et al., addresses problems for design and validation of force control loops for a 3-DOF parallel manipulator in drilling applications. In particular, the control design has been investigated for a built prototype of CaPaMan2bis at LARM (Laboratory of Robotics and Mechatronics of Cassino). Two control loops have been developed, each one with two types of controllers. The first one is a Constrained Control Loop, which limits the force that is applied to an object to stay below a given value. The second one is a Standard Control Loop with external force feedback, which keeps the force at a given value. The control loops have been implemented on CaPaMan2bis by a Virtual Instrument in LABVIEW Software. CaPaMan2bis has been attached to a serial robot to make dynamic tests. The results of the experimental tests show the effectiveness and quick response of both algorithms after a careful calibration process.

Next, "High Performance Control of Stewart Platform Manipulator Using Sliding Mode Control with Synchronization Error" by Dereje Shiferaw et al. presents the design and analysis of a high performance robust controller for the Stewart platform manipulator. The controller is a variable structure controller that uses a linear sliding surface which is designed to drive both tracking and synchronization errors to zero. In the controller, the model based equivalent control part of the sliding mode controller is computed in task space and the discontinuous switching controller part is computed in joint space; hence, it is a hybrid of the two approaches. The hybrid implementation helps to reduce computation time and to achieve high performance in task space without the need to measure or estimate 6DOF task space positions. Effect of actuator friction, backlash, and parameter variation due to loading have been studied, and simulation results confirmed that the controller is robust and achieves better tracking accuracy than other types of sliding mode controllers and simple PID controller.

Hamoon Hadian et al.'s chapter, "Kinematics and Dynamics Modeling of a New 4-DOF Cable-Driven Parallel Manipulator," addresses the kinematics and dynamics modeling of a 4-DOF cable-driven parallel manipulator with new architecture, and a typical Computed Torque Method (CTM) controller is developed for dynamic model in SimMechanics. The novelty of kinematic architecture and the closed loop formulation is presented. The workspace model of the mechanism's dynamic is obtained in an efficient and compact form by means of Natural Orthogonal Complement (NOC) method which leads to the elimination of the nonworking kinematic-constraint wrenches and also to the derivation of the minimum number of equations. To verify the dynamic model and analyze the dynamical properties of novel 4-DOF cable-driven parallel manipulator, a typical CTM control scheme in joint-space is designed for dynamic model in SimMechanics.

In Chapter 18, "Kinematic Isotropic Configuration of Spatial Cable-Driven Parallel Robots," Hamoon Hadian and Abbas Fattah study the kinematic isotropic configuration of spatial cable-driven parallel robots by means of four different methods, namely (1) symbolic method, (2) geometric workspace, (3) numerical workspace and Global Tension Index (GTI), and (4) numerical approach. The authors apply the mentioned techniques to two types of spatial cable-driven parallel manipulators to obtain their isotropic postures. These are a 6-6 cable-suspended parallel robot and a novel restricted three-degree-of-freedom cable-driven parallel robot. Eventually, the results of isotropic conditions of both cable robots are compared to show their applications.

G. Satheesh Kumar and T. Nagarajan continue with "Experimental Investigations on the Contour Generation of a Reconfigurable Stewart Platform." Reconfiguration of Stewart platform for varying tasks accentuates the importance for determination of optimum geometry catering to the specified task. They propose a solution using the variable geometry approach through the formulation of dimensionless parameters in combination with generic parameters like configuration and joint vector. The methodology proposed provides an approach to develop a complete set of design tools for any new reconfigurable Stewart platform for two identified applications, contour generation and vibration isolation. This chapter details the experimental investigations carried out to validate the analytical results obtained on a developed Stewart platform test rig and error analysis is performed for contour generation. The experimental natural frequency of the developed Stewart platform has also been obtained.

Finally, there is considerable scientific and commercial interest in understanding the mechanics of mastication. In "Parallel Architecture Manipulators for Use in Masticatory Studies," Madusudanan Sathia Narayanan et al. develop quantitative engineering tools to enable this process by (1) designing a general purpose mastication simulator test-bed based on parallel architecture manipulator capable of producing the requisite motions and forces and (2) validating this simulator with a range of test-foods, undergoing various mastication cycles under controlled and monitored circumstances. Such an implementation provides a test bed to quantitatively characterize the mastication based on "chewability index." Due to the inherent advantages of locating actuators at the base (ground) in terms of actuator efforts and structural rigidity as well as benefits of using prismatic sliders compared to revolute actuators, the 6-P-U-S system was chosen. A detailed symbolic kinematic analysis was then conducted. For the practical implementation of the test-bed, the analytical Jacobian was examined for singularities and the design was adapted to ensure singularity free operation. A comprehensive parametric study was undertaken to obtain optimal design parameters for desired workspace and end effector forces. Experiments captured jaw motion trajectories using the high speed motion capture system which served as an input to the hardware-in-the-loop simulator platform.

Shahin Sirouspour
McMaster University, Canada

Chapter 1
Open Loop Force Control of Piezo–Actuated Stick–Slip Drives

Christoph Edeler
University of Oldenburg, Germany

Sergej Fatikow
University of Oldenburg, Germany

ABSTRACT

In this paper a new method to generate forces with stick-slip micro drives is described. The forces are generated if the runner of the stick-slip drive operates against an obstacle. It is shown that the generated force can be varied selectively without additional sensors and that virtually any force between zero and a limiting force given by certain parameters can be generated. For the investigated micro actuator this force is typically in the range up to hundreds of mN. For this reason, the method has the potential to expand the application fields of stick-slip positioners. After the presentation of the testbed containing the measured linear axis, measurements showing the principle and important parameters are discussed. Furthermore, it is shown that the force generation can be qualitatively simulated using state-of-the-art friction models. Finally, the results are discussed and an outlook is given.

INTRODUCTION

Stick and slip micro-drives or micro-actuators (SSA) are well-known and have been under investigation for at least two decades. One of the first SSA was presented by Pohl in 1987. It is driven by a piezoceramic tube, which operates on a slider, carrying the object of interest. With control frequencies of up to 500 Hz, a slider's velocity of 0.2 mm/s can be reached. With the

development of piezoceramics, control electronics and fabrication techniques and the rising interest on small, vacuum capable ultra-fine positioners a lot of developments came up. In 1995, Zesch presented the locomotion platform Abalone based on the stick-slip principle (Zesch, Buechi, Codourey, & Siegwart, 1995). It is driven by three piezoceramic stack-actuators and can move in three degrees of freedom (DOF) on a working surface. This allows a flexible positioning of specimen,

DOI: 10.4018/978-1-4666-3634-7.ch001

e.g. for light- and electron-based microscopes. Other research groups came up with similar approaches. The Piezowalker is a design close to that of Zesch, but with a driven rod in the middle of the structure (Mariotto, D'Angelo, & Shvets, 1999). The group around Breguet and Bergander presented several mobile micro robots for different applications (Bergander, Driesen, Varidel, & Breguet, 2003a; Bergander, Driesen, Varidel, & Breguet, 2003b; Breguet, 1998). Munassypov et al. (1996) presented a mobile platform driven by bending piezoceramic tubes (Munassypov, Grossmann, Magnussen, & Fatikow, 1996). Each of the three tubes can be displaced in two DOF similar to a leg. Thus, motions in three DOF are possible. Another mobile robot driven by piezoceramic tubes is the Nanowalker (Martel et al., 2001). It is designed in such a way that the miniaturized control electronics including an energy source is carried by the robot aiming towards an untethered mobile nanohandling robot. However, several problems such as the influence of the robot's mass on the working principle, wear caused by friction or overheating electronics prevented extensive applications.

It can be noted that the most important advantages of SSAs are their simple design (a SSA rarely consists of more than a handful of mechanical parts), the small piezoelectric coefficient which allows positioning in the nanometer scale, the stick-slip working principle itself which combines fine positioning with large travels, and the high potential of miniaturization. Nevertheless, the function of SSAs is intimately connected with friction characteristics and therefore, results such as the performed step lengths can vary. For this reason, measurement and control of the generated displacements is necessary in most cases.

Today, SSAs are established in research as well as in the commercial field. Linear and rotary positioners are commercially available, e.g. by the companies SmarAct, Kleindiek Nanotechnik, Klocke Nanotechnik or Attocube. Applications performed with these actuators can be found in

(Rabenorosoa, Clévy, Lutz, Bargiel, & Gorecki, 2009; Peng et al., 2004; Noyong, Blech, Rosenberger, Klocke, & Simon, 2007; Meyer, Sqalli, Lorenz, & Karrai, 2005; Vogel, Stein, Pettersson, & Karrai, 2001). Since early 2010, Imina Technologies, a spin-up of the Ecole Polytechnique Fédérale de Lausanne, Switzerland, firstly offers mobile nanohandling robots for nanomanipulation applications. The robots are the result of several research projects. Descriptions of the robots can be found in (Bergander et al., 2004; Canales et al., 2008). Commercial applications of the robots have not been documented yet. In research, mobile, multi-DOF nanohandling robots are still in the focus of investigation in contrast to single-DOF actuators. An almost all-embracing classification and description of mobile nanohandling robots can be found in Driesen (2008). Appreciable examples were published by the group of Fatikow (Edeler, 2008; Jasper & Edeler, 2008). The center of gravity is on the automation of the robots as part of nanohandling scenarios. The mobile nanohandling robots offer a wide velocity range combined with high resolution and good open- and closed-loop control characteristics. Another approach of mobile robots using the stick-slip principle was presented by Das, Zhang, Popa, and Stephanou (2007) (Murthy, Das, & Popa, 2008; Murthy & Popa, 2009). A stick and slip crawling motion with electro-thermal actuators is used to drive mobile robots with dimensions in the sub-mm scale. General examples for the application of SSAs in research are cell manipulation (Trüper, Kortschack, Jähnisch, Hülsen, & Fatikow, 2004; Hagemann, Krohs, & Fatikow, 2007; Brufau et al., 2005), the handling of carbon-nano-tubes (Eichhorn, Carlson, Andersen, Fatikow, & Bøggild, 2007; Eichhorn et al., 2009) or applications in material science (Breguet, Driesen, Kaegi, & Cimprich, 2007; Fatikow et al., 2007). It can be concluded that SSAs are exclusively used to position objects in present literature. This is in contrast to investigations concerning the generated forces of piezo-driven actuators not using the stick-slip principle. An example for those is (Ronkanen, Kallio, & Koivo, 2007). Apparently, there seems to be low interest in the force-generating capabilities of micro-SSAs at the moment.

In this paper, a method to generate defined forces with a small SSA is presented. Not only that a gap in research is lit up, in fact the method could offer attractive applications in micro- and nanohandling. Due to the fact that no force sensor is needed and SSAs can be miniaturized very well, several applications are thinkable, such as metrology (measurements and calibrations of elastic moduli without force sensor), surface analysis and "safe" micro- and nanohandling processes without deformation of objects. However, several disadvantages come up. As SSAs are always "vibrating" actuators, the method will not be suitable for ultra-sensitive applications. A detailed discussion of the ad- and disadvantages will be given at the end of this paper.

The organization of the paper is as follows: The method to generate defined forces is presented with the help of measurements of a linear SSA. The used testbed and the drive are described. Furthermore, basic terms are introduced. After that, the influences of important parameters are shown via according measurements. It is estimated to what extend the measured force can be simulated using state-of-the-art friction models. Finally, the results are discussed and an outlook is given.

METHOD

As indicated in the introduction, the method presented in this paper is based on the idea that the runner of a SSA runs against an obstacle. In this very moment a force appears. This force can be controlled by the control signal's amplitude of the SSA. Thus, after a calibration process, the dependency of the amplitude on the generated force is known and can be applied. At the time of contact, the drive changes its function from a positioner to a force generator.

Linear Stick-Slip Micro Drive and Testbed

In this section the stick-slip drive used for the investigations is described. The piezoactuators are fabricated out of piezoceramic plates (material: PZT-5H) with a thickness of 500 μm (see Figure 1 a).

As the runner of the axis is driven and hold in position by six ruby hemispheres, two types of piezoactuators are fabricated. The first type of piezoactuator drives and carries two ruby hemispheres (Figure 1 a, left). It is further referred to as A-type actuator. The type holding one ruby hemisphere will be called B-type actuator. The ruby hemispheres are in contact with the runner surface. They offer beneficial friction characteristics and low wear. Each two actuator types are arranged with an angle of ninety degrees to each other (Figure 1 b for the B-type). This allows a beneficial guiding of the runner. In Figure 1 c, the runner with ruby hemispheres and both types of actuators is shown. The motion of direction is perpendicular to the shown drawing. An image of the runner in the testbed can be seen in Figure 1 d. The runner is equipped with a mirror to measure its position via laser interferometry. Every ruby hemisphere is glued to two segments of the piezoceramic. Each segment can be controlled separately. With symmetrical control voltages (e.g., +100 V and -100 V), a rotation of the ruby hemispheres can be achieved. Synchronous rotations of the six hemispheres lead to a linear displacement of the runner. For this reason, the static displacement of the runner (stick-phase) is - in a first approximation - proportional to the control voltage. Very fast hemisphere rotations can be used to turn the hemispheres without runner displacement. The ruby hemispheres just slip over the runner's surface (slip-phase). Alternation of both phases is used to displace the runner with travels larger than the static displacement. For detailed descriptions of the principle the reader

Figure 1. In a) the two types of piezoactuators for the linear axis are shown. The left one actuates two ruby hemispheres (A-type), the right one a single hemisphere (B-type). b) Two B-type piezoactuators are equipped with ruby hemispheres (diameter 1mm) and are joined together with an angle of 90° to form one half of the guiding. The other half of the guiding is made by two A-type actuators in the same way, carrying four ruby hemispheres. c) CAD-based sketch of the runner hold by the guiding. Two of the six ruby hemispheres are occluded. d) Image of the runner in the testbed. A mirror is mounted for position measurements.

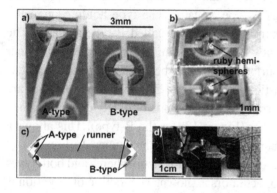

is referred to literature (Breguet, 1998; Edeler, 2008; Driesen, 2008).

In Figure 2, a sketch of the testbed containing the stick-slip drive is shown.

Three of the six ruby hemispheres are drawn in the sketch. The B-type actuators are sketched at the top of the runner; two A-type actuators at the bottom. This allows for displacements of the runner to the left and right. The position of the runner is measured via laser interferometry. The laser interferometer is a SIOS SP-S120. With additional actions (vibration isolation table and acoustic isolation chamber), the measurement resolution is approximately 5 nm. In opposite direction to the laser interferometer, a miniature load cell (Honeywell load cell 31) is mounted. As shown in later measurements, the surface condition between runner and load cell is of importance. For this reason, the load cell carries a piece of a

silicon wafer and the runner is equipped with a ruby hemisphere. The load cell is capable to measure +/-500 mN with an internal full bridge of strain gauges. The amplifier signal is afflicted with noise in the range of 5 µN. For every stick-slip drive it is necessary to create a normal force for the friction contacts between the actuator and the runner. This normal force, further called pre-load, is created by a mechanical spring with known compliance. By setting up the length of the spring, preload is defined. The length of the spring was chosen to be approximately 50-60 mm, thus improving the accuracy of preload and decreasing the influence of the runner's surface tolerances. The spring affects a carriage holding the A-type actuators. The carriage is guided by low-friction linear ball bearings. For the measurements in this

Figure 2. Sketch of the testbed for the force measurements. The runner is hold in position by six ruby hemispheres, carried by two A-type and two B-type actuators (compare to Figure 1). The A-type actuator is mounted to a linear bearing so that the normal force between the hemispheres and the runner is defined by a preload spring. The length of the spring can be modified to gather different preloads. Furthermore, the spring is exchangeable to achieve different preload ranges. The position of the runner is measured via laser interferometer. The runner operates against an obstacle in form of a load cell. The generated force and the position are measured simultaneously.

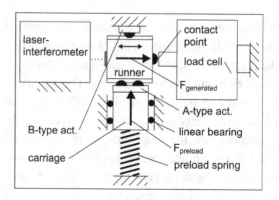

paper, the runner runs horizontally. Consequently, gravity takes no effect in the direction of travel. However, an influence on preload is thinkable, as the ruby hemisphere guidance also gathers the forces caused by gravity. Particularly if preload was very low, reaction could be a question of substance. The testbed is designed in such a way that it can be rotated to investigate runner displacements with respect to different gravity angles.

Preload is one of the most important parameters for stick-slip drives. If preload is too low, the axis is not driven properly and can become mechanically unstable. If preload is too high, the axis vibrates, but remains in the stick-phase without further displacement. This empirical experience has been known for a long time, but has rather been investigated systematically. Last but not least preload plays an important role for the piezoactuators. Piezoactuators offer a characteristic in such a way that displacements can only be generated below a certain force, further called piezo-blocking-force. To get an overview of the piezo-blocking-force of the used piezoactuators, a Finite-element analysis was adducted. Figure 3 shows the static displacements of the A-type

actuator for averaged preload of each ruby hemisphere of 35 mN.

The value was calculated for a typical preload of approximately 0.1 N. According to experience, the static rotation of the ruby hemispheres and therefore the runner's displacement is calculated to be 60 nm. This correlates with measurements of single stick-slip steps, where the step length is 130 nm. It is twice the static displacement due to the fact, that two stick-phases are applied for a single step. Thus, the step length is at the nominal value for low preloads. Table 1 shows the static runner displacements for increasing preloads.

The step length decreases a bit up to preloads in the range of 25-50 N. With a preload of 100 N, virtually no displacement of the hemispheres can be simulated. This indicates that, from the piezo-actuators point of view, preloads up to 20 N and more should be possible. In practice this is not true, as will be shown later on.

Edeler et al. (2010) presented the influence of preload on the zero-step amplitude. The zero-step amplitude is the minimum control signal's amplitude which is needed to generate stick-slip steps. Breguet shortly mentions a similar value (Breguet, 1998), but in his investigations it is

Figure 3. Finite element simulation of the A-type piezoactuator (static case). With a control voltage of 300 Vpp, a rotation of the ruby hemispheres is generated which leads to a displacement of the runner of approximately 60 nm. For this reason, the full step length can be calculated to 120 nm.

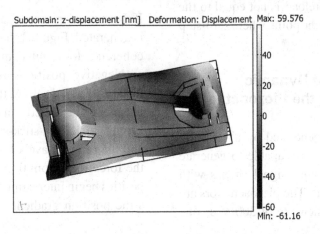

Table 1. Simulated static displacements of the runner with 300 Vpp (Volts peak-peak). The middle normal force per ruby hemisphere in the linear axis is calculated to 0.035 N. The simulated displacement is 60 nm (compare to Figure 3). With rising normal force, the displacement decreases. Values near the blocking force of approximately 2 N still lead to displacements of 45 nm. Very high normal forces larger than 50 N can limit the displacement of the piezoactuator to zero. However, this will likely cause damage of the actuator during usage.

Normal force on single ruby hemisphere [N]	Runner displacement derived from Z-displacements [nm]
0.035	60
0.5	55
1	55
10	45
25	40
50	25
100	5

considered as a constant. As a matter of fact, the zero-step amplitude rises with preload. In other words, if the drive exhibits large preload, most of the amplitude is spent to get the runner to the slip-phase and the step lengths decrease. The extreme case is reached if zero-step amplitude reaches 100%. Then the step length is zero and the runner just oscillates at its current position. It will be shown that preload is not equal to the piezo-blocking-force at the point, where zero-step amplitude is 100%.

Control Signal and Dynamic Characteristics of the Piezoactuators

The control signal is generated by a specially built signal generator. It is capable to generate sawtooth signals and other signal shapes with slewrates up to 600 V/µs. The piezoactuators are usually driven with a slewrate of 300 V/µs due

to technical reasons. This allows the slip phase to be easily reached in such a way that the parasitic motion during the slip phase is minimized and large steps are performed. However, the abrupt change in the signal affects the actuator's and drive's dynamics. For safe operation of stick-slip drives, the eigenfrequencies should be as high as possible. This can only be reached by a stiff design (Breguet, 1998; Edeler, 2008). But even if this is done, the slip signal is almost a kind of unit step and therefore the whole frequency spectrum is theoretically excited (Edeler, 2010). Few papers deal with techniques to reduce stick-slip vibrations of micro actuators. A detailed discussion would be too extensive for this paper. The reader is referred to Bergander (2003c). Therefore, a certain amount of noise or respectively vibrations is unavoidable during operation.

The sawtooth control signal is defined by polarity, frequency, amplitude and slewrate. Polarity defines the direction of motion. Amplitude and frequency together determine the runner's velocity and slewrate is supposed to be high enough to enable a reliable change from stick to slip. Investigations on the influence of the slewrate can be found in Edeler, Meyer, & Fatikow, 2010. In the following section it is shown that the control amplitude can be used to modulate the generated force.

Force Generation

If the runner operates against an obstacle, a force is generated. Figure 4, from 0 to 1.8 s, shows the coherency for rough contact surfaces.

Negative positions represent the direction towards the load cell. At the beginning the runner is positioned directly in front of the load cell. From 0 to 1.25 s, stick-slip steps are performed and the runner travels 1.2 µm. At the same time the force rises from 0 to 2 mN. The surface asperities begin interacting. Between 1.25 and 1.75 s the position gradient decreases and the force increases. Every single step can be identified in the force signal. At the point of time 1.4 s the position has reached a static condition. No steps are performed. A drift in the force can be identified until the runner retracts at 1.75 s. The drift is likely caused by the piezoactuators' relaxation. The large travel and the low force gradient in the beginning are indicators for a high level of surface roughness within the range of 1 μm. Nevertheless, the force adjusts to a constant value dependent on the amplitude. The right part of the figure represents the same procedure for flat and hard surfaces (ruby hemisphere and silicon wafer). In contrast to the rough surfaces, the force is generated much faster within a small travel. But it also adjusts to a certain level. Thus, the force generation is independent from surface condition of the contact between runner and obstacle.

RESULTS

Parameter Variation

The level of force reached with measurements as in Figure 4 can be recorded for different control amplitudes. The resulting graphs can be taken from Figure 5.

Each point represents the averaged value of ten measurements. The error bars show the minimum and maximum force. At first, the graph for a preload of 0.1 N is analyzed. Force generation starts with an amplitude of 20%. Below the amplitude of 20%, no steps are performed due to the zero-step amplitude. Therefore, the runner does not travel and no force can be generated. With rising amplitude, the generated force also rises until the amplitude reaches 60%. For higher

Figure 4. Generated forces and positions with different surface roughness between runner and load cell are shown exemplarily. In case "rough" (0 – 1.8 s), the untreated runner surface meets a screw head mounted to the load cell. The force slowly increases from 0 to 3 mN until 1.3 s. After that, the force increases much faster to 25 mN. A static drift of the force can be observed between 1.45 and 1.75 s. A travel of 1.3 μm is needed to generate the force because of the surface roughness. The flat contact – a ruby hemisphere in contact to a piece of a flat silicon wafer mounted to the load cell – needs only 150 nm travel to generate a force of 39 mN. The force levels are generated with different runners and amplitudes and cannot be compared directly. It can be noted that the force generation functions with different contact conditions. However, the contact conditions strongly influence the gradient of force.

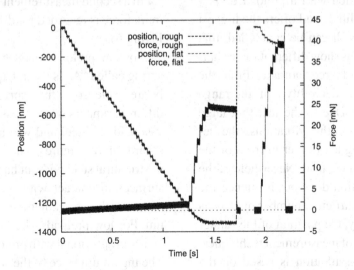

Figure 5. Generated maximum forces dependent on amplitude and preload. Different characteristics can be identified. With low preload of 0.1 N, force generation starts with an amplitude of 20%. Force increases with amplitude up to approximately 60% amplitude. For larger amplitudes, the force rises marginal. A kind of saturation is reached, likely due to the sliding friction limit of the drive. For higher preloads, the start amplitude and the saturation limit rise (preload 0.25 N). With large preload of 1.0 N, start amplitude reaches almost 50%. Large forces are generated, but saturation is not reached. Overall the repeatability is in the range of 10-25%.

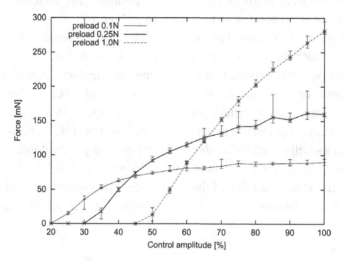

amplitudes the force does not increase further. A force-saturation at 85 mN is reached. The graph representing the preload of 0.25 N does not start until the amplitude of 30% is reached. Again, the force rises, but the force gradient decreases with rising amplitude. The level of the generated forces is higher, but there is no distinct saturation effect. Force generation with a preload of 1 N starts with 50% amplitude and offers the highest forces up to 270 mN. It can be concluded, that different preload levels show different force characteristics. The level of force-saturation limits the generated force. Reproducibility is in the range between 5 and 15%. Due to the fact that large preloads cause noteworthy vibrations, the sequence of positioning the runner in front of the load cell is a challenging task. Nevertheless, the runner's initial position does not influence the generated force if a sufficient number of steps is performed. Generally, the position of the runner is of interest because of measurement techniques. At first sight, force-saturation is based on the kinetic friction coefficient. However, as far as

literature values of approximately 0.1 - 0.2 for steel on ruby are concerned, the large force level in Figure 5, preload 0.1 N, cannot be explained with kinetic friction. This is the case even if influence of gravity is included. The high level of force points to a new value, based on the dynamic force generation process.

In a second measurement the generated maximum force is recorded for different preloads (see Figure 6).

For low preloads between 0.05 and 0.3 N, a spring called "A" is used, larger preloads up to 2 N are achieved with a spring called "B". Two different runners were investigated. The first one was made of steel and was hardened to 55 HRc ("hard" in the figure). The other one was made of structural steel without hardening ("soft"). All surfaces in contact with the ruby hemispheres were polished to gather a quality better than 0.3 μm. For low preloads, the maximum force rises nearly proportional with preload for both runners. The maximum force of the soft runner decreases after a maximum at 260 mN and 0.7 N preload.

Figure 6. Maximum generated force about preload for two different runner surfaces. The first surface (hard) was hardened to HRc 55, the second is structural steel, not hardened (soft). The identifiers A and B stand for different preload springs (compare to Figure 2) and the identifiers 1 and 2 stand for two different measure runs due to technical reasons. Both graphs (hard and soft) show an almost linear rise with low preloads, after that they reach maximums and finally decrease in the range before blocking force. The decreasing is, because steps cannot be performed anymore. Thus, no force can be generated.

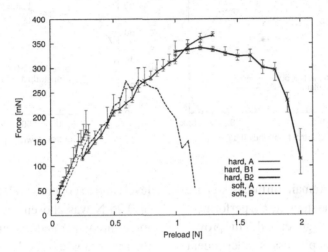

At 1.2 N preload, the runner barely generates a force. The hardened runner reaches its maximum force at this very preload. However, the characteristic is the same. Again, with rising preload the maximum force decreases and collapses after a preload of 2 N. The larger preloads for the hard runner had to be recorded with the same spring in two sessions "B1 and B2". The force levels at the overlap of B1 and B2 are not exactly the same because it is mechanically difficult for large preloads to set up preload. It can be concluded that preload mainly influences the maximum force and also force characteristics. With the fact of interacting asperities between runner and ruby hemispheres in mind, the issue is discussed more detailed in the last section. During measurements it was observed that parasitic vibrations of the runner rise with preload. Especially in the case of the largest preload, vibrations are significant. With even more preload, step generation stops. The runner just oscillates at its initial position. This is a clear indicator for the fact that the function of this SSA is not limited by the piezo-blocking force, but by a certain stick-slip blocking-

force. This blocking force is likely defined by the friction characteristics. More precisely, if friction characteristics prevent entering the slip-phase (dependent on preload), blocking-force is reached.

Previous measurements clarify that the force generation process is dependent on control amplitude and preload. Now there is the question to what extend other parameters influence the force generation process? Figure 7, left, shows the influence of the control frequency on the generated force.

Figure 7, right, shows how the control slewrate during the slip affects the force. Both figures represent typical characteristics of different preload regimes. It can be seen that the force is largely independent from the control frequency. Only for control frequencies higher than 10 kHz the forces decrease. A possible reason could be the different friction characteristic for very fast stick-phases. Such frequencies can cause the friction state to remain in slip without stick phase. Control slewrate also has a marginal influence in the forces. For very low slewrates, the forces generation does not come about. This can be eas-

Figure 7. Left: the dependency of the control frequency on the force is shown. The forces (for two different amplitudes) remain constant as far as possible. Dynamical aspects cause decreasing forces with very high control frequencies. Right: the dependency of control slewrate on the force is shown. Except at very low slewrates - where the drive does not generate steps anyway - there is virtually no dependency.

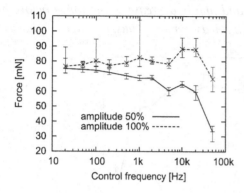
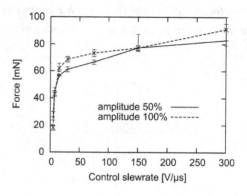

ily explained. The runner remains in the stick-phase for very low slewrates and steps are not performed. Therefore, no force can be generated. This proves that the force generation process is independent from the control frequency and the control slewrate for a wide parameter range.

Another important parameter of stick-slip drives is the mass of the runner. It is known for stick-slip positioners especially that the mass of the runner has great influence on the function of the drive, in particular on the step size. Interestingly, the mass does not have any weight for the force generation process if the positioner works well. Because of the fact that the runner is virtually in a standstill during force generation, neither inertia forces nor the mass play an important role. Therefore, the force generation is independent from the runner's mass.

Generated Force as Target Value

For applications of the force generation it will be of interest to set different force levels one after another. To test this, a set-value as in Figure 8 was defined.

Four different forces are supposed to be generated, 50, 150, 100 and 200 mN. The sequence was chosen to test not only rising, but also falling force

levels and to cover the full force range. A preload of 0.25 N was chosen. In the first instance, the forces are generated one after another by setting the according amplitude. The control amplitude was derived after modeling the characteristic of the force generation (similar to that of preload 0.25 N in Figure 5) using a simple logarithmic approximation. This approach leads to the run labeled "force without retraction". The first force level is only covered by 50%. After that, the second force level, higher than the first one, is reached quite well. The run of the set value fits "force with retraction," therefore it does not match "force without retraction" very well. The set value then falls to 0 mN. The measured force level does rather decrease. The runner cannot change its state to a lower force level. This means that some sort of initial condition has to be established before the force generation. Again, this indicates a clear hysteresis characteristic. The last force level, 200 mN, is reached quite well. Again, this is a change with rising level. Consequently, a modified force generation sequence was measured. The new sequence includes a single retraction step with full amplitude after every force level. The result is shown in the run "force with retraction". As a result, the force decreases abrupt after every level. After the level of 150 mN, force does not

Figure 8. Generating predefined force levels. Without retraction after a completed force level ("force without retraction"), a lower level cannot be reached. At 1.6 s, the force does not decrease, but remains at the level of 145 mN. Inserting a retraction sequence also allows reaching lower force regimes ("force with retraction"). The set value is fulfilled as far as possible. However, error is about 20% at the levels of 50 and 150 mN. Furthermore, at 1.45 s, zero is not reached because there were not enough retract steps. It can be seen that the retraction travel is not sufficient at 1.45 s.

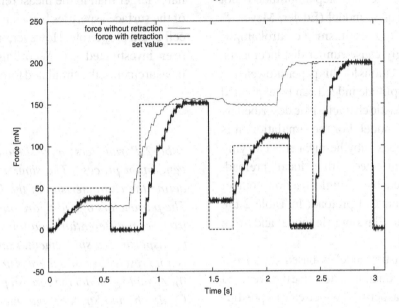

reach zero, but remains at 30 mN. This effect is founded by the fact that the load cell is afflicted with certain compliance. By generating a force level, the runner has to travel the way the load cell is buckled by the runner. If, as in this case, the distance of retraction is less than the distance the load cell is indented, the resulting force level is unequal to zero. Overall, the force's set-value is fulfilled with an accuracy of approximately 25%. It has to be analyzed why the other measurements were performed with better accuracy? Maybe the order of measurements or the time of detention plays a role. This could be connected with the time-dependent characteristic of the piezoceramic.

Simulation of the Force Generation Process

It is assumed that the function of the force generation is strongly connected with the stick-slip principle and the interaction of the actuator with the runner. In particular, the interacting asperities of the surfaces are supposed to play an essential role dependent on changing control amplitudes. This assumption should be provable if the friction which causes the variable forces can be simulated. Preliminary simulations show, that the force generation does not happen without a stick-slip control signal. Theoretically, effects such as piezohysteresis and piezo-caused nonlinearity, exact friction characteristics including material conditions and external conditions such as gravity have to be modeled for an all-embracing simulation. However, many of these parameters are unknown or difficult to determine. Hence, only friction between ruby hemispheres and runner and the load cell's compliance are modeled.

The force generating process is simulated via a single friction contact represented by the Elastoplastic model. The model is a well-developed friction model to describe not only the Stribeck effect and Coulomb's friction but also presliding and friction hysteresis (Dupont, Hayward, Armstrong, & Altpeter, 2002). Recent simulations denote that the step length of a stick-slip drive can be simulated very well using the LuGre-model (Wit, Olsson, Astrom, & Lischinsky, 1995). It is a forerunner of the Elastoplastic model. However, the effect of the zero-step amplitude is not covered by the LuGre model (Edeler, Meyer, & Fatikow, 2010). As a result, small control amplitudes misleadingly cause runner displacements. Preliminary simulations for this paper showed that the zero-step-amplitude indeed can be explained with the elasto-plastic characteristic described by the Elastoplastic model. For the simulation, it is assumed that the six ruby hemispheres carrying the runner, each affected with a "local" preload, can be substituted by a single friction contact with the full amount of preload. In Table 2 the parameters for the Elastoplastic model and drive data are outlined.

The Elastoplastic model is based on a first-order differential equation to describe the averaged deflection of the asperities, which causes presliding and oscillations after abrupt changes. The parameters σ_0 and σ_1 define the frequency and damping of such oscillations. As a matter of fact, a friction lubricant is not present. Therefore, σ_2 equals zero. The characteristic of the zero-step amplitude is defined by z_{ss} and z_{ba}. The Coulomb' and Stribeck's friction parameters μ_{static}, $\mu_{kinetic}$ and $v_{stribeck}$ are uncritical. They are derived by static friction measurements. The drive's parameters such as mass, preload and the control signal's parameters are set equal to the measurements. The load cell is modeled as a spring with the according compliance. For initial condition the runner is placed to the load cell with a distance of 200 nm. This was done in order to ensure a fair transition from positioning to force generation. The result can be obtained from Figure 9.

Position and force signal start at zero. Until the runner reaches the load cell shortly before 10 ms, the force equals zero. Almost two steps are performed before the force generation begins.

This matches the step size of approximately 120 nm. Equal to the force signal in Figure 4, the force rises until a force-saturation is reached. As a matter of fact, the force generation can be simulated in principal. However, the noise in the force signal is larger than in the measurements. Damping of the surface asperities between runner and load cell could play a role. However, this issue has not been investigated yet. According to the testbed measurements, the simulated force-saturation can

Table 2. Parameters for the simulation of the force generation process. The simulation parameters define the characteristic of the friction contact. The parameters σ_0 and σ_1 are parameters for the first-order differential equation system modeling the asperities' displacement. Lubricated friction is not present, because there is no lubricant ($\sigma_2 = 0$). Z_{ss} and z_{ba} model the zero-step amplitude. The Coulomb' and Stribeck' parameters (μ_{static}, $\mu_{kinetic}$ and $v_{stribeck}$) are less crucial and are derived from several preliminary measurements. A preload of 0.1 N is chosen for simulation. The mass of the runner, slewrate, maximum actuator amplitude and frequency are equal to the testbed conditions.

Name	Value	Unit
σ_0	$1.8 * 10^6$	N/m
σ_1	10	Ns/m
σ_2	0	Ns/m
z_{ss}	40	nm
z_{ba}	20	nm
μ_{static}	0.14	-
$\mu_{kinetic}$	0.14	-
$v_{Stribeck}$	0.001	m/s
preload	0.1	N
runner's mass	3	g
slewrate	2	µs
amplitude$_{max}$	160	nm
frequency	200	Hz

Figure 9. Simulation of the force generation process. The Elastoplastic friction model was used to describe the friction between the ruby hemispheres and the runner. The load cell (modeled as a spring with the same stiffness) is positioned with a distance of 200 nm to the runner. This is the reason why the force initially remains zero. At the time when the position exceeds 200 nm, a force is generated. Analogue to the measured data, the force rises until saturation occurs. Thus, the force generation process can be simulated in principal.

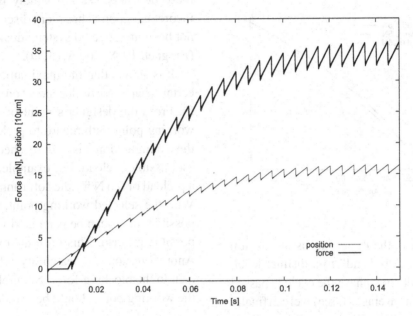

be measured in dependency to control amplitude and preload. Figure 10 shows the coherency.

Distinct and constant zero-step amplitude is simulated for all preload sets. Thus, the Elastoplastic model is not able to cover the dependency of the zero-step amplitude dependent on preload yet. Nevertheless, increased preload levels cause increased generated forces. If the characteristics of the generated forces are compared to those in Figure 5, similarities are rare. There is no pronounced force-saturation, the characteristic does not change with different preloads, and the level of the generated forces is by far lower. It is difficult to justify simulation results. The preconditioned simplification of the friction contact, influence of gravity on the runner or different contact characteristics between runner and ruby hemispheres, but also between runner and load cell, could be of importance.

CONCLUSION

In this paper a new method to generate controlled forces with a micro stick-slip drive was described. The concept was proven and important parameters were identified and investigated. Ultimately, the question arises, how the force generation can be explained.

Theoretical Explanation of the Force Generation Method

It is conceivable that, during force generation, the surface asperities are deflected by the actuator, dependent on the actuator's amplitude. This results in a force defined by the asperity compliance and the deflection during stick-phase. A continuous actuator displacement preconditioned, the generated force rises proportionally to a certain value.

Figure 10. Simulated force against amplitude and preload. It can be seen that different preloads generate different force levels, but the rising zero-step amplitude and the saturation effect (compare to Figure 5) cannot be simulated yet. As a matter of fact, the simulated force levels are low.

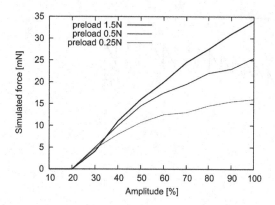

During slip-phase, the runner keeps in position and the asperities "re-bend" to a distinct level. In the subsequent stick-phase the cycle is passed through again. The averaged force level can finally be measured. The dependency on the control amplitude can be also explained with this approach. With larger control amplitude, the maximum deflection of the asperities rises and as a result the averaged forces rise. Whereas this explanation is supported by simulation results, measurements (Figure 4) rather show explicit vibrations. This can have several reasons. The force signal is afflicted with noise, which complicates identification of dynamic effects. Furthermore, the compliant part of the load cell weights approximately 15 g and represents an inertia damper. Measurements with more dynamic, lighter sensors will be necessary for verification. Another reason can be found in the contact characteristics between load cell and runner. The interacting surface asperities stay in contact during force generation and may also act as a damper. However, a stiff contact such as the pair silicon wafer/ruby hemisphere should barely show pronounced damping. It seems to be sure that the function of the surface asperities is very important. Furthermore, asperities are assumed to be essential for the function of the force generation method. From this point of view, surface quality and the chosen material of runner and hemisphere strongly influence characteristics. At the moment, these dependencies are widely unknown. Few theoretical approaches were discussed, but have not been investigated systematically for SSA yet (Breguet, 1998; Popov, 2009).

It is shown that the maximum force shows a certain characteristic dependent on preload (Figure 5). From the designer's point of view, the best working point with regard to preload is given if the maximum force is even reached with the lowest possible preload. For example, this could be a preload of 0.6 N for the soft runner in Figure 5. With the selected working point, the maximum possible forces can be generated with the lowest possible preload. This prevents excessive wear. Another reason is repeatability, which decreases near to the blocking-force. So, preloads larger as the working point should be avoided.

Advantages

The great advantage of the method is the fact that force generation is an open-loop process, which is not in the need of an additional sensor. The known methods to generate forces in a comparable scale are closed-loop approaches and make at least use of a position sensor. In fact, the performance of this method can be improved using a position sensor, e.g., to optimize the distance of retraction to detect the maximum force. However, the idea is to get along without sensors to benefit from the simple stick-slip design and the miniaturization potential. These are exactly the properties which make the method attractive for applications.

It is an important fact that the level of vibration scales with preload. In other words: Drives with large preload exhibit more vibrations than drives with less preload. Due to technical reasons, preload is closely linked with miniaturization. Drives with small dimensions will exhibit less preload and therefore a lower generated force level, but will also show less vibrations. It will be difficult to use a "large" drive for very low forces and small vibrations. But, a smaller drive could fulfill the requirements. The force range of the drive also plays a role. However, the individual application defines the outline data. Generally, smaller structures use smaller forces. This method can adapt to different scenarios primarily by selecting the adequate preload.

Disadvantages

An evident disadvantage of the force generating method is the fact that the force is arranged in distinct portions, respectively steps. This does not allow for true continuous force establishing. In the measurement in Figure 4, the steps are performed with a control frequency of 1 kHz. If the control frequency was lower, the rise of the force could be less abrupt due to the lower asperity bending gradient. However, the discontinuity given by the slip-phase is at least there. In simulations (Figure 9), slip leads to abrupt force discontinuities of approximately 2.5 mN. In Figure 4 there is barely such a discontinuity. Possible reasons have already been discussed. It will be a matter of investigation how vibrations can be minimized and how far a semi-continuous force generation can be established.

The aspect of force drift can be observed in Figure 4. Drift, or relaxation of piezoactuators, is well-known. Several approaches have been proposed to face it. Because of the fact that it appears within the stick-phase, known methods could be used for compensation. Only if the maximum actuator amplitude was reached, a slip-phase could be inserted to provide compensation travel.

Wear plays an important role for SSAs. During the measurements, a continuously rising force level was observed due to worn out runner surfaces. This effect is distinct for the soft runner, but was also seen for the hard runner. Therefore, wear should be integrated into a model which describes the generated forces. Hertzian' contacts in relation to the number of interacting asperities could be a theoretical basis.

Outlook

The investigation of the stick-slip drive is motivated by the design of a linear axis for a mobile nanohandling robot. It can be assumed that the results of this paper can be extrapolated to the residual three DOF of the robot or to other common SSAs. For the robot, this would enable a three-dimensional (3-D) force generation. Three control amplitudes for the SSA in X, Y, and Z would define an arbitrary force vector. It is imaginable to drive the prospect of such robots from "simple" 4-D positioners to calibrated force probers. And this can be done without the use of sensors. Nevertheless, the circumstances are much more difficult in the robot. If other SSAs can also be supported, it can be assumed that force generation is independent from the design of the SSA. Finally, virtually any SSA could be used to generate defined forces after calibration. This would push the limit of possible applications.

It will be important in further investigations to see the role of gravity in the force generation process. The measurements in this paper were performed with horizontal axis adjustment and preload is comparatively large to a gravitational force of 30 mN. Obviously, if the aspect ratio between preload and gravitational force rises, force characteristics is by far defined by preload. For vertical and intermediate adjustments the characteristics will be different. Non-symmetrical, direction-dependent characteristics can be expected. Unfortunately, measurement effort rises heavily. Therefore, simulations will play an important role in the future.

The characteristic of the surface asperities is one of the important parameters for the friction simulation. Another issue is the integration of the preload-dependent zero-step amplitude and blocking force characteristic into a friction model, preferably the Elastoplastic model. In preliminary simulations, the preload-dependent zero-step amplitude could be modeled with the preload-dependent parameters zss and zba, who define the elastic and plastic regimes of the asperities. This could be a start for an all-embracing model of the force generation process.

REFERENCES

Bergander, A. (2003c). *Control, wear testing & integration of stick-slip micropositioning.* Unpublished doctoral dissertation, Ecole Polytechnique Fédérale de Lausanne, Lausanne, Switzerland.

Bergander, A., Driesen, W., Lal, A., Varidel, T., Meizoso, M., Bleuler, H., et al. (2004). Position Feedback for Microrobots based on Scanning Probe Microscopy. In *Proceedings of the International Conference on Intelligent Robots and Systems* (Vol. 2, pp. 1734-1739).

Bergander, A., Driesen, W., Varidel, T., & Breguet, J.-M. (2003a). Development of Miniature Manipulators for Applications in Biology and Nanotechnologies. In *Proceeding of Microrobotics for Biomanipulation Workshop, IEEE/RSJ International Conference on Intelligent Robots and Systems* (pp. 11-35).

Bergander, A., Driesen, W., Varidel, T., & Breguet, J.-M. (2003b). Monolithic piezoelectric push-pull actuators for inertial drives. In *Proceedings of 2003 International Symposium on Micromechatronics and Human Science.* doi:10.1063/1.1139566

Breguet, J.-M. (1998). *Actionneurs "stick and slip" pour micro-manipulateurs.* Unpublished doctoral dissertation, Ecole Polytechnique Fédérale de Lausanne, Lausanne, Switzerland.

Breguet, J.-M., Driesen, W., Kaegi, F., & Cimprich, T. (2007). Applications of piezo-actuated micro-robots in micro-biology and material science. In *Proceedings of the IEEE International Conference on Mechatronics and Automation* (pp. 57-62).

Brufau, J., Puig-Vidal, M., Lapez-Sanchez, J., Samitier, J., Driesen, W., Breguet, J.-M., et al. (2005). MICRON: Small Autonomous Robot for Cell Manipulation Applications. In *Proceedings of the IEEE International Conference on Robotics and Automation* (pp. 844-849).

Canales, C., Kaegi, F., Groux, C., Breguet, J.-M., Meyer, C., Zbinden, U., et al. (2008). A nanomanipulation platform for semi automated manipulation of nano-sized objects using mobile microrobots inside a Scanning Electron Microscope. In *Proceedings of the 17th World Congress International Federation of Automatic Control* (pp. 13737-13742).

Das, A., Zhang, W. P., Popa, D., & Stephanou, H. (2007). μ³: Multiscale, Deterministic Micro-Nano Assembly System for Construction of On-Wafer Microrobots. In *IEEE International Conference on Robotics and Automation* (pp. 461-466). doi:10.1109/ROBOT.2007.363829

de Wit, C. C., Olsson, H., Aström, K., & Lischinsky, P. (1995). A new model for control of systems with friction. *IEEE Transactions on Automatic Control, 40,* 419–425. doi:10.1109/9.376053.

Driesen, W. (2008). *Concept, modeling and experimental characterization of the modulated friction inertial drive (MFID) locomotion principle.* Unpublished doctoral dissertation, Ecole Polytechnique Fédérale de Lausanne, Lausanne, Switzerland.

Dupont, P., Hayward, V., Armstrong, B., & Altpeter, F. (2002). Single state elastoplastic friction models. *IEEE Transactions on Automatic Control, 47,* 787–792. doi:10.1109/TAC.2002.1000274.

Edeler, C. (2008). Simulation and Experimental Evaluation of Laser-Structured Actuators for a Mobile Microrobot. In *Proceedings of the IEEE International Conference on Robotics and Automation (ICRA 2008)* (pp. 3118-3123). doi:10.1109/ROBOT.2008.4543685

Edeler, C. (2010). Dynamic-mechanical Analysis of Piezoactuators for mobile Nanorobots. In *Proceedings of the International Conference on New Actuators (ACTUATOR2010)*, Bremen, Germany (pp. 1003-1006).

Edeler, C., Meyer, I., & Fatikow, S. (in press). Simulation and Measurements of Stick-Slip-Microdrives for Nanorobots. In *Proceedings of the European Conference on Mechanism Science (EUCOMES 2010)*, Cluj-Napoca. *Romania.*

Eichhorn, V., Carlson, K., Andersen, K. N., Fatikow, S., & Bøggild, P. (2007). Nanorobotic Manipulation Setup for Pick-and-Place Handling and Nondestructive Characterization of Carbon Nanotubes. In *Proceedings of the IEEE International Conference on Intelligent Robots and Systems* (pp. 291-296). doi:10.1109/IROS.2007.4398979

Eichhorn, V., Fatikow, S., Wortmann, T., Stolle, C., Edeler, C., Jasper, D., et al. (2009). NanoLab: A Nanorobotic System for Automated Pick-and-Place Handling and Characterization of CNTs. In *Proceedings of the IEEE International Conference on Robotics and Automation* (pp. 1826-1831). doi:10.1109/ROBOT.2009.5152440

Fatikow, S., Eichhorn, V., Krohs, F., Mircea, I., Stolle, C., & Hagemann, S. (2007). Development of automated microrobot-based nanohandling stations for nanocharacterization. *Microsystem Technologies*, *14*, 463–474. doi:10.1007/s00542-007-0471-5.

Hagemann, S., Krohs, F., & Fatikow, S. (2007). *Automated Characterization and Manipulation of Biological Cells by a Nanohandling Robot Station.* Poster presented at Nanotech Northern Europe Conference and Exhibition, Helsinki, Finland.

Jasper, D., & Edeler, C. (2008). Characterization, Optimization and Control of a Mobile Platform. In *Proceedings of the International Workshop on Microfactories* (pp. 143-148).

Mariotto, G., D'Angelo, M., & Shvets, I. V. (1999). Dynamic behavior of a piezowalker, inertial and frictional configurations. *The Review of Scientific Instruments*, *70*, 3651–3655. doi:10.1063/1.1149972.

Martel, S., Sherwood, M., & Helm, C. de, W. G., Fofonoff, T., Dyer, R., et al. (2001). Three-legged wireless miniature robots for mass-scale operations at the sub-atomic scale. In *Proceedings of the IEEE International Conference on Robotics and Automation* (pp. 3423-3428).

Meyer, C., Sqalli, O., Lorenz, H., & Karrai, K. (2005). Slip-stick step-scanner for scanning probe microscopy. *The Review of Scientific Instruments*, *76*. doi:10.1063/1.1927105.

Munassypov, R., Grossmann, B., Magnussen, B., & Fatikow, S. (1996). Development and Control of piezoelectric actuators for a mobile micromanipulation system. In *Proceedings of the 5th International Conference on New Actuators (ACTUATOR1996)* (pp. 213-216).

Murthy, R., Das, A., & Popa, D. (2008). ARRIpede: A stick-slip micro crawler/conveyor robot constructed via 2 ½D MEMS assembly. In *Proceedings of the IEEE/RSJ International Conference on Intelligent Robots and Systems* (pp. 34-40). doi:10.1109/IROS.2008.4651181

Murthy, R., & Popa, D. O. (2009). A four degree of freedom microrobot with large work volume. In *Proceedings of the IEEE International Conference on Robotics and Automation* (pp. 1028-1033). doi:10.1109/ROBOT.2009.5152812

Noyong, M., Blech, K., Rosenberger, A., Klocke, V., & Simon, U. (2007). In situ nanomanipulation system for electrical measurements in SEM. *Measurement Science & Technology*, *18*(84).

Peng, L. M., Chen, Q., Liang, X. L., Gao, S., Wang, J. Y., & Kleindiek, S. et al. (2004). Performing probe experiments in the SEM. *Micron (Oxford, England), 35*, 495–502. doi:10.1016/j.micron.2003.12.005.

Pohl, D. W. (1987). Dynamic piezoelectric translation devices. *The Review of Scientific Instruments, 58*, 54–57. doi:10.1063/1.1139566.

Popov, V. L. (2009). Kontaktmechanik und Reibung. Berlin: Springer. doi: doi:10.1007/978-3-540-88837-6.

Rabenorosoa, Clévy, Lutz, Bargiel, & Gorecki, (2009). A micro-assembly station used for 3D reconfigurable hybrid MOEMS assembly. In *Proceedings of the IEEE International Symposium on Assembly and Manufacturing* (pp. 95-100).

Ronkanen, P., Kallio, P., & Koivo, H. N. (2007). Simultaneous Actuation and Force Estimation Using Piezoelectric Actuators. In *Proceedings of the 2007 IEEE International Conference on Mechatronics and Automation* (pp. 3261-3265).

Trüper, T., Kortschack, A., Jähnisch, M., Hülsen, H., & Fatikow, S. (2004). Transporting Cells with Mobile Microrobots. *IEEE Proceedings of Nanobiotechnology, 151*, 145–150. doi:10.1049/ip-nbt:20040839.

Vogel, M., Stein, B., Pettersson, H., & Karrai, K. (2001). Low-temperature scanning probe microscopy of surface and subsurface charges. *Applied Physics Letters, 78*, 2592–2594. doi:10.1063/1.1360780.

Zesch, W., Buechi, R., Codourey, A., & Siegwart, R. Y. (1995). Inertial drives for micro- and nanorobots: two novel mechanisms. *SPIE, 2593*, 80–88. doi:10.1117/12.228638.

Chapter 2
Optimal Robot Path Planning with Cellular Neural Network

Yongmin Zhong
Curtin University of Technology, Australia

Bijan Shirinzadeh
Monash University, Australia

Xiaobu Yuan
University of Windsor, Canada

ABSTRACT

This paper presents a new methodology based on neural dynamics for optimal robot path planning by drawing an analogy between cellular neural network (CNN) and path planning of mobile robots. The target activity is treated as an energy source injected into the neural system and is propagated through the local connectivity of cells in the state space by neural dynamics. By formulating the local connectivity of cells as the local interaction of harmonic functions, an improved CNN model is established to propagate the target activity within the state space in the manner of physical heat conduction, which guarantees that the target and obstacles remain at the peak and the bottom of the activity landscape of the neural network. The proposed methodology cannot only generate real-time, smooth, optimal, and collision-free paths without any prior knowledge of the dynamic environment, but it can also easily respond to the real-time changes in dynamic environments. Further, the proposed methodology is parameter-independent and has an appropriate physical meaning.

INTRODUCTION

Real-time collision-free motion planning in a non-stationary environment is an important and challenging issue in many autonomous systems including robotics and intelligent systems. It pro-vides intelligent robotic systems with an ability to plan motions and to navigate autonomously. This ability becomes critical particularly for robots which operate in dynamic environments, where unpredictable and sudden changes may occur. Whenever the robot's sensory system detects a

DOI: 10.4018/978-1-4666-3634-7.ch002

dynamic change, its planning system has to adapt and modify its paths accordingly. Prominent examples include real world environments that involve interaction with people, such as museums, shops, and households.

Based on the analogy between cellular neural network (CNN) and robot path planning, this paper presents a new neural dynamics based methodology for optimal collision-free robot path generation in an arbitrarily varying environment. The real-time collision-free robot trajectory is formulated as the dynamic CNN activity. The target activity is treated as an energy source injected into the neural system, and is propagated in the state space by neural dynamics. By formulating the local connectivity of cells as the local interaction of harmonic functions, an improved CNN model is established to propagate the target activity within the state space in the manner of heat conduction, which guarantees the target and obstacles remain at the peak and bottom of the activity landscape, respectively. The proposed neural network model cannot only generate real-time, smooth and optimal robot paths without learning procedures, prior knowledge of target or barrier movements, optimizations of any cost functions, and explicitly searching over the free work space or collision paths, but it can also easily respond to real-time changes in dynamic environments. Further, the proposed neural network model is parameter-independent and has an appropriate physical meaning.

RELATED WORK

The earliest work on robot path planning was by Lozano-Perez and Wesley (1979), who presented a path planning algorithm to avoid polyhedral obstacles based on a visibility graph. Since then, various approaches have been studied, especially during the past twenty years. Plenty of global approaches such as decomposition, road-map, distance transform and retraction methods, ran-domized approaches and genetic algorithms were reported to search the possible paths in the work space (Latombe, 1991; Hwang & Ahuja, 1992; Zelinsky, 1994; Henrich, 1997). A number of local methods such as potential field methods were also reported (Khatib, 1986; Barraquand & Latombe, 1991; Glasius, Komada, & Gielen, 1994; Li & Bui, 1998). Oriolo et al. (1998) proposed a method by combining global and local approaches for robot path planning, in which a global path planning plus a local graph search algorithm and several cost functions are used. Seshadri and Ghosh (1993) presented a path-planning model by using an iterative approach. This method is computationally complicated, especially in a complex work space. Ong and Gilbert (1998) presented a method for robot path planning with penetration growth distance. This method searches over collision paths instead of the free space, and it can generate optimal and continuous robot paths only in a static environment. In general, although most of the above methods can generate the accessible path with free collision in the work space, they can only deal with the static environment. A moving object or introduction of new objects requires that the whole work environment be constructed dynamically. In addition, with the increase of obstacles, the complexity of the algorithms increases exponentially.

Neural network based methods have received considerable attention for generating real-time robot paths. There is a number of learning based neural networks reported for mobile robot path planning (Svestka & Overmars, 1997; Zalama et al., 1995, 2002; Quoy, Moga, & Gaussier, 2003; Lebedev, Steil, & Ritter, 2005). These learning based approaches are time-consuming in operation, and the generated paths are not optimal, particularly during the initial learning phase (Willms & Yang, 2006). To avoid the time-consuming learning process Glasius et al. (1995, 1996) proposed a Hopfield-type neural network model for real-time path generation with obstacle avoidance in a non-stationary environment. It is rigorously proven that the generated path does not suffer from undesired local minima and is globally optimal in a stationary environment. However, these models require that the robot dynamics be faster than the dynamics of the target and obstacles, and have difficulty in accommodating fast changing environments. The shunting equation (Hodgkin & Huxley, 1952; Grossberg, 1988) is also a neural network approach to collision-free robotic path generation in dynamic environments (Yang & Meng, 2003; Yuan & Yang, 2003; Bhattacharya & Talapatra, 2005). However, this approach cannot guarantee the elimination of unexpected attractive points. The generated paths are also affected by the relationships between the model parameters. Pashkevich et al. (2006) reported a topologically ordered neural network. However, this neural network allows search path without the connectivity between neighbouring nodes, and thus reducing the high resolution or bitmap sampling of the configuration space. However, only simple static cases are considered, and the neural network also works in an off-line way. Sadati et al. (2008) also studied the use of the Hopfield neural network to find the shortest path among multiple targets. However, this method focuses on shortest path finding and only simple results on shortest path finding are reported.

There are several investigations reported to carry out robot path planning by using a CNN. Kanaya and Tanaka (1998) reported a three-layer neural network for generating collision-free paths of multi-robots. However, this method was not fully based on a CNN, since a CNN is only used as the first layer to produce the local currents of edges. Kim et al. (2002) reported a multi-layer CNN to find the optimal path. In this study, the neural connections are symmetric and only a static environment is discussed without any discussions on a changing environment. In addition, the computational time is increased due to the multi-layers of the CNN model. In contrast, the proposed neural network is a single-layer CNN model for robot path planning in both static and changing environments. The contribution of the proposed methodology incorporates the elegant properties of harmonic functions in a neural system to carry out real-time robot path planning. The novelty of the proposed neural network model is that local connectivity of neurons is a nonlinear harmonic function rather than a linear distance function in the existing neural network models.

ORIGINALITY AND ANALOGY

The CNN was first developed by Chua and Yang (1988), and was primarily intended for image processing and pattern recognition. A CNN is a dynamic nonlinear circuit composed by locally coupled, spatially recurrent circuit units called cells, which contain linear capacitors, linear resistors, and linear/nonlinear current sources. A CNN can be applied to different grid types. Without loss of generality, we consider a CNN on a rectangular grid with M rows and N columns. Each node on the grid is occupied by a cell. The dynamics of the array of M×N cells are described by the equation and conditions (Chua & Yang, 1988) shown in Box 1, where (i, j) denotes a cell on the ith row and the jth column. C is the capacitance of a linear capacitor. Since the capacitance is a scalar coefficient, we can set $C = 1$ without loss of generality. R_x is the resistance of a linear resistor, and I_{ij} is the independent current source at cell (i, j). A is the feedback template and B is the control template, whose values depend only on the relative positions of cells (i, j) and cells (k, l). $v_{uij}(t)$, $v_{xij}(t)$ and $v_{yij}(t)$ are the input, state and output of cell (i, j) at time t, respectively. $v_{yij}(t)$ is a nonlinear sigmoid function of $v_{xij}(t)$, and it is bounded by a constant K, which is equal to or greater than one. $N_r(i, j)$ is the neighbourhood of cell (i, j) within a radius r, and r is the size of the neighbourhood defining the interaction range of cells, which can be different integer values from one. For the sake of convenience, the smallest neighbourhood, i.e., $r = 1$, is considered in the following derivation.

One significant feature of CNNs, as well as the basic difference from other neural networks, is the local connectivity of cells, i.e., any cell in a CNN is connected only to its neighbouring cells. Adjacent cells directly interact with each other. Cells not directly connected to each other have indirect effect due to the propagation effects of the continuous-time dynamics of a CNN. The activity of a cell is propagated to other cells through the local connectivity of cells with the evolution of time. Therefore, the CNN has both local and global dynamics. The global property can be used by the target to globally attract the robot in the whole state space through neural activity propagation, while the local property can be used by obstacles to have only local effect in a small region for avoiding collisions and achieving the clearance from obstacles.

A CNN is a stable neural network, and thus it ensures the existence of the optimal path to find the target. In addition, given the initial state and the external input, the activity of a CNN is only determined by the local connectivity of cells. Therefore, by appropriate design of the local connectivity of cells, the CNN activity can be used for propagating the target activity in the state

space to guarantee the target and the obstacles remain at the peak and the bottom of the activity landscape, respectively.

Further, a CNN offers an incomparable speed advantage due to the collective and simultaneous activity of all cells (Roska et al., 1995; Reljin et al., 2004). The computation advantage of a CNN is very suitable for the real-time computation requirement of robot path planning.

CNN MODEL FOR PATH PLANNING

The neural network architecture of the proposed model is a discrete topographically organized map, which has been used in many neural network models (Willms & Yang, 2006; Glasius, Komoda, & Gielen, 1995, 1996; Yang & Meng, 2003; Yuan & Yang, 2003; Bhattacharya & Talapatra, 2005; Kohonen, 1982). The location of the *i*th neuron at the grid in the finite dimensional state space represents a position in the work space or a configuration in the joint space. Each cell has only local connectivity to its neighbouring cells that constitute the neighbourhood of the cell. The cell responds only to the stimulus within its neighbourhood. The proposed CNN model is an autonomous CNN derived from Equation (1):

Box 1.

$$C \frac{dv_{xij}(t)}{dt} = -\frac{1}{R_x} v_{xij}(t) + \sum_{(k,l) \in N_r(i,j)} A(i,j;k,l) v_{ykl}(t) + \sum_{(k,l) \in N_r(i,j)} B(i,j;k,l) v_{ukl} + I_{ij} \qquad (1a)$$

$$v_{yij}(t) = \frac{1}{2}(|v_{xij}(t) + K| - |v_{xij}(t) - K|), \quad K \geq 1 \qquad (1b)$$

$$|v_{xij}(0)| \leq K, \quad |v_{uij}| \leq K \qquad (1c)$$

$$N_r(i,j) = \{(a,b) \mid \max\{|a - i|, |b - j|\} \leq r, \ 1 \leq a \leq M, \ 1 \leq b \leq N\}$$

$$(1 \leq i \leq M; \ 1 \leq j \leq N) \qquad (1d)$$

$$\frac{dv_{xij}(t)}{dt} = -\frac{1}{R_x} v_{xij}(t) + \sum_{(k,l) \in N_r(i,j)} A(i,j;k,l)v_{ykl}(t) + I_{ij}$$

(2)

where

$$I_{ij} = \begin{cases} E & \text{if there is a target} \\ 0 & \text{otherwise} \end{cases}$$

(3)

$$v_{xij} = F \text{ if there is an obstacle}$$

(4)

where E is a positive constant. F is an arbitrary constant bounded by the constant K, and it can be simply set to zero. Since the CNN model has no inputs, the CNN constraint conditions Equation (1c) can be easily satisfied by:

$$v_{xij}(0) = F$$

(5)

Formulation for Local Connectivity of Cells

Due to the elegant properties of harmonic functions, such as the complete elimination of local minima by nature, the extrapolation of obstacle geometry, the derivation of smooth and safe paths, and the capacity for complete coverage of a working environment (Connolly, Burns, & Weiss, 1990; Connolly, 1993, 1994), the local connectivity of cells is defined as a harmonic form, i.e. Laplace operator that encodes each node relative to its neighbourhood.

To formulate the local connectivity of cells, Laplace operator has to be discretized on a grid map. The discretization of Laplace operator on a grid map can be easily obtained by a finite difference scheme (Kamiadakis & Kirby II, 2003) or a finite volume scheme (Versteeg & Malalasekera, 1995), and thus the local connectivity of cells can be subsequently obtained. For example, for the point $\mathbf{P}_{i,j}$ shown in Figure 1, Laplace operator discretized at point $\mathbf{P}_{i,j}$ by using a finite

difference scheme is shown in Equation (6) (see Box 2), where ∇ represents Laplace operator, $y_{i,j}$ is the activity at point $\mathbf{P}_{i,j}$, and $\left\| \overrightarrow{\mathbf{P}_{i-1,j}\mathbf{P}_{i,j}} \right\|$ and other similar terms represent the magnitudes of vector $\overrightarrow{\mathbf{P}_{i-1,j}\mathbf{P}_{i,j}}$ and other similar vectors.

From Equation (6), it can be easily seen that the local interaction of Laplace operator has the property that the sum of the weights at each node and its neighbourhood is zero.

Thus, the local connectivity of the cell at point $\mathbf{P}_{i,j}$ can be obtained as Equation (7) (see Box 3) where h is a positive constant and is called the control coefficient.

It is not difficult to see that the property of the local interaction of Laplace operator is also inherited by the locally connected cells, i.e., the sum of the connection weights for each cell and its neighbouring cells is zero.

Physical Meaning

It must be emphasized that the proposed CNN model is stable and convergent since it satisfies all the constraint conditions of a general CNN. In addition, the proposed neural network has an appropriate physical meaning.

Figure 1. A grid map

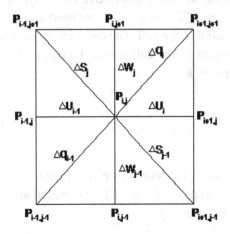

Box 2.

$$
(\nabla y)_{\mathbf{P}_{i,j}} =
$$

$$
\frac{2y_{i+1,j}}{\Delta u_i(\Delta u_{i-1} + \Delta u_i)} + \frac{2y_{i-1,j}}{\Delta u_{i-1}(\Delta u_{i-1} + \Delta u_i)} + \frac{2y_{i,j+1}}{\Delta w_j(\Delta w_{j-1} + \Delta w_j)}
$$

$$
+ \frac{2y_{i,j-1}}{\Delta w_{j-1}(\Delta w_{j-1} + \Delta w_j)} + \frac{2y_{i+1,j+1}}{\Delta q_i(\Delta q_{i-1} + \Delta q_i)} + \frac{2y_{i-1,j-1}}{\Delta q_{i-1}(\Delta q_{i-1} + \Delta q_i)}
$$

$$
+ \frac{2y_{i-1,j+1}}{\Delta s_j(\Delta s_{j-1} + \Delta s_j)} + \frac{2y_{i+1,j-1}}{\Delta s_{j-1}(\Delta s_{j-1} + \Delta s_j)}
$$

$$
- \frac{2y_{i,j}}{\Delta v_{i-1}\Delta v_i} - \frac{2y_{i,j}}{\Delta w_{j-1}\Delta w_j} - \frac{2y_{ij}}{\Delta q_{i-1}\Delta q_i} - \frac{2y_{ij}}{\Delta s_{j-1}\Delta s_j}
$$

$$
\Delta u_{i-1} = \left\|\overrightarrow{\mathbf{P}_{i-1,j}\mathbf{P}_{i,j}}\right\| \quad \Delta u_i = \left\|\overrightarrow{\mathbf{P}_{i,j}\mathbf{P}_{i+1,j}}\right\| \quad \Delta w_{j-1} = \left\|\overrightarrow{\mathbf{P}_{i,j-1}\mathbf{P}_{i,j}}\right\| \quad \Delta w_j = \left\|\overrightarrow{\mathbf{P}_{i,j}\mathbf{P}_{i,j+1}}\right\|
$$

$$
\Delta q_{i-1} = \left\|\overrightarrow{\mathbf{P}_{i-1,j-1}\mathbf{P}_{i,j}}\right\| \quad \Delta q_i = \left\|\overrightarrow{\mathbf{P}_{i,j}\mathbf{P}_{i+1,j+1}}\right\| \quad \Delta s_{j-1} = \left\|\overrightarrow{\mathbf{P}_{i+1,j-1}\mathbf{P}_{i,j}}\right\| \quad \Delta s_j = \left\|\overrightarrow{\mathbf{P}_{i,j}\mathbf{P}_{i-1,j+1}}\right\|
$$

(6)

Box 3.

$$
A = \left(\begin{array}{ccc} \dfrac{2h}{\Delta s_j(\Delta s_{j-1} + \Delta s_j)} & \dfrac{2h}{\Delta w_j(\Delta w_{j-1} + \Delta w_j)} & \dfrac{2h}{\Delta q_i(\Delta q_{i-1} + \Delta q_i)} \\[12pt] \dfrac{2h}{\Delta u_{i-1}(\Delta u_{i-1} + \Delta u_i)} \quad \dfrac{1}{R_x} - \dfrac{2h}{\Delta u_{i-1}\Delta u_i} - \dfrac{2h}{\Delta w_{j-1}\Delta w_j} - \dfrac{2h}{\Delta q_{i-1}\Delta q_i} - \dfrac{2h}{\Delta s_{j-1}\Delta s_j} & \dfrac{2h}{\Delta u_i(\Delta u_{i-1} + \Delta u_i)} \\[12pt] \dfrac{2h}{\Delta q_{i-1}(\Delta q_{i-1} + \Delta q_i)} & \dfrac{2h}{\Delta w_{j-1}(\Delta w_{j-1} + \Delta w_j)} & \dfrac{2h}{\Delta s_{j-1}(\Delta s_{j-1} + \Delta s_j)} \end{array} \right)
$$

(7)

Since the local connectivity of cells in the proposed CNN model is formulated as Laplace operator and the cell state can be represented as a function of the cell output, Equation (2) may be written as:

$$
\frac{dv_{xij}}{dt} = h\nabla v_{xij} + I_{ij}
$$

(8)

It is not difficult to see that Equation (8) is similar to the well-known heat conduction equation described by:

$$
\frac{\partial T}{\partial t} = k\nabla T + q
$$

(9)

where T is the temperature, k is the thermal conductivity, and q is the heat rate.

By comparing Equation (8) with Equation (9), it can be easily seen that the proposed neural network is actually a discrete form of the continuous heat conduction in the space. The cell state v_x corresponds to the temperature T, the control coefficient h corresponds to the thermal conductivity k, the current source I corresponds to the heat rate q, and the local connectivity of cells corresponds to the local interaction of heat conduction. Therefore, the proposed neural network has the similar energy propagation behaviour as heat conduction.

Robot Movement

The proposed CNN model guarantees that the target activity can be propagated to the whole state space through local connectivity of cells, while the activities of the obstacles remain locally only. Therefore, the target globally influences the whole state space to attract the robot, while the obstacles have only local effect to avoid collisions. In addition, the activity propagation from the target is blocked when it hits the obstacles. Such a property is very important for a maze-solving type of problem.

The positions of the target and obstacles may vary with time. As shown in Equation (2), the activity of each cell dynamically changes according to the current source, the local connectivity of the cell and the states of the cell's neighbouring cells. As shown in Equations (3) and (4), the target corresponds to the current source, and the obstacles have a fixed state and the position change of an obstacle corresponds to the change of the state of a cell. Therefore, each cell can respond to the real-time changes of the target and obstacles, and there is no prior knowledge of the varying environment in the proposal model. The real-time path is generated from the dynamic activity landscape by a steepest ascent rule. For a given present position $\mathbf{P}_{i,j}$ in the state space of the neural network (i.e., a position in the Cartesian work space), assuming the corresponding cell is *(i, j)*, the next position $\mathbf{P}_{m,n}$ is obtained by:

$$\mathbf{P}_{m,n} \Leftarrow v_{ymn} = \max\{v_{ykl}, (k,l) \in N_r(i,j)\} \quad (10)$$

After the present position gets to its next position, the next position becomes a new present position. If the found next position is the same as the present position, the robot remains there without any movement. The present position adaptively changes according to the varying environment.

The dynamic activity of the topologically organized CNN is used to determine where the next robot position will be. However, when the next robot position is generated, it is determined by the robot moving speed. In a static environment, the CNN activity will reach a steady state. In most cases, the robot reaches the target earlier before the CNN activity reaches the steady state. When a robot is in a changing environment, the CNN activity will never reach the steady state. With the evolution of the CNN activity, the target always keeps remaining at the top and the obstacles always keep remaining at the bottom. The robot keeps moving toward the target with obstacle avoidance according to the steepest ascent rule till the designated objective is obtained.

With the harmonic local connectivity of cells and the heat conduction propagation behaviour of the target activity, the proposed methodology is able to generate optimal robot paths with obstacle avoidance in both static and dynamic environments. The generated path in a static environment is globally optimal in the sense of a shortest path from the starting position to the target, and the robot always reaches the target along a shortest path. In a dynamic environment, the optimality is in the sense that the robot travels a continuous and smooth route to the target.

SIMULATION STUDIES AND DISCUSSIONS

The proposed methodology for real-time robot path planning has been evaluated comprehensively in both static and dynamic environments. In a static environment, experiments have been conducted to examine the path generation with avoidance of concave U-shaped obstacles, and a maze-solving type of problem. In a dynamic environment, tests have been carried out to investigate robot path planning for tracking a moving target with avoidance of static obstacles, and tracking a static target or a moving target with avoidance of moving obstacles. In addition, the comparison with the existing neural network models is also discussed.

Path Planning in Static Environments

The obstacle avoidance for U-shaped obstacles has been achieved by the proposed methodology, while potential field based methods and other strictly local avoidance schemes cannot handle this type of problem. Figure 2 shows an example for robot path planning with avoidance of concave U-shaped obstacles. The CNN has 30×30 topologically organized cells. A set of concave U-shaped obstacles are represented by green solid squares. The model parameters are $E = 1$ and $h = 1$. The generated path is shown in Figure 2 by white hollow circles connected with white lines. A solid blue circle indicates that the robot reaches the target, and it also indicates the position of the target in a static environment. The start position of the robot is indicated by the first hollow circle, which is the farthest from the solid blue circle along the white line. It can be seen from Figure 2 that the generated path is a continuous and smooth route from the start point to the target with obstacle avoidance. The stable activity landscape of the neural network model is shown in Figure 3, where the peak remains at the target location and the bottom remains at the locations of the obstacles.

The solution to the maze-solving type of problem can be treated as a special case of the path planning problem, in which a mobile robot reaches the target from a given start position with obstacle avoidance. Figure 4 shows an example of the well-known beam robot competition micromouse maze, where a green solid square represents a typical quarter of the maze. The CNN model has the same size and model parameters

as the CNN model shown in Figure 2. The generated globally optimal solution is illustrated in Figure 4, where the robot path is indicated by white hollow circles connected with white lines.

It can be seen from these two examples that the proposed methodology does not suffer from undesired local minima, i.e., the robot will not be trapped in the situation with concave U-shaped obstacles or with complex maze-like obstacles.

Path Planning in Changing Environments

- **Moving target:** Real-time tracking of a moving target has been achieved by the proposed methodology. Experiments have been conducted by setting different robot moving speeds to investigate the effect of the relative moving speed between the target and the robot on the generation of the tracking path and obstacle avoidance when tracking a moving target. A CNN model with 40×40 cells and the same parameters as that in Figure 2 is chosen in these experiments. As shown in Figures 5-8, the moving target is displayed by a hollow green box and the path that the target moves along is displayed by the red line. The target starts from point (10, 10), moves along the path at a speed of 50 blocks/min, and stops at point (30, 30) after 40 steps. The initial position of the robot is at point (5, 5). The generated robot path is displayed by hollow circles. Figure 5 shows an example for tracking a moving target without any obstacles, in which the robot moves at a speed of 25 blocks/min and takes 22 steps to reach the moving target.

Figure 6 and Figure 7 illustrate two examples for real-time path generation by increasing the speed of the robot. The robot speed in Figure 6 is 40 blocks/min, which is faster than that in Figure 5 but still slower than the target speed. As shown

Figure 2. Path planning for tracking a static target with avoidance of concave U-shaped obstacles

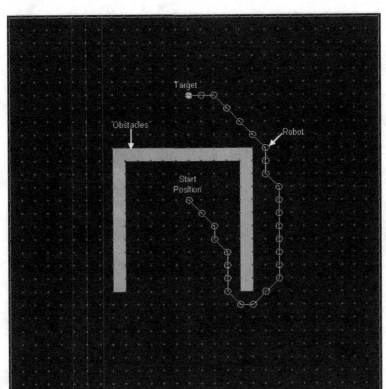

Figure 3. Stable activity landscape of the neural network model in Figure 2

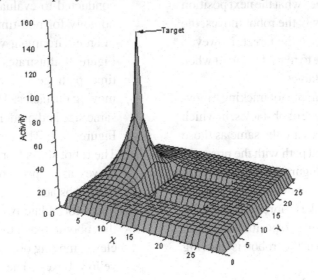

in Figure 6, the robot takes 30 steps to reach the target from the target's start position. The robot speed in Figure 7 is 62.5 blocks/min, which is faster than the target speed. It can be seen that the robot takes 6 steps to catch the target before the target goes through the path indicted by the red line. With respect to Figures 5-7, it can be seen that the robot moving slower than the target takes

Figure 4. Path planning with avoidance of maze-like obstacles

less steps to reach the target, since the robot has more time to "wait and see" what the next position is. Further, the more slowly the robot moves, the fewer steps it takes to reach the target. However, the robot spends less time to reach the target when it moves faster than the target.

Figure 8 shows an example for tracking a moving target with the existence of obstacles, in which the target and robot speeds are the same as those in Figure 6. The generated path with the presence of obstacles is shown in Figure 8. Compared with Figure 6, it can be seen that due to the existence of obstacles, the robot takes more steps to reach the target and it jumps up to two positions (as seen in the yellow box) when the robot is moving closely to the obstacles.

With reference to the above examples, it can be seen that the generated robot paths are continuous and smooth, and the robot's traveling path is generally shorter than the target's traveling path. The proposed CNN model also responds to obstacles sensitively.

• **Moving obstacles:** Experiments have been conducted to evaluate the proposed methodology for real-time robot path planning in an environment with moving obstacles. Figure 9 illustrates an example for real-time path generation with avoidance of moving obstacles. The CNN model has the same size and model parameters as those in Figures 5-8. The target is at point (14, 6). The robot starts from position (19, 32) and moves at a speed of 25 blocks/min. The static obstacles displayed by solid green boxes formulate two possible channels for the robot to reach the target. There are also eleven moving obstacles displayed by solid yellow boxes. The moving obstacles are initially located at positions from (9, 13) to (20, 14) to completely obstruct the left channel. They move towards the right channel at a speed of 20 blocks/min, and finally stop at positions from (19, 13) to (30, 14) to completely block the right channel and

Figure 5. Path planning for tacking a moving target when the robot moves at 25 blocks/min

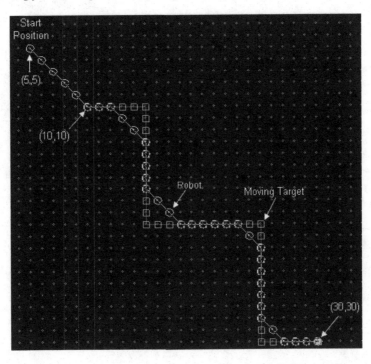

Figure 6. Path planning for tracking a moving target when the robot moves at 40 blocks/min

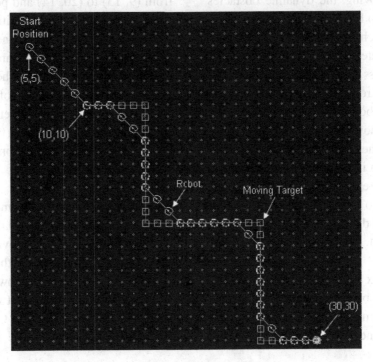

Figure 7. Path planning for tracking a moving target when the robot moves at 62.5 blocks/min

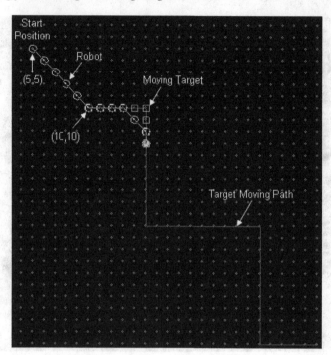

free the other. The robot initially moves to the target through the right channel since the left channel is blocked. While the robot starts moving, the dynamic obstacles are moving to block the right channel and leave the left channel open. The right channel is completely blocked before the robot is able to pass through the right channel, and thus the robot is turning and moving towards the bottom of the right channel and finally moving through the left channel to reach the target. The generated robot path is shown in Figure 9, where the robot reverses and reaches the target via the left channel due to a rapid change of the CNN activities and the direction of the CNN activity gradient.

A more complex case is the presence of continuously moving obstacles. Figure 10 shows an example for real-time generated path to reach a target with avoidance of continuously moving obstacles. The CNN model is the same as that in Figure 9 except that the obstacles are continuously moving back and forth between positions from (9, 13) to (20, 14) and positions from (19, 13) to (30, 14). Compared with Figure 9, the robot in Figure 10 is turning and moving towards the bottom of the right channel since the right channel is being closed, and then is turning again and moving towards the top of the right channel since the right channel is being opened again. Accordingly, there is a loop (the robot moves in the counterclockwise direction once in the loop) generated in the robot path.

- **Moving target and moving obstacles:** Real-time robot path planning with both a moving target and moving obstacles has also been achieved by the proposed methodology. Figure 11 shows an example for tacking a moving target with avoidance of moving obstacles. The CNN model is the same as that in Figure 10 except for the

Figure 8. Path planning for tracking a moving target with avoidance of static obstacles

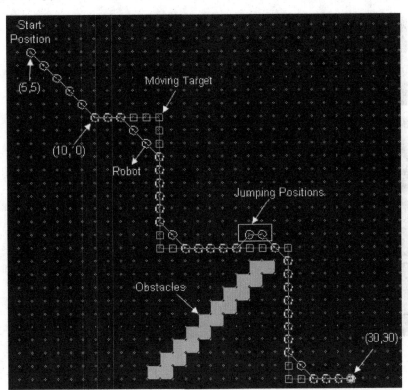

moving target. The target starts from position (9, 6) and moves back and forth between position (9, 6) and position (29, 6) at the same speed (20 blocks/min) as the moving obstacles (shown in Fig. 11 by hollow green boxes). The real-time generated robot path is shown in Figure 11. Although there still is a turning loop generated in the generated path since the right channel is being closed and then opened again, the turning loop becomes smaller since the target is remaining around the right channel. After passing through the right channel, the robot moves in a zigzag fashion to reach the target. It can be also seen that the proposed CNN model responds to the real-time movements of the target and obstacles without the prior knowledge of the varying environment.

Comparison Analysis

The fundamental difference of the proposed methodology from the existing neural network methods for robot path planning is that the proposed methodology incorporates the elegant properties of harmonic functions in neural networks to carry out robot path planning in real time. Besides the advantages of the existing neural network models, such as real-time path generation without learning procedures, prior knowledge of target or barrier movements, optimizations of any cost functions, and explicitly searching over the free work space or collision paths, the proposed neural network model also inherits the advantages of harmonic functions, such as the avoidance of local minima, smooth and optimal path generation, and sensitive respond to the sudden changes in dynamic environments.

Figure 9. Path planning for tracking a static target with avoidance of moving obstacles

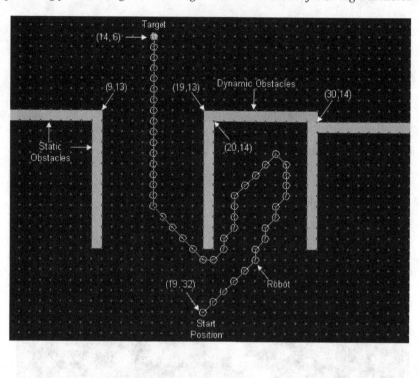

Figure 10. Path planning for tracking a static target with avoidance of back and forth moving obstacles

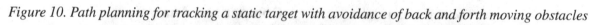

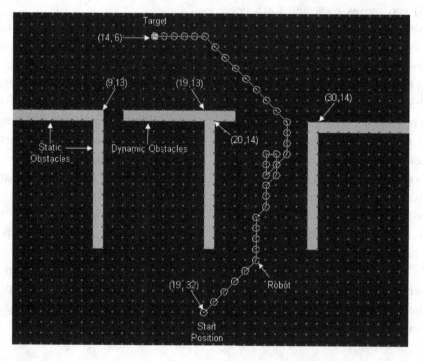

Figure 11. Path planning for tracking a moving target with avoidance of back and forth moving obstacles

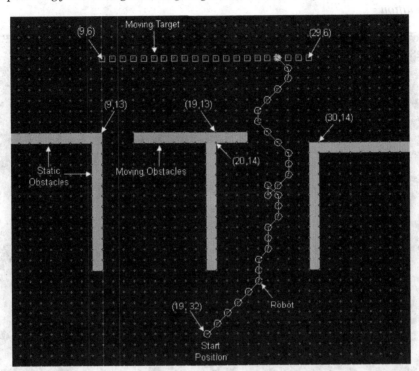

The performance of the proposed neural network model has been evaluated in comparison with the shunting equation (Yang & Meng, 2003; Yuan & Yang, 2003; Bhattacharya & Talapatra, 2005) for robot path planning. The local connectivity of neurons is a nonlinear harmonic function in the proposed model, while the local connectivity of neurons is a linear distance function in the shunting equation. The neighbourhood of a neuron in the proposed model is simply defined as the locally and naturally connected neurons, while the neighbourhood of a neuron in the shunting equation is defined by a given Euclidean distance and the neighbourhood definition is more complex in the case of an asymmetric grid map. The proposed model has an appropriate physical meaning, while the shunting equation does not have an explicit physical meaning.

Although both methods can generate collision-free trajectories in both static and dynamic environments, the shunting equation is less likely

to find the optimal path. Considering the path planning problem as shown in Figure 2, the path generated by the shunting equation is shown in Figure 12. Comparing Figure 2 with Figure 12, it can be seen that there is a jump point on the path generated by the shunting equation (marked by the red circle), although the robot is not too close to the obstacles. In addition, the proposed model solves the local minima problem by using the theory of harmonic functions, while there is a lack of theoretical evidence in the shunting equation to support the avoidance of local minima.

Further, the shunting equation is less sensitive to respond to a variable environment than the proposed neural network model. Let us consider the shunting equation for the path planning problem as shown in Figure 10. The path generated by the shunting equation is shown in Figure 13. As shown in Figure 13, although there is a turning loop (the robot moves in a counterclockwise direction once in the loop) since the right channel is

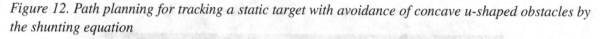

Figure 12. Path planning for tracking a static target with avoidance of concave u-shaped obstacles by the shunting equation

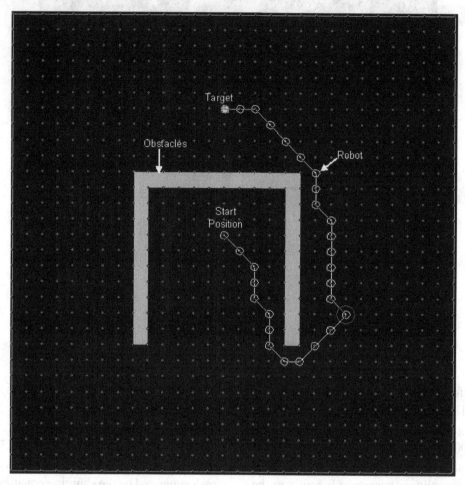

being closed, the robot does not move forward to the top of the right channel any more even if the right channel is being opened again, and finally it is going through the left channel to reach the target. In contrast, the robot in Figure 10 makes a turning loop and then moves forward to the top of the right channel since the right channel is being closed and then opened again, and finally it is going through the right channel to reach the target. Comparing Figure 10 with Figure 13, it can be seen that the proposed neural network model is more efficient to find the optimal robot path than the shunting equation, and the proposed model responds to the variable environment more sensitively than the shunting equation.

In addition, the shunting equation has a set of parameters, and the generated paths depend on the relationships between the model parameters. For example, with the above two examples, if the parameters are not set appropriately, the shunting equation may not even find a path to reach the target. In contrast, the proposed neural network model only has two parameters, i.e., the current source I and the control coefficient h. The proposed model is also independent of these two parameters, which means that the two parameters can be chosen as arbitrary positive constants and they do not affect the generated paths. For example, with the above examples, the generated robot paths do

Figure 13. Path planning for tracking a static target with avoidance of continuously moving obstacles by the shunting equation

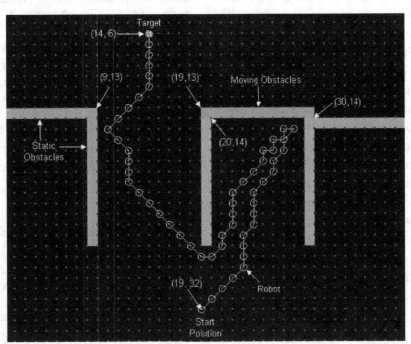

not change whenever both the parameters are set to an arbitrary positive constant.

It is noted that the formulation for the local connectivity of neurons in the proposed model is more complex than that in the shunting equation. Similar to the shunting equation, the neural connections are only related to the grid map and do not change with variations of the target and obstacles. Under a given grid map, the extra computational load on the formulation of neural connections can be avoided by the pre-computation of neural connections. Accordingly, the formulation for the local connectivity of neurons in the proposed model only increases the pre-computational time, and the real-time computational complexity linearly depends on the size of the neural network, which is the same as the shunting equation. The computational performance can be further improved by digital hardware implementation (Chua & Yang, 1988).

CONCLUSION

This paper presents a new cellular neural network based methodology for optimal robot path generation with obstacle avoidance in an arbitrarily dynamic environment. The contribution of the paper is that a cellular neural network model is developed by incorporating the elegant properties of harmonic functions in a neural system for real-time path generation in both static and dynamic environments. The local connectivity of neurons in the proposed neural network model is a nonlinear harmonic function rather than a linear distance function in the existing neural network models. Similar to the existing neural network models, the proposed neural network model works in real time and does not require prior knowledge of target or barrier movements, learning procedures, and explicit searching over the global work space or searching collision paths. On the other hand, with the properties inherited from harmonic functions, the proposed model cannot only generate smooth and optimal robot paths, but it can also easily respond to the real-time changes in dynamic environments, as well as is suitable for robot path planning with safety consideration. Further, the proposed neural network model is parameter-independent and has an appropriate physical meaning.

Future research work will focus on the extension of the proposed methodology to the complex path planning problem such as the path generation of multiple robots for tracking multiple moving targets and the path generation for complete coverage of a working environment. The proposed neural network model will be extended to simultaneous tracking of multiple targets by setting multiple current sources to multiple targets. It will also be extended to the path generation for complete coverage of a working environment.

ACKNOWLEDGMENT

This research is supported by Australian Research Council (ARC) Linkage grant (ARC Linkage: LP0668052).

REFERENCES

Barraquand, J., & Latombe, J.-C. (1991). Robot motion planning: A distributed representation approach. *The International Journal of Robotics Research*, *10*(6), 628–649. doi:10.1177/027836499101000604.

Bhattacharya, S., & Talapatra, S. (2005). Robot motion planning using neural networks: a modified theory. *International Journal of Lateral Computing*, *2*(1), 9–13.

Chua, L. O., & Yang, L. (1988). Cellular neural network: Theory. *IEEE Transactions on Circuits and Systems*, *35*(10), 1257–1272. doi:10.1109/31.7600.

Connolly, C. I. (1993). The application of harmonic functions to robotics. *Journal of Robotic Systems*, *10*(7), 931–946. doi:10.1002/rob.4620100704.

Connolly, C. I. (1994). Harmonic functions and collision probabilities. In *Proceedings of the IEEE International Conference on Robotics and Automation*, San Diego (pp. 3015-3019).

Connolly, C. I., Burns, J. B., & Weiss, R. (1990). Path planning using Laplace's equation. In *Proceedings of the IEEE International Conference on Robotics and Automation*, Cincinnati, OH (pp. 2101-2106).

Glasius, R., Komada, A., & Gielen, S. (1994). Population coding in a neural net for trajectory formation. *Network (Bristol, England)*, *5*(4), 549–563. doi:10.1088/0954-898X/5/4/007.

Glasius, R., Komoda, A., & Gielen, S. (1996). A biologically inspired neural net for trajectory formation and obstacle avoidance. *Biological Cybernetics*, *74*(6), 511–520. doi:10.1007/BF00209422.

Glasius, R., Komoda, A., & Gielen, S. C. A. M. (1995). Neural network dynamics for path planning and obstacle avoidance. *Neural Networks*, *8*(1), 125–133. doi:10.1016/0893-6080(94)E0045-M.

Grossberg, S. (1988). Nonlinear neural networks: principles, mechanisms, and architectures. *Neural Networks*, *1*(1), 17–61. doi:10.1016/0893-6080(88)90021-4.

Henrich, D. (1997). Fast motion planning by parallel processing-a review. *Journal of Intelligent & Robotic Systems*, *20*(1), 45–69. doi:10.1023/A:1007948727999.

Hodgkin, A. L., & Huxley, A. F. (1952). A quantitative description of membrane current and its application to conduction and excitation in nerve. *The Journal of Physiology*, *117*(4), 500–544.

Hwang, Y. K., & Ahuja, N. (1992). Gross motion planning-a survey. *ACM Computing Surveys*, *24*(3), 219–291. doi:10.1145/136035.136037.

Kanaya, M., & Tanaka, M. (1998). Path planning method for multi-robots using a cellular neural network. *Electronics and Communications in Japan-Part 3, 81*(3), 335-345.

Karniadakis, G. E., & Kirby, R. M. II. (2003). *Parallel scientific computing in C++ and MPI: a seamless approach to parallel algorithms and their implementation*. New York: Cambridge University Press.

Khatib, O. (1986). Real-time obstacle avoidance for manipulators and mobile robots. *The International Journal of Robotics Research, 5*(1), 90–98. doi:10.1177/027836498600500106.

Kim, H., Son, H., Roska, T., & Chua, L. O. (2002). Optimal path finding with space- and time-variant metric weights via multi-layer CNN. *International Journal of Circuit Theory and Applications, 30*(2-3), 247–270. doi:10.1002/cta.199.

Kohonen, T. (1982). Self-organized formation of topologically correct feature maps. *Biological Cybernetics, 43*(1), 59–69. doi:10.1007/BF00337288.

Latombe, J. C. (1991). *Robot motio planning*. Boston: Kluwer.

Lebedev, D. V., Steil, J. J., & Ritter, H. J. (2005). The dynamic wave expansion neural network model for robot motion planning in time-varying environment. *Neural Networks, 18*(3), 267–285. doi:10.1016/j.neunet.2005.01.004.

Li, Z. X., & Bui, T. D. (1998). Robot path planning using fluid model. *Journal of Intelligent & Robotic Systems, 21*(1), 29–50. doi:10.1023/A:1007963408438.

Lozano-Perez, T., & Wesley, M. A. (1997). An algorithm for planning collision-free paths among polyhedral obstacles. *Communications of the ACM, 22*(10), 560–570. doi:10.1145/359156.359164.

Ong, C. J., & Gilbert, E. G. (1998). Robot path planning with penetration growth distance. *Journal of Robotic Systems, 15*(2), 57–74. doi:10.1002/(SICI)1097-4563(199802)15:2<57::AID-ROB1>3.0.CO;2-R.

Oriolo, G., Ulivi, G., & Vendittelli, M. (1998). Real-time map building and navigation for autonomous robots in unknown environments. *IEEE Transactions on Systems, Man, and Cybernetics. Part B, Cybernetics, 28*(3), 316–333. doi:10.1109/3477.678626.

Pashkevich, A., Kazheunikau, M., & Ruano, A. E. (2006). Neural network approach to collision free path-planning for robotic manipulators. *International Journal of Systems Science, 37*(8), 555–564. doi:10.1080/00207720600783884.

Quoy, M., Moga, S., & Gaussier, P. (2003). Dynamical neural networks for planning and low-level robot control. *IEEE Transactions on Systems, Man, and Cybernetics. Part A, Systems and Humans, 33*(4), 523–532. doi:10.1109/TSMCA.2003.809224.

Reljin, B., Krstic, I., Kostic, P., Reljin, I., & Kandic, D. (2004). CNN applications in modelling and solving non-electrical problems. In Slavova, A., & Mladenov, V. (Eds.), *Cellular neural networks: Theory and Applications* (pp. 135–172). New York: Nova Science Publishers.

Roska, T., Chua, L. O., Wolf, D., Kozek, T., Tetzlaff, R., & Puffer, F. (1995). Simulating nonlinear waves and partial differential equations via CNN-Part I: basic techniques. *IEEE Transactions on Circuits and Systems, 42*(10), 807–815. doi:10.1109/81.473590.

Sadati, S. H., Alipour, K., & Behroozi, M. (2008). A combination of neural network and ritz method for robust motion planning of mobile robots along calculated modular paths. *International Journal of Robotics and Automation, 23*(3), 187–198.

Seshadri, C., & Ghosh, A. (1993). Optimum path planning for robot manipulators amid static and dynamic obstacles. *IEEE Transactions on Systems, Man, and Cybernetics*, *23*(2), 576–584. doi:10.1109/21.229471.

Svestka, P., & Overmars, M. H. (1997). Motion planning for carlike robots using a probabilistic approach. *The International Journal of Robotics Research*, *16*(2), 119–145. doi:10.1177/027836499701600201.

Versteeg, H. K., & Malalasekera, W. (1995). *An introduction to computational fluid dynamics: the finite volume method*. Harlow, UK: Longman Scientific & Technical.

Willms, A. R., & Yang, S. X. (2006). An efficient dynamic system for real-time robot-path planning. *IEEE Transactions on Systems, Man, and Cybernetics. Part B, Cybernetics*, *36*(4), 755–766. doi:10.1109/TSMCB.2005.862724.

Yang, S. X., & Meng, M. (2003). Real-time collision-free motion planning of a mobile robot using a neural dynamics-based approach. *IEEE Transactions on Neural Networks*, *14*(6), 1541–1552. doi:10.1109/TNN.2003.820618.

Yuan, X., & Yang, S. X. (2003). Virtual assembly with biologically inspired intelligence. *IEEE Transactions on Systems, Man and Cybernetics. Part C, Applications and Reviews*, *33*(2), 159–167. doi:10.1109/TSMCC.2003.813148.

Zalama, E., Gaudiano, P., & Coronado, J. L. (1995). A real-time, unsupervised neural network for the low-level control of a mobile robot in a nonstationary environment. *Neural Networks*, *8*(1), 103–123. doi:10.1016/0893-6080(94)00063-R.

Zalama, E., Gomez, J., Paul, M., & Peran, J. R. (2002). Adaptive behavior navigation of a mobile robot. *IEEE Transactions on Systems, Man, and Cybernetics. Part A, Systems and Humans*, *32*(1), 160–169. doi:10.1109/3468.995537.

Zelinsky, A. (1994). Using path transforms to guide the search for findpath in 2D. *The International Journal of Robotics Research*, *13*(4), 315–325. doi:10.1177/027836499401300403.

This work was previously published in the International Journal of Intelligent Mechatronics and Robotics, Volume 1, Issue 1, edited by Bijan Shirinzadeh, pp. 20-39, copyright 2011 by IGI Publishing (an imprint of IGI Global).

Chapter 3
Impedance Control of a Spherical Parallel Platform

Luca Carbonari
Polytechnic University of Marche, Italy

Luca Bruzzone
University of Genova, Italy

Massimo Callegari
Polytechnic University of Marche, Italy

ABSTRACT

This article describes the impedance control of an in-parallel actuated orientation platform. The algorithm is based on a representation of platform orientation which exploits the equivalent axis of rotation: this approach is more intuitive and easier to visualize than conventional methods based on Cardan or Euler angles. Moreover, since for small angular displacements the Mozzi's axis lies very close to angular velocity, impedance control algorithms based on such representation provides better performances and smoother motions. Results of numerical simulations and experimental tests are shown and commented with reference to the spherical parallel machine.

INTRODUCTION

Parallel kinematics machines, PKMs, are now used in many different application fields, both in industrial environments and in advanced robotics developments. Such machines are given by a moving platform that is actuated in-parallel by several limbs, each one driven by actuators that are fixed with the ground.

On the other hand, due to the complex kinematics of 6-dof full mobility PKMs, research community is addressing the studies towards simpler

DOI: 10.4018/978-1-4666-3634-7.ch003

mechanisms that are able to perform elementary motions, like pure translations, pure rotations or even planar displacements. In fact some important applications may well be accomplished by these kinds of reduced mobility PKMs, while in other cases hybrid machines can be designed (e.g., a conventional "serial" wrist on top of a parallel shoulder) or mini-maxi architectures can be experimented.

Alternatively, a full-mobility task may be decomposed into elemental sub-tasks, to be performed by separate minor mobility machines, as already done in conventional machining operations (Callegari & Suardi, 2004). In this case a proper mechatronic design allows exploiting, at least partially, the advantages of both architectures, while the disadvantages can be minimized. In this way it is possible to realize hybrid cooperative systems with many degrees of freedom, leading to a modular and reconfigurable system architecture.

This is the case of the research being developed at the Department of Mechanics of the Polytechnic University of Marche, which aims at assessing the feasibility of complex assembly tasks (e.g., 6 axes operations) by means of two cooperating parallel robots, both characterized by a simple mechanical and control architecture. The two machines are based on the same 3-CPU kinematic topology but are capable of pure translation and pure rotation motions respectively. The control systems of the two machines will be equipped with an impedance controller, so that the relative stiffness of the system can be varied during parts' mating to allow an effective accomplishment of the task; on the other hand, the complexity of the hybrid position/force algorithms (needing proper force sensors and the availability of real time robots' inverse dynamics models) is avoided.

The present article describes the studies performed at the University of Genova and at the Polytechnic University of Marche in order to develop a well performing impedance control for the orientation platform: it is anticipated that few realizations are reported in scientific literature about the control of the stiffness for machines of pure rotation, let alone applied to parallel robots. This is the reason that addressed Bruzzone and Molfino (2006) to introduce the application of the natural invariants of the rotation matrix to the impedance control of PKM's. Bruzzone and Callegari (2010) studied the control of the present robot by computer simulation; the present article presents the experimental results that are now available after its implementation on the prototype (see Figure 1).

REPRESENTATION OF ORIENTATION

There are many methods to describe the orientation of a rigid body, each of them having pros and cons that are well known to the research community since a long time (Tandirci et al., 1992; Bonev & Ryu, 2001; Bonev et al., 2002; Selig, 2005).

The most commonly used method in robotics is making reference to three rotations around the axes of the fixed or mobile frame: the order of

Figure 1. The spherical parallel machine used in the experiments

the rotations is relevant but, on the other hand, the rotations around the axes of the mobile frame provide the same final orientation of the same rotations, performed in the reverse order, around the axes of the fixed frame. A total number of 12 different angles set can be found: 6 sets, that are called *Cardan angles*, are obtained through rotations around the *x, y* and *z* axes, taken in any possible order; if the rotation around the same axis is repeated twice (not consecutively), 6 more sets of angles are obtained, which are called *Euler angles*. The drawbacks of such method are well known too:

- There are 12 different sets of angles among which select one, and such choice influences the resulting equations; furthermore, in case the orientation of the robot has to be controlled in the task space, the control algorithm is arbitrarily affected by such choice;
- Each angle set has representation singularities that correspond to task space orientations where it is not possible to describe robot's attitude orientation or angular velocity.

An alternative to the use of the Euler or Cardan angles is to describe the rotation of a rigid body by an axis of rotation (identified by a unit vector **e**) and the amount of the rotation ϑ about this axis according to the right hand rule (Craig, 2005). The vector **e** and the angle ϑ represent the natural invariants of the rotation matrix that describes the orientation of the body and can be gathered into the *equivalent angle-axis* vector E:

$$\mathbf{E} = \vartheta \mathbf{e} \tag{1}$$

which provides a minimal representation of orientation and is independent of the reference frame. The advantages of this approach derive from the

strict relation of the vector E with the angular velocity ω (Angeles, 2007)

$$\omega = \sin \vartheta \dot{e} + \left(1 - \cos \vartheta\right) e \times \dot{e} + \dot{\vartheta} e \tag{2}$$

In particular, it is easy to demonstrate that when the angle ϑ tends to 0, the angular velocity vector tends to the time derivative of the natural invariant:

$$\lim_{\vartheta \to 0} \omega = \vartheta \dot{e} + \dot{\vartheta} e = \frac{d}{dt} E \tag{3}$$

Such close relationship can also be seen by expressing the rotation matrix R as a function of the angle-axis parameters as follows:

$$R = e^{\vartheta \begin{bmatrix} 0 & -e_z & e_y \\ e_z & 0 & -e_x \\ -e_y & e_x & 0 \end{bmatrix}} \tag{4}$$

where the matrix at the exponent has the same structure of the angular velocity matrix.

The mentioned properties of this representation are important when the attitude of a body must be controlled in 3D space: in fact in this case both the information of orientation and angular velocity are used to plan the motion and smoother trajectories can therefore be obtained.

Caccavale et al. (1999) studied several methods for the impedance control of the attitude, including the screw axis representation, but developed further only the unit quaternion technique; Bruzzone and Molfino (2006) proposed the use of the equivalent angle-axis of rotation for the impedance control of robots while Bruzzone and Callegari (2010) showed by simulation that control algorithms based on this representation provide better performances: in fact in this case the information of orientation and angular velocity are used to compute the stiffness and damping forces respectively.

The present article applies such algorithms for the control of a 3 dof's spherical parallel machine and a comparison is made between the simulation results and the experimental tests.

ARCHITECTURE AND KINEMATICS OF THE ROBOT

The 3-CPU spherical parallel machine under study is made of three identical serial chains connecting the moving platform to the fixed base, as shown in Figure 2; each leg is composed by two links: the first one is connected to the frame by a cylindrical joint (C) while the second link is connected to the first one by a prismatic joint (P) and to the end-effector by a universal joint (U). In order to constraint the end effector to a spherical motion, the axes of the joints must respect a few (geometrical) *manufacturing conditions*:

- The axes of the cylindrical joints $(\mathbf{a}_i, i = 1,2,3)$ are aligned on the x, y, z axes of a reference base frame and intersect at the centre O of the spherical motion;
- The axis $\mathbf{b}_i$ of each prismatic pair is perpendicular to the axis of the respective cylindrical joint $\mathbf{a}_i$;

- The first axis of each universal joint is perpendicular to the plane of the corresponding leg (plane identified by the axes $\mathbf{a}_i$ and $\mathbf{b}_i$);
- The second axis of the 3 universal joints are aligned along the v, w, u axes of a local frame centered in P and attached to the mobile platform.

For a successful operation of the mechanism, a *mounting condition* must be satisfied too: assembly should be operated in such a way that the two frames $O(x,y,z)$ and $P(u,v,w)$ come to coincide.

If these conditions are verified, the points P and O remain coincident during the motion and the moving platform performs a spherical motion. In the initial configuration the three displacements a_i are equal to the length of the second link c and the displacements of the prismatic joints b_i are equal to the constant distance d.

The platform is actuated by driving the strokes of the 3 cylindrical joints, therefore joint space displacements are gathered into the following vector q:

$$\mathbf{q} = \begin{bmatrix} a_1 \\ a_2 \\ a_3 \end{bmatrix} \qquad (5)$$

Figure 2. Kinematic schemes of the 3-CPU robot (a) and geometry of the legs (b)

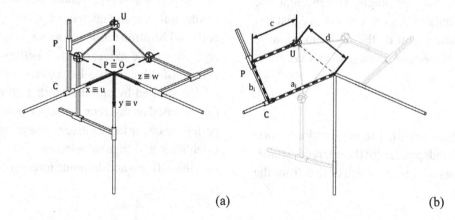

(a)   (b)

Box 1.

$$
{}_P^O R(\alpha, \beta, \gamma) = R_x(\alpha) R_y(\beta) R_z(\gamma) = \begin{bmatrix} c\beta c\gamma & -c\beta s\gamma & s\beta \\ s\alpha s\beta c\gamma + c\alpha s\gamma & -s\alpha s\beta s\gamma + c\alpha c\gamma & -s\alpha c\beta \\ -c\alpha s\beta c\gamma + s\alpha s\gamma & c\alpha s\beta s\gamma + s\alpha c\gamma & c\alpha c\beta \end{bmatrix} \tag{6}
$$

The position kinematics of the robot expresses the relation between the orientation of the and the displacements of the actuators; the attitude of the mobile platform machine in space is fully provided by the rotation matrix ${}_P^O R$, that can also be conveniently expressed as a composition of elemental rotations. In the development of robot's kinematics, the Cardan angles set displayed in Box 1 is used.

Callegari (2008) worked out the following resolving system for position kinematics:

$$
\begin{cases} r_{12} = -c\beta s\gamma = \dfrac{a_1 - c}{d} \\ r_{23} = -s\alpha c\beta = \dfrac{a_2 - c}{d} \\ r_{31} = -c\alpha s\beta c\gamma + s\alpha s\gamma = \dfrac{a_3 - c}{d} \end{cases} \tag{7}
$$

where r_{ij} is the element at the ith row and j^{th} column of rotation matrix ${}_P^O R$. The solution of the direct position kinematics (DPK) problem for the robot

Table 1. Main data used by the numerical model

Geometrical data			Mass data		
c	210	mm	slider	7,15	kg
d	490	mm	link 1	1,90	kg
h	280	mm	link 2	2,21	kg
$a_{i\,min}$	319	mm	platform	11,73	kg
$a_{i\,max}$	661	mm			
$b_{i\,min}$	130	mm			
$b_{i\,max}$	210	mm			

requires the computation of the rotation matrix ${}_P^O R$ as a function of internal coordinates q: appendix A shows how such relation can be derived starting from (7) and by taking advantage of the mentioned Cardan angles set (Gabrielli, 2009). According to Innocenti and Parenti-Castelli (1993), a maximum number of 8 different configurations can be worked out; however, a single feasible solution is found when the real workspace of the robot is considered, i.e., the actual mobility of the joints is introduced.

Once the rotation matrix is known, it is easy to translate this orientation information in terms of the natural invariant $\mathbf{E}$:

$$
\theta = \cos^{-1}\left(\frac{r_{11} + r_{22} + r_{33} - 1}{2}\right) \tag{8a}
$$

$$
e = \frac{1}{2\sin\theta}\begin{bmatrix} r_{32} - r_{23} \\ r_{13} - r_{31} \\ r_{21} - r_{12} \end{bmatrix} \tag{8b}
$$

Of course, inverse position kinematic (IPK) problem admits just one solution and it is easily solved by working out joint displacements q in (7).

Turning to differential kinematics, the expression of the analytic Jacobian $\mathbf{J}_A$ is obtained as a function of the Cardan angles and their rates (see Box 2).

By taking into account the following relation (10) between the derivatives of the Cardan angles and the angular velocity ω, the geometric Jacobian $\mathbf{J}_G$ is obtained too:

Table 2.

Motors properties			
M_s	2,95	kg	*Stator mass*
K_t	58		*Torque constant*
I_n	3	A	*Nominal supply current*
T_n	184	N	*Nominal thrust*
v_n	6		*Nominal speed*

$$\begin{bmatrix} \omega_x \\ \omega_y \\ \omega_z \end{bmatrix} = \begin{bmatrix} 1 & 0 & s\beta \\ 0 & c\alpha & -s\alpha c\beta \\ 0 & s\alpha & c\alpha c\beta \end{bmatrix} \begin{bmatrix} \dot{\alpha} \\ \dot{\beta} \\ \dot{\gamma} \end{bmatrix} = T \begin{bmatrix} \dot{\alpha} \\ \dot{\beta} \\ \dot{\gamma} \end{bmatrix} \qquad (10)$$

$$\begin{bmatrix} \dot{a}_1 \\ \dot{a}_2 \\ \dot{a}_3 \end{bmatrix} = J_A \begin{bmatrix} \dot{\alpha} \\ \dot{\beta} \\ \dot{\gamma} \end{bmatrix} = J_A T^{-1} \begin{bmatrix} \omega_x \\ \omega_y \\ \omega_z \end{bmatrix} = J_G \begin{bmatrix} \omega_x \\ \omega_y \\ \omega_z \end{bmatrix} \qquad (11)$$

with Equation 12 in Box 3.

It has been noted already by Bruzzone and Callegari (2010) that the use of different sets of external coordinates affects the implementation of the impedance control law and yields different system behaviors. In this work the tree components of the vector $\mathbf{E}_P$, natural invariant of the rotation

matrix ${}^O_P\mathbf{R}$, are assumed as set of external coordinates; therefore, the analytical Jacobian matrix assumes in this case a different shape and relates the vector $\dot{\mathbf{q}}$ to the time derivative of $\mathbf{E}_P$:

$$\frac{d}{dt}\mathbf{q} = \mathbf{J}_E \frac{d}{dt}\mathbf{E}_P \qquad (13)$$

Bruzzone and Molfino (2006) demonstrated that the matrices $\mathbf{J}_E$ and $\mathbf{J}_G$ are equal in the case of small angles of rotation ϑ :

$$\lim_{\vartheta \to 0} \mathbf{J}_G = \lim_{\vartheta \to 0} \mathbf{J}_E \qquad (14)$$

In order to better define the spatial compliance and to be able to easily visualize and understand the assigned tasks and the obtained results, it is necessary to choose a different set of reference frames, as shown in Figure 3. The fixed frame $O*$ has been defined as follows:

- The origin is located at the centre of the moving platform when it assumes its initial configuration (on the plane identified by the three points B_1, B_2 and B_3).

Box 2.

$$\begin{bmatrix} \dot{a}_1 \\ \dot{a}_2 \\ \dot{a}_3 \end{bmatrix} = d \cdot \begin{bmatrix} 0 & -s\beta s\gamma & c\beta c\gamma \\ c\alpha c\beta & -s\alpha s\beta & 0 \\ -s\alpha s\beta c\gamma - c\alpha s\gamma & c\alpha c\beta c\gamma & -c\alpha s\beta s\gamma - s\alpha c\gamma \end{bmatrix} \begin{bmatrix} \dot{\alpha} \\ \dot{\beta} \\ \dot{\gamma} \end{bmatrix} = J_A \begin{bmatrix} \dot{\alpha} \\ \dot{\beta} \\ \dot{\gamma} \end{bmatrix} \qquad (9)$$

Box 3.

$$J_G = d \cdot \begin{bmatrix} 0 & -c\alpha s\beta s\gamma - s\alpha c\gamma & c\alpha c\gamma - s\alpha s\beta s\gamma \\ c\alpha c\beta & 0 & -s\beta \\ -s\alpha s\beta c\gamma - c\alpha s\gamma & c\beta c\gamma & 0 \end{bmatrix} \qquad (12)$$

Table 3. Steady state values of the Cardan angles

	Angle [deg]	φ_x	φ_y	φ_z
Test number	test 1	-0,347	-6,242	0,203
	test 2	-0,203	-5,489	0,050
	test 3	-0,188	-4,857	0,075
	test 4	-0,332	-5,968	0,190
	test 5	-0,259	-6,184	0,080

- The z axis has opposite direction to the vector $\mathbf{g}$ of gravity acceleration.
- The x axis lies on the plane identified by the three points B_1, B_2 and B_3, directed as the axis of the cylindrical joint of the first leg $\mathbf{a}_1$.
- The y axis is determined by the first two to obtain a right-handed triad.

The frame attached to the end-effector P^* is coincident to the fixed frame when the robot is in its initial configuration. Of course, the kinematics of the robot has been rewritten according to the newly defined frames.

Once the location of the new frame O^* has been defined by means of the ${}^O_O\mathbf{R}$ rotation matrix, the orientation of the mobile platform can be described in the new frames by:

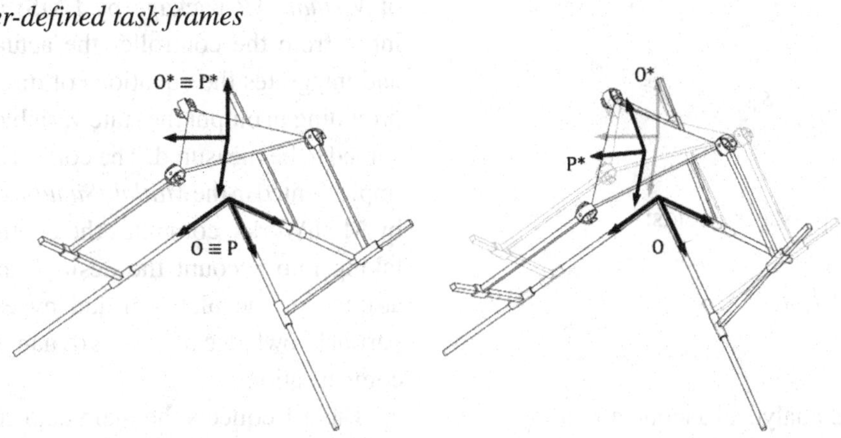

$$
{}^{O^*}_{P^*}\mathbf{R} = {}^O_{O^*}\mathbf{R}^T\, {}^O_P\mathbf{R}\, {}^O_{O^*}\mathbf{R} \tag{15}
$$

where it has been used the identity ${}^O_{O^*}\mathbf{R} = {}^P_{P^*}\mathbf{R}$. Of course, having changed the mobile and fixed frames, also the Cardan angles $\varphi_x, \varphi_y, \varphi_z$ that yield the rotation matrix ${}^{O^*}_{P^*}\mathbf{R}$ are different from the previously described set (α, β, γ):

$$
{}^{O^*}_{P^*}R\left(\varphi_x, \varphi_y, \varphi_z\right) = R_{x^*}\left(\varphi_x\right) R_{y^*}\left(\varphi_y\right) R_{z^*}\left(\varphi_z\right) \tag{16}
$$

Henceforth these angles are used to describe the orientation of the manipulator and to assign the stiffness and damping of the mobile platform; since they are assumed as external coordinates for the computation of the differential kinematics, the analytic and the geometric Jacobians are worked out again as previously described, providing similar but more complex relations.

CONTROL ALGORITHM

When an impedance control law is used, the end effector, if subject to external forces, moves according to a predefined spatial compliance given by a *stiffness matrix* $\mathbf{K}$ and a *damping matrix* $\mathbf{C}$ (Hogan, 1985; Siciliano & Villani, 2000; Park &

Figure 3. User-defined task frames

Lee, 2004). When a reference configuration, described by frame D, is assigned, it represents a *virtual equilibrium state*; in fact it corresponds to an equilibrium configuration only if no force is exerted by the environment.

Using the virtual work principle and Equations (11) and (14), the impedance control law can be expressed as:

$$f = J_G^{-T}\left[K_r\left(E_D - E_P\right) + C_r\left(\omega_D - \omega_P\right)\right] + f_g = f_{imp} + f_g \tag{17}$$

where:

- f is the vector of actuation forces along the axis a_1, a_2, a_3 of the three cylindrical joints;
- J_G is the geometrical Jacobian matrix;
- E_D and E_P define respectively the desired and the actual orientation of the manipulator;
- ω_D and ω_P are the desired and the actual angular velocities;
- K_r and C_r are the rotational stiffness and damping;
- f_g is the vector of actuation efforts needed to compensate the gravity force acting on each member of the robot.

The vector f_g is computed using once again the virtual works principle (Tsai, 2000; Yang, 2010):

$$f_g^T \delta q = -\sum_{i=1}^{n}\left(m_i g\right)^T \delta x_i \tag{18}$$

that, after a few passages, yields:

$$f_g = -J_A^{-T}\sum_i m_i J_i^T g \tag{19}$$

where J_A is the analytical Jacobian matrix, m_i is the mass of the *i-th* member, J_i is the jacobian

matrix that links the velocity of the centre of gravity of the *i-th* member to the vector $\dot{q}$; g is the gravity acceleration.

The stiffness (K_r) and damping (C_r) matrices in general are not diagonal. However, in order to simplify the comprehension of their geometrical meaning, it is possible to use a task frame (T) in which they are diagonal; of course the orientation ${}^O_T R$ of the task frame with respect to the reference frame of the robot is known. The value assumed by K_r and C_r in the reference frame O is readily computed as:

$$K_r = {}^O_T R\,{}^T K_r\,{}^O_T R^T \tag{20}$$

$$C_r = {}^O_T R\,{}^T C_r\,{}^O_T R^T \tag{21}$$

Figure 4 shows a scheme of the control loop that has been implemented for the 3-CPU spherical robot.

TEST RIG

The control law (17) has been tested by running several dynamic simulations on a numerical model and then it has been finally implemented on the 3-CPU parallel wrist. Figure 5 shows the scheme of the virtual prototyping environment: the multibody model is implemented by means of *VirtualLAB* package by LMS: it receives in input from the controller the actuation torques and integrates the equations of direct dynamics, providing in output the state variables that are assumed to be measured. The control system that is implemented in the *Matlab/Simulink* environment by MathWorks, computes the control actions by taking into account the desired and the actual attitude of the platform and by exploiting the partial knowledge of robot's dynamics for gravity compensation.

Table 1 collects the main data characterizing the virtual model that has been used for the

Figure 4. Scheme of the impedance control system

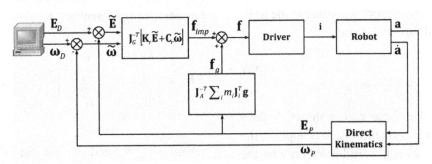

simulations: the meaning of the geometrical data can be understood from the sketch in Figure 6; moreover, it is pointed out that the mass of the platform includes the mass of the lever arm too.

The robot is actuated by three brushless linear motors by *Phase* and controlled by a *National Instrument* box based on the *PXI/FlexMotion* hardware. The force developed by the sliders is obtained by directly setting the current loop of the drivers, according to the usual relation between the current i and the thrust F:

$$F = K_t i \tag{22}$$

with $K_t [N / A]$, usually called *torque constant*, is a parameter characterizing the performances of the motor. The described impedance control loop has been implemented in *LabView* on the *PXI* card.

TEST RESULTS

In order to test an impedance control law, one of the possible approaches is to exert on the manipulator a known external wrench and then observe the response of the robot. In the case of a pure rotational machine, if known forces F_{ext} and torques M_{ext} are applied at the mobile platform in the centre of rotation O, the first kind of action is

Figure 5. Virtual prototyping environment

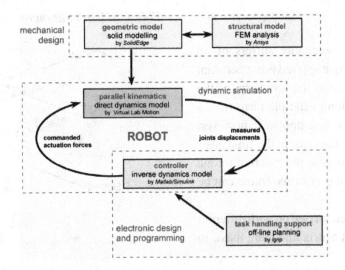

entirely born by frame bearings while the torque is reflected to the actuators according to the well-known cardinal formula of statics:

$$\mathbf{f}_{ext} = \mathbf{J}_G^{-T} \mathbf{M}_{ext} \qquad (23)$$

Therefore, if a force whose direction does not intercept the centre of motion is applied, the platform moves as if it were loaded only by the resulting torque, while the force is totally born by the structure. The main trouble in such type of tests is to apply a torque of known intensity around a known direction. For this reason a simple fixture has been realized, which uses the gravity force to load the end-effector: it consists of an arm of known length anchored to the moving platform by a plate of metal sheet, see Figure 7, that is assembled in such a way that its centre of mass is located in the centre O of the spherical motion; if a known mass is hung at one end of the bar, the active torque acting on the manipulator at point O is only due to the known weight of the lumped mass.

The procedure followed for testing the robot consists of the following steps:

• Drive the robot in its initial configuration (*virtual equilibrium state*) and then brake each motor in this position;
• Load the bar with the hung mass;
• Release brakes and record the evolution of the motion.

Unfortunately this procedure suffers of an intrinsic defect: the torque applied is not constant because of the variation of the lever arm of the force during the rotation of the platform (i.e., a perfect step moment is not applied); however, for limited oscillations the error in the final position is small and in any case the external torque actually applied at O at every time frame can be easily computed.

Some tests have been carried on with different values for the stiffness matrix and hung mass. In particular, it has been chosen to use a diagonal but non isotropic matrix $^T\mathbf{K}_r$:

$$^T\mathbf{K}_r = \begin{bmatrix} K_x & 0 & 0 \\ 0 & K_y & 0 \\ 0 & 0 & K_z \end{bmatrix} \left[Nm/rad \right] \qquad (24)$$

where K_x, K_y and K_z are the values of rotational stiffness around the axes x, y and z respectively of the user-defined frame $O*$ (i.e. they correspond to the $\varphi_x, \varphi_y, \varphi_z$ Cardan angles). In the test cases studied in this work, it has been chosen to load only one angle per time by the application of the torque along the corresponding direction. To better observe the motion of the manipulator, the stiffness of the loaded direction is suitably set at a low value; on the other hand it is useful to maintain higher values of the stiffness on the other two directions to avoid undesired rotations around them. Thereby the moving platform moves with a defined compliance so that the only possible rotation is around the axis with the lower value of stiffness. In theory the platform should not bend around the directions that are not loaded, as is actually found in simulation trials, but in real world experiments friction, noise and other unmodeled effects cause a coupling of the

Figure 6. Explanation of main geometrical data

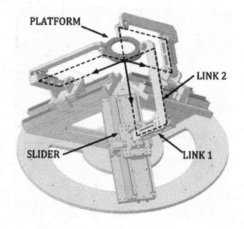

Figure 7. Fixtures realized for the experiment (a) and variation of the lever arm of the external load (b)

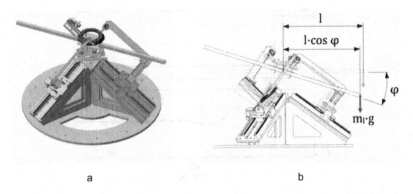

a   b

response of the machine that shows unwanted side-motions.

In the implementation of robot's controller, it has been neglected the damping term $^{T}\mathbf{C}_{r}$ in law (17): in fact, the presence of significant friction in kinematic pairs makes the dynamical system already damped and a further contribution of dissipation coming from the control law would lead to an over-damped system. On the other hand, the analytical model (as well as the multibody virtual mock-up) should provide some kind of dissipation; otherwise a step external load would cause an indefinite oscillation of the platform around the home configuration.

A few results are now presented for different load conditions and values of the stiffness matrix.

First case of study: The external torque has been applied around the y axis of the frame P^{*} by means of a *3 kg* lumped mass (m_{l}). The stiffness has been set to

$$^{T}\mathbf{K}_{r} = diag\begin{bmatrix}2;0.2;2\end{bmatrix}kNm / rad,$$

while for the numerical model the damping matrix $^{T}\mathbf{C}_{r} = diag\begin{bmatrix}30;30;30\end{bmatrix}kg / (s\,rad)$ has been used.

As shown in Figure 8, the input torque can be roughly assumed as a step signal of $-22\,Nm$, even if simulation shows that, due to machine's dynamics, the steady state is achieved through small oscillations; therefore the response expected by the robotic system is that of an undamped

system loaded by a step moment, with a steady state value for the angle φ_{y} given by:

$$\varphi_{y} = \frac{M_{y}}{K_{y}} = -0,1097\,rad = -6,28\ \text{deg}$$

During the motion of the platform the reading of the encoders have been recorded as well as the currents developed by motors' drivers at a *15 ms* sampling rate: such information allowed to compute platform's orientation (in terms of angles φ_{x}, φ_{y}, φ_{z}) by means of direct kinematics (7), the forces developed by the three linear motors by using (22) and the restoring torque available at the end-effector (if system's dynamics is neglected) by inverting (23).

Figure 9 plots the time-response of the system in 5 different tests: the thick black lines represent the angular values computed in simulation while actual angles of rotations are plot with thin colored lines: it can be noted that the real system is affected by a more damped transient because of the dissipations which it is subject to. The steady-state values of the Cardan angles in the 5 tests are collected in Table 3, with an optimal agreement between the experimental results and the pre-computed values.

Figure 10 shows the actuation forces computed by simulation (thick black line) or measured in the 5 experimental tests (thin colored lines). It can be observed that the experimental values are

Figure 8. Plot of external torque applied at the mobile platform at different scales

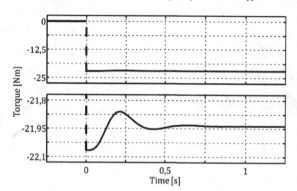

subject to high variability with respect to the expected and to the computed ones, even if the values of the Cardan angles was in close agreement with theoretical data: this dispersion of experimental values is probably due to the high friction in prismatic pairs.

The actuation forces correspond to a moment in the workspace, which is applied at the mobile platform: its static value can be evaluated through the geometric Jacobian of the platform and acts as a restoring effect which tries to bring the platform back to its initial equilibrium configuration. Figure 11 shows the time history of this moment either for the simulation or for the 5 experimental tests.

It can be seen that at the initial time the actuation torque available at the platform is not null, due to the gravity compensation term; therefore it is interesting to separate the two terms of the control law in order to understand which part of the actuation effort comes from gravity compensation and which one is due to the impedance control law, as shown in Figure 12.

It should be noted that the impedance control is actually due only to the forces shown in Figure 12 b or the torques in Figure 12d; therefore the previously mentioned "restoring effect" lies in the difference between the applied external wrench (Figure 8) and this part of the actuation effort (a plot is shown in Figure 13). The expected behav-

Figure 9. Time history of the 3 Cardan angles (a) and difference between computed and measured angles (b) in 5 different tests

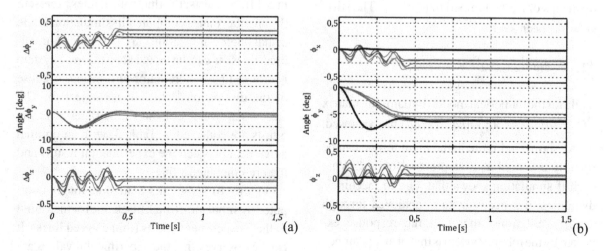

Figure 10. Actuation forces developed by the 3 motors

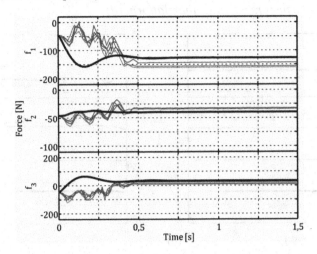

Figure 11. Actuation torque available at the platform

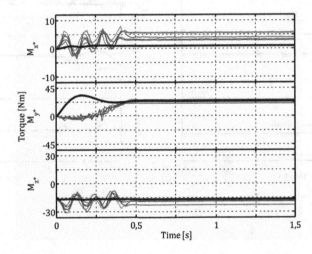

ior is the trend to the null value of difference in torque when the new equilibrium configuration is reached. The curves show that the multibody simulation is consistent with this hypothesis, while the experimental results deviate slightly. This difference is probably due once again to the effects of friction in kinematic pairs (mainly the prismatic ones).

Second case of study: This time the torque has been applied around the x axis of the frame P^*, by keeping all other parameters as in previous test case. Figure 14 a shows the rotational behavior of the platform and Figure 14 b the forces exerted by the three motors, with a good agreement

between the numerical model and the experimental tests.

In order to evaluate the isotropic behavior of the closed-loop system, it is useful to compare the results obtained in the two cases. The blue line in Figure 15 plots the average rotation of the platform around the bending axis y during the experiments performed in the first test case, while the red line is referred to the second case (rotation around x axis): it is shown that comparable results are obtained in the two cases, with steady state rotations given by: $Æ_y = -5.748 \; deg$ and $Æ_x = -5.478 \; deg$.

Figure 12. Gravity component of the actuation effort at the motors (a) or at the platform (c) and impedance control law term at the motors (b) or at the platform (d)

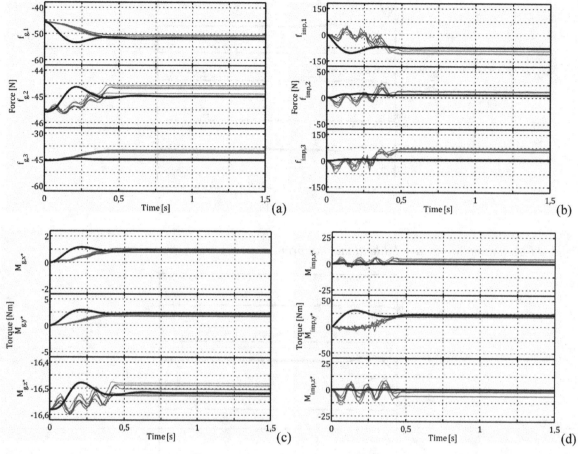

Figure 13. Difference between the external torque and the actuation torque available at the platform

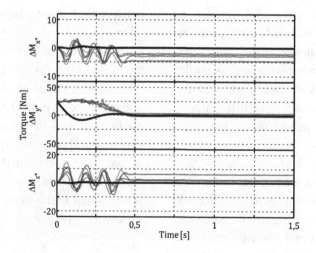

Figure 14. Time history of the three Cardan angles (a) and forces developed by the motors (b) in the numerical model (solid black lines) and in 5 different tests (colored lines)

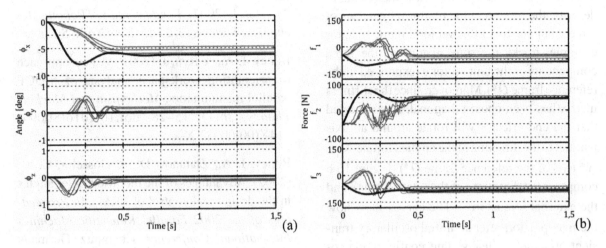

Figure 15. Rotations around the loaded directions in the first case of study (φ_y, blue line) and in the second case (φ_x, red line)

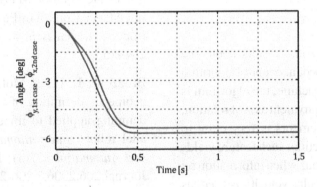

Figure 16. Components of the unit vector e (direction of the axis of rotation) (a) and difference between unit vector e and direction of the external torque

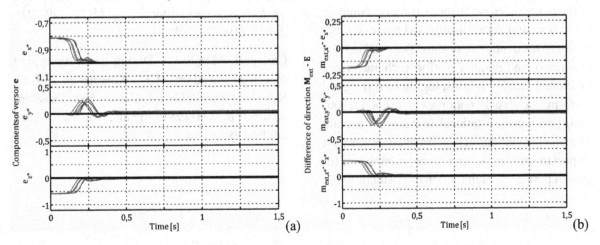

Since the *equivalent angle-axis vector* E is a key entity for the resent work, it is interesting to describe the motion of the platform by means of three components of this vector. Figure 16 a, for example, tracks the time history of the three components of the unit vector e (expressed in the reference frame O^*). Moreover, since the stiffness matrix has been chosen diagonal, it is expected that the end-effector will rotate around an axis parallel to the direction of the applied torque $\mathbf{M}_{ext}$ (y^* axis of the reference frame O^*): Figure 16 b compares the directions of the rotation axis e and the external torque $\mathbf{m}_{ext}$. The curves show how the manipulator, after an initial oscillatory transient, assumes a steady state configuration rotated around the only y^* axis.

CONCLUSION

The article has shown some results about the implementation of an impedance control algorithm on a parallel spherical machine; the algorithm is based on the use of the equivalent axis of rotation, which for small displacements is very close to the angular velocity direction; such feature yields smooth and natural motions when information on both orientation and angular velocity is used, as is the case of the stiffness control.

The algorithm has been tested first on a virtual prototyping environment, then several experimental trials have been performed in order to compare theoretical and actual results and assess the behavior of the control algorithm. The good inertial features of the parallel platform, characterized by limited moving masses, highlighted the good performance of such control law, with good decoupling between the rotational degrees-of-freedom. The small deviations of experimental results with respect to simulation outputs are probably due for greater part to the high friction of prismatic pairs.

REFERENCES

Angeles, J. (2007). *Fundamentals of Robotic Mechanical Systems* (3rd ed.). New York: Springer.

Bonev, I. A., & Ryu, J. (2001). A new approach to orientation workspace analysis of 6-DOF parallel manipulators. *Mechanism and Machine Theory*, *36*(1), 15–28. doi:10.1016/S0094-114X(00)00032-X.

Bonev, I. A., Zlatanov, D., & Gosselin, C. M. (2002). Advantages of the modified Euler angles in the design and control of PKMs. In *Proceedings of the 2002 Parallel Kinematic Machines International Conference*, Chemnitz, Germany (pp. 171-188).

Bruzzone, L., & Callegari, M. (2010). Application of the Rotation Matrix Natural Invariants to Impedance Control of Purely Rotational Parallel Robots. *Advances in Mechanical Engineering*, 284976.

Bruzzone, L. E., & Molfino, R. M. (2006). A geometric definition of rotational stiffness and damping applied to impedance control of parallel robots. *International Journal of Robotics and Automation*, *21*(3), 197–205. doi:10.2316/Journal.206.2006.3.206-2838.

Caccavale, F., Natale, C., Siciliano, B., & Villani, L. (1999). Six-DOF impedance control based on angle/axis representations. *IEEE Transactions on Robotics and Automation*, *15*(2), 289–300. doi:10.1109/70.760350.

Callegari, M. (2008). Design and Prototyping of a SPM Based on 3-CPU Kinematics. In J.-H. Ryu (Ed.), Parallel Manipulators: New Developments (pp. 171-198). Vienna, Austria: I-Tech publications.

Callegari, M., & Suardi, A. (2004). Functionally-Oriented PKMs for Robot Cooperation. In [Amsterdam, The Netherlands: IOS Press.]. *Proceedings of the Intelligent Autonomous Systems Conference, IAS-8*, 263–270.

Craig, J. J. (2005). *Introduction to Robotics: Mechanics & Control* (3rd ed.). Upper Saddle River, NJ: Pearson.

Gabrielli, A. (2009). *Mini- robotic applications for miniaturized assembly tasks*. Unpublished doctoral dissertation, Polytechnic University of Marche, Ancona, Italy.

Hogan, N. (1985). Impedance Control: An Approach to Manipulation: Part I, II, III. *Journal of Dynamic Systems, Measurement, and Control, 107*(1), 1–24. doi:10.1115/1.3140702.

Innocenti, C., & Parenti-Castelli, V. (1993). Echelon Form Solution of Direct Kinematics for the General Fully-Parallel Spherical Wrist. *Mechanism and Machine Theory, 28*, 553–561. doi:10.1016/0094-114X(93)90035-T.

Park, H., & Lee, J. M. (2004). Adaptive impedance control of a haptic interface. *Mechatronics, 14*(3), 237–253. doi:10.1016/S0957-4158(03)00040-0.

Selig, J. M. (2005). *Geometric fundamentals of robotics* (2nd ed.). New York: Springer.

Siciliano, B., & Villani, L. (2000). *Robot force control*. Boston: Kluwer.

Tandirci, M., Angeles, J., & Darcovich, J. (1992). The Role of Rotation Representations in Computational Robot Kinematics. In *Proceedings of the IEEE Intl. Conf. Robotics and Automation*, Nice, France (pp.344-349). Washington, DC: IEEE Press.

Tsai, L.-W. (2000). Solving the inverse Dynamics of a Stewart-Gough Manipulator by the Principle of Virtual Work. *ASME J. Mech. Design, 122*(1), 3–9. doi:10.1115/1.533540.

Yang, C., Huang, Q., Jiang, H. O., Ogbobe, P., & Han, J. (2010). PD control with gravity compensation for a hydraulic 6-DOF parallel manipulator. *Mechanism and Machine Theory, 45*(4), 666–677. doi:10.1016/j.mechmachtheory.2009.12.001.

APPENDIX

In robot's control it is sometimes required to be able to calculate the entire rotation matrix once three elements are known: such problem, for instance, is often encountered in the direct kinematics of spherical parallel machines, such as wrists, pointing devices, space orienting mechanisms, etc. For some applications like artificial vision a numerical approach can be deployed, leading to a recursive building up of the complete matrix; in other cases, however, such as robot's control or machine kinematics design, a closed form solution is needed, providing all the 8 admissible attitudes of the machine for the assigned values of rotation matrix elements.

The described problem is easily solved when the three known elements lie on matrix diagonal or when two of them are lined up on the same row or column; otherwise the solution can still be found (Gabrielli, 2009) but the calculations are rather arduous and need the handling of many singular cases. This is indeed the case of the 3-CPU mechanism described in the article, where the r_{12}, r_{23}, r_{31} elements of the rotation matrix R are known and the whole matrix is needed for kinematics analysis and also for wrist control. It is just noted that similar results hold in case different (sparse) elements are known.

The values of the six missing elements of the rotation matrix must be worked out from the knowledge of the three known ones: obviously it is possible to set up a system of 6 solving equations imposing the 3 conditions of orthogonality and the 3 conditions of unitary modulus of the columns (or rows), however it is simpler to parametrize the matrix by means of a proper set of Cardan angles.

If the orientation is described by using three ordered rotations α, β and γ around the body axes x, y and z respectively, the rotation matrix R can be written as follows in (A.1), with $-\pi < \alpha \leq \pi$, $-\pi/2 < \beta \leq \pi/2$, $-\pi < \gamma \leq \pi$; it is noted that the chosen parametrization degenerates when $\beta = \pm \pi/2$ and it is not possible to determine all the three angles but only the value of β and the sum or the difference of the other two angles. From (1) it is obtained:

$$R = \begin{bmatrix} r_{11} & \mathbf{r_{12}} & r_{13} \\ r_{21} & r_{22} & \mathbf{r_{23}} \\ \mathbf{r_{31}} & r_{32} & r_{33} \end{bmatrix} = \begin{bmatrix} c\beta c\gamma & -c\beta s\gamma & s\beta \\ s\alpha s\beta c\gamma + c\alpha s\gamma & -s\alpha s\beta s\gamma + c\alpha c\gamma & -s\alpha c\beta \\ -c\alpha s\beta c\gamma + s\alpha s\gamma & c\alpha s\beta s\gamma + s\alpha c\gamma & c\alpha c\beta \end{bmatrix} \tag{A.1}$$

$$\begin{cases} -c\beta s\gamma = r_{12} \\ -s\alpha c\beta = r_{23} \\ -c\alpha s\beta c\gamma + s\alpha s\gamma = r_{31} \end{cases} \tag{A.2}$$

that is a non linear algebraic system of 3 equations in the 3 unknowns α, β, γ. By considering separately the singular cases, from the first two equations in (A.2) it is obtained:

$$s\alpha = \frac{r_{23}}{r_{12}} s\gamma \tag{A.3}$$

and by considering the first and the last equation in (A.2) it results:

$$\frac{r_{12}^2}{s^2\gamma} + \frac{r_{31}^2 + s^2\alpha s^2\gamma - 2r_{31}s\alpha s\gamma}{c^2\alpha c^2\gamma} = 1 \tag{A.4}$$

Equation (A.4) can be elaborated so as to obtain that displayed in Equation (A.5) and taking into consideration (A.3) it is obtained:

$$r_{12}^2 \cdot \left(1 - s^2\gamma - s^2\alpha + s^2\alpha s^2\gamma\right) + s^2\gamma \cdot \left(r_{31}^2 + s^2\alpha s^2\gamma - 2r_{31}s\alpha s\gamma\right) = s^2\gamma \cdot \left(1 - s^2\gamma - s^2\alpha + s^2\alpha s^2\gamma\right) \tag{A.5}$$

$$A \cdot s^4\gamma + B \cdot s^2\gamma + C = 0 \tag{A.6}$$

with

$$A = -2\frac{r_{31}r_{23}}{r_{12}} + \frac{r_{23}^2}{r_{12}^2} + r_{23}^2 + 1 > 0$$

$$B = r_{31}^2 - r_{23}^2 - r_{12}^2 - 1 \leq 0$$

$$C = r_{12}^2 \geq 0$$

It can be observed that, if the discriminant of (A.6) is positive, the signs of A, B and C guarantee that the two solutions of (A.6) are positive real, so that 4 real values for $sin\gamma$ and 8 values for the angle γ can be found. The values of the three parameters r_{12}, r_{23} and r_{31} which cause negative values of the previous discriminant correspond to reflections matrices (i.e. the determinant is -1): since this kind of matrices cannot be found using the parametrization of the Cardan angles, in this case the solutions of (A.6) become complex.

Now from the first equation in (A.2) the value of $cos\beta$ can be obtained and then two values for the angle β:

$$c\beta = -\frac{r_{12}}{s\gamma} \tag{A.7}$$

It can be noted that angle β varies in the range $[-\pi/2, \pi/2]$: since four of the eight previously obtained values of γ are certainly in quadrants I or IV whereas the other four are in quadrants II or III, there are at the same time 4 positive values of $cos\beta$ (hence acceptable) and 4 negative values, which cannot be considered any further. Hence, up to now, 8 acceptable solutions for the pair of angles (γ, β) have been found.

In the end, in order to find the value of α the last two equations in (A.2) are considered; since only the parameters $cos\alpha$ and $sen\alpha$ are now unknown, it is worked out:

$$\begin{cases} s\alpha = -\dfrac{r_{23}}{c\beta} \\ c\alpha = -\dfrac{r_{31}c\beta + r_{23}s\gamma}{s\beta c\beta c\gamma} \end{cases} \tag{A.8}$$

Therefore the function *atan2* gives just one value of α for each pair (γ, β) previously determined. Hence the problem, and also the direct kinematics of the 3-CPU mechanism, admits at most eight solutions, according to Innocenti and Parenti-Castelli (1993). The problem is comprehensively solved when all the following singular cases are separately treated:

- $r_{12} = r_{23} = r_{31} = 0 \ (c\beta = 0 \text{ or } s\gamma = 0)$
- $r_{12} = r_{23} = 0 \text{ and } r_{31} \neq 0 \ (c\beta = 0 \text{ or } s\gamma = 0)$
- $r_{12} = r_{31} = 0 \text{ and } r_{23} \neq 0 \ (s\gamma = 0)$
- $r_{23} = r_{31} = 0 \text{ and } r_{12} \neq 0 \ (s\alpha = 0)$
- $r_{12} = 0, \ r_{23} \neq 0 \text{ and } r_{31} \neq 0 \ (s\gamma = 0)$
- $r_{23} \cdot r_{31} = r_{12} \ (c\alpha = 0)$
- $r_{12} \cdot r_{31} = r_{23} \ (c\gamma = 0)$
- $r_{12} \cdot r_{23} = r_{31} \ (s\beta = 0)$

This work was previously published in the International Journal of Intelligent Mechatronics and Robotics, Volume 1, Issue 1, edited by Bijan Shirinzadeh, pp. 40-60, copyright 2011 by IGI Publishing (an imprint of IGI Global).

Chapter 4
Stiffness Modeling and Analysis of Passive Four-Bar Parallelogram in Fully Compliant Parallel Positioning Stage

X. Jia
Tianjin University, China

D. Zhang
Tianjin University, China

Y. Tian
Tianjin University, China

J. Liu
Hebei University of Technology, China

ABSTRACT

In order to investigate the influence of the stiffness of the compliant prismatic pair, a planar four-bar parallelogram, in a fully compliant parallel mechanism, the stiffness model of the passive compliant prismatic pair in a compliant parallel positioning stage is established using the compliant matrix method and matrix transformation. The influences of the constraints and the compliance of the connecting rods on the flexibility characteristics of the prismatic pair are studied based on the developed model. The relative geometric parameters are changed to show the rules of the stiffness variation and to obtain the demands for simplification in the stiffness modeling of the prismatic pair. Furthermore, the finite element analysis has been conducted to validate the analytical model.

INTRODUCTION

As a combination of parallel manipulator and compliant mechanism, compliant parallel mechanisms can provide the merits of both mechanisms, such as high stiffness, wide response band, vacuum

compatibility, clean room compatibility, zero error accumulation, no backlash, friction free and no need for lubrication. Consequently, compliant parallel mechanisms have a potential in various fields where an ultra-precision manipulation system is first and foremost required (Ryu & Gweon,

DOI: 10.4018/978-1-4666-3634-7.ch004

1997a; Ryu & Gweon, 1997b; Yi, Na, & Chung, 2002; Tian, Shirinzadeh, & Zhang, 2008; Tian, Shirinzadeh, & Zhang, 2009).

The stiffness characteristic of the flexure-based parallel mechanism plays an important role in the practical applications. It will affect the workspace, load-carrying capacity, dynamic behavior and the positioning accuracy of the entire mechanism. A spatial compliant parallel mechanism with high stiffness can provide a more dexterous and precise motion and possess high natural frequency. Currently, the research on the stiffness of flexure-based mechanism mainly focuses on the modeling and analysis of whole mechanism or individual flexure hinges including elliptical flexure hinges, corner-filleted hinges, and right circular hinges, etc. Lobontiu and Garcia (2003) provided an analytical method for stiffness calculations of planar compliant mechanism with single-axis flexure hinges based on the strain energy and Castigliano's displacement theorem. By discussing about the relationship between the input/output stiffness and mechanism geometric parameters, a stiffness optimization procedure was proposed. Koseki, Tanikawa, and Koyachi (2000) and Yu, Bi, and Zong (2002) obtained the stiffness model based on the matrix method and the equilibriums of displacements and forces. Pham and Chen (2005) analyzed the stiffness of a six-DOF flexure parallel mechanism based on the simplified modeling method. Xu and Li (2006) introduced a more straightforward approach, which employed only one kind of transformation matrix, to derive the stiffness matrix of an orthogonal compliant parallel micromanipulator. Dong, Sun, and Du (2008) established the stiffness equation of a 6-DOF high-precision parallel mechanism via assembling stiffness matrices and formulating constraint equations. By simplifying the flexure hinge as an ideal revolution joint with a linear torsional spring and the equation equilibriums of forces, Tian, Shirinzadeh, & Zhang (2008) derived the stiffness model of a compliant five-bar

mechanism. Based on the model, the influences of the position on the stiffness and the position of the end-effector point are discussed.

Stiffness modeling of individual flexure hinge is the premise to analyze the compliance of spatial flexure-based mechanism. In general, the single axis flexure hinge is utilized as the compliant revolute joint and the flexure-based planar four-bar parallelogram as compliance prismatic pair. Paros and Weisbord (1965), Smith, Badami, and Dale (1997), and Tian, Shirinzadeh, and Zhang (2010) have done a number of works on the stiffness modeling of flexure hinge with different notches, and established the analytic formulations. However, when the stiffness of compliant prismatic pair was involved in the above literatures, the simplifications have commonly been adopted. The flexure hinges and the right/left links in the parallelogram were regarded as elastic bodies and the compliance of the connecting rods on the upper/lower terminal of the parallelogram was neglected. In addition, the research was primarily focused on the stiffness along the translation direction of compliant prismatic pair, although Yang, Yin, and Ma (2005) formulated the stiffness equation of compliant prismatic pair using the energy method.

In order to solve above mentioned problems, the mechanical structure of a 3-DOF flexure-based parallel positioning stage is described briefly in this paper. The stiffness of the passive compliant prismatic pair is modeled analytically and the influences of the geometric dimensions, restriction condition and the compliance of the connecting rod on the stiffness are investigated and discussed. Finally, the proposed analytical model is validated by FEM simulation.

MECHANICAL DESIGN

A modified Delta parallel mechanism which consists of only revolute joints, firstly presented by Tsai and Stamper (1996), is chosen as the prototype of compliant mechanism. The revolute joints only offer rotation about one axis that is similar to the single-axis flexure hinge. Therefore, replacing the revolute joints in the modified Delta mechanism with single-axis right circular flexure hinges and adjusting the length of the links, a novel precision positioning stage whose kinematics performance is equivalent to the primary mechanism can be achieved and shown in Figure 1. The material of the compliant stage is chosen as spring steel 65Mn. Compared with other types of flexure hinges, circular flexure hinge with rectangular cross-section has high precision for rotational accuracy and is the optimal choice as the revolute joint for precision mechanism design and manufacture (Tian, Shirinzadeh, & Zhang, 2009). Thus, the right circular flexure hinge is used in the stage, and the flexure hinges on the passive four-bar parallogram have the same dimensions. The mechanism has a simple, compact and symmetrical structure to improve its performance in terms of precision, stiffness and dynamics. Piezoelectric actuators are chosen to supply the force/displacement on the middle of the active arms of the mechanism.

COMPLIANCE ANALYSIS OF THE PRECISION POSITIONING STAGE

Compliance Matrix and Compliance Transformation

As shown in Figure 1, the developed 3-DOF compliant precision positioning stage consists of a mobile platform, a fixed base, and three limbs with identical kinematic structure. Each limb connects the fixed base to the mobile platform by two revolute pair in sequence and one flexure prismatic pair (passive four-bar parallelogram) followed by one flexure hinge. Thereby, each limb can be regarded as the basic compliant elements of a cantilever and a right circular flexure hinge (as shown in Figure 2) with different structure forms.

Referring to the work of Koseki, Tanikawa, and Koyachi (2000), the compliance matrices about local frames of two types of basic compliant elements are expressed as:

$$C = \begin{bmatrix} c_1 & 0 & 0 & 0 & c_7 & 0 \\ 0 & c_2 & 0 & -c_8 & 0 & 0 \\ 0 & 0 & c_3 & 0 & 0 & 0 \\ 0 & -c_8 & 0 & c_4 & 0 & 0 \\ c_7 & 0 & 0 & 0 & c_5 & 0 \\ 0 & 0 & 0 & 0 & 0 & c_6 \end{bmatrix} \qquad (1)$$

where the factors c_i ($i=1,...,8$) are the same as in Koseki, Tanikawa, and Koyachi (2000), E and G are modulus of longitudinal elasticity (Young's modulus) and modulus of transverse elasticity, respectively (Figure 2).

The compliance transformation from local frame to reference frame has been described in Xu and Li (2006) as follows:

$$\bar{C}_i = J_i C_i J_i^T \qquad (2)$$

where $\bar{C}_i$ is the compliance matrix of the compliant element expressed in the reference frame, C_i

Figure 1. 3-DOF precision positioning stage

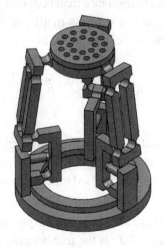

Figure 2. Basic compliant element

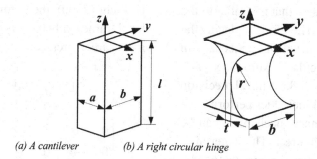

(a) A cantilever   *(b) A right circular hinge*

is the compliance matrix expressed in the local frame. J_i represents a 6×6 transformation matrix and can be written as

$$J_i = \begin{bmatrix} R_i & -R_i S(r_i) \\ 0 & R_i \end{bmatrix} \qquad (3)$$

where R_i denotes the rotation matrix of the local frame with respect to reference frame, $r_i = [\ r_{ix}\ r_{iy}\ r_{iz}]^T$ is a vector expressed in the reference frame, $S(r_i)$ represents the skew-symmetric operator with the notation of

$$S(r_i) = \begin{bmatrix} 0 & -r_{iz} & r_{iy} \\ r_{iz} & 0 & -r_{ix} \\ -r_{iy} & r_{ix} & 0 \end{bmatrix} \qquad (4)$$

Thus, the compliance matrix of compliant serial chain and the stiffness matrix of compliant parallel structure can be given as

$$\bar{C} = \sum_{i=1}^{m} \bar{C}_i \text{ and } \bar{K} = \sum_{i=1}^{n} J_{ci}^{-T} \bar{K}_{ci} J_{ci}^{-1} \qquad (5)$$

where $\bar{C}$ and $\bar{K}$ denote the compliance matrix for a serial chain and the stiffness matrix for a parallel structure expressed in the reference frame, respectively. $\bar{K}_{ci}$ is the stiffness matrix of the ith limb of the parallel structure expressed in the local frame, and J_{ci} is the transformation matrix

from the local frame of the ith limb to the reference frame of the parallel structure.

Compliance of the Passive Four-Bar Parallelogram

As shown in Figure 1, the planar four-bar parallelogram places aslant and intersects with the platform in each chain. Therefore, the force and moment will be transferred and acted on the four-bar parallelogram when the external operation is applied on the platform. Hence, it is necessary to investigate the influence of the stiffness of the compliant prismatic pair on the output stiffness of the compliant parallel stage. As shown in Figure 3, the four-bar parallelogram consists of three parts: four flexure hinges, the left/right links between the upper and lower flexure hinges, and the upper and lower connecting rods. In general, the compliance of the connecting rods is neglected because of their large rigidity, and only the flexibilities of the flexure hinges and the left/right links are taken into account in the stiffness modeling of the four-bar parallelogram. However, it is unclear whether this kind of simplification is feasible or not. Therefore, during the stiffness analysis of the spatial flexure-based stage, the effect of the compliance of the connecting rods should be discussed in detail to make sure that the simplification can be adopted sequentially.

Based on the above analyses, the planar four-bar parallelogram will be modeled and discussed analytically in three cases: (1) neglecting the

Figure 3. The geometry parameters and coordinate systems of the planar four-bar parallelogram

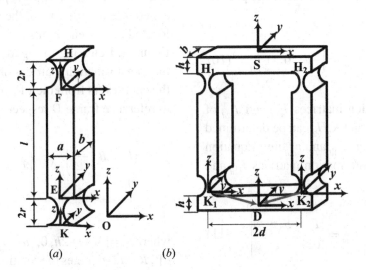

(a) (b)

compliance of the upper and lower connecting rods; (2) restricting completely the lower connecting rod and calculating the compliance of the upper connecting rod; (3) applying the constraints on the middle of the lower connecting rod and calculating the compliance of the two connecting rods.

Neglecting the Compliance of the Connecting Rods

The four-bar parallelogram can be considered as two symmetric parallel compliant limbs K_1H_1 and K_2H_2 as shown in Figure 3(a) if the connecting rods are regarded as rigid bodies and the compliance is neglected. Each limb K_iH_i ($i=1, 2$) is a serial chain with two flexure hinges H_i, K_i and one compliant link E_iF_i. Thus, the transformation matrices from the compliance from their local frames F, E and K to frame K respectively are:

$$J_{fk} = \begin{bmatrix} I & -S(r_{fk}) \\ 0 & I \end{bmatrix} J_{ek} = \begin{bmatrix} I & -S(r_{ek}) \\ 0 & I \end{bmatrix} J_k = \begin{bmatrix} I & 0 \\ 0 & I \end{bmatrix}$$

$$(6)$$

where $r_{ek}=[0, 0, -2r]^T$, $r_{fk}=[0, 0, -(2r+l)]^T$, I is a 3×3 identity matrix.

The compliance matrices C_{hingeH}, C_{hingeK}, C_{prisml} of the flexure hinges H_i, K_i and compliant link E_iF_i are assembled to an 18×18 matrix:

$$C_0 = diag\begin{pmatrix} C_{hingeH} & C_{prisml} & C_{hingeK} \end{pmatrix} \quad (7)$$

where the expressions of the compliant matrices C_{hingeH}, C_{hingeK} and C_{prisml} are same as ones listed in Koseki, Tanikawa, and Koyachi (2000).

The transformation matrices in Equation (6) can be assembled to a matrix as follows:

$$J_0 = \begin{bmatrix} J_{fk} & J_{ek} & J_k \end{bmatrix} \quad (8)$$

Thus, according to Equation (2), the compliance matrix of the limb is determined as:

$$C_{limb} = J_0 C_0 J_0^T \quad (9)$$

For the symmetric structure of the planar four-bar parallelogram, the middle point D of the connecting rod K_1K_2 is defined as the origin of reference frame. Hence, vectors pointing from the tips K_i of limbs K_iH_i to the reference point D are denoted as:

$$r_{k1d} = \begin{bmatrix} d & 0 & -h \end{bmatrix}^T r_{k2d} = \begin{bmatrix} -d & 0 & -h \end{bmatrix}^T \quad (10)$$

The transformation matrices J_{k1d} and J_{k2d} of the two limbs K_1H_1 and K_2H_2 can be determined by substituting vectors r_{k1d} and r_{k2d} into Equation (3). Hence, the transformation matrices J_{k1d} and J_{k2d} yield:

$$J_{k1d} = \begin{bmatrix} I & -S(r_{k1d}) \\ 0 & I \end{bmatrix} J_{k2d} = \begin{bmatrix} I & -S(r_{k2d}) \\ 0 & I \end{bmatrix} \quad (11)$$

Considering the location of two flexure limbs as shown in Figure 3(b), then the compliance matrix of the planar four-bar parallelogram is described as:

$$C_1 = \left(\sum_{i=1}^{2} \left(J_{kid} C_{limb} J_{kid}^T \right)^{-1} \right)^{-1} \quad (12)$$

Restricting completely the lower connecting rod and calculating the compliance of the upper connecting rod.

Due to the completely fixed constraint, the lower connecting rod should be regarded as rigid body and only the flexibility of the upper connecting rod is calculated during formulizing the stiffness of the parallelogram. In this case, the planar four-bar parallelogram can be developed as an equivalent system exactly like the crank and follower of the four-bar mechanism. From Figure 3, it can be observed that the upper connecting rod connect with the two compliant limbs at its two endpoints, and therefore the connecting rod can be considered as two cantilevers with length d composing serial chains with two compliant limbs after connecting with each other in parallel at middle point S.

To facilitate the stiffness analysis of the parallelogram, the upper connecting rod is detached

from the compliant limbs separately, and the local coordinate systems of the two cantilevers are located at two endpoints H_1 and H_2 as shown in Figure 4. Referring to Figure 4 and Figure 3(b), the transformation matrices of the compliance of the cantilevers from their local frames H_1 and H_2 to reference frame D respectively are:

$$J_{h1} = \begin{bmatrix} I & -R_y\left(\dfrac{-\pi}{2}\right)S(r_{h1}) \\ 0 & I \end{bmatrix} J_{h2} = \begin{bmatrix} I & -R_y\left(\dfrac{\pi}{2}\right)S(r_{h2}) \\ 0 & I \end{bmatrix} \quad (13)$$

where $r_{h1}=[4r+l+2h, 0, -d]^T$, $r_{h2}=[-(4r+l+2h), 0, d]^T$, $R_y(\pi/2)$ denotes the rotation of $\pi/2$ around the y-axis and the others are similar.

Based on Equation (2), the compliance matrices of the cantilevers expressed in reference frame D can be derived as:

$$C_{h1} = J_{h1} C_{prismd} J_{h1}^T \text{ and } C_{h2} = J_{h2} C_{prismd} J_{h2}^T \quad (14)$$

where C_{prismd} is the compliance matrix of the cantilever.

Considering that two cantilevers connect with each other in parallel and then forming a serial structure with the two parallel compliant limbs, and thus, referring to Equation (5), the compliance matrix of the planar four-bar parallelogram yield:

$$C_2 = \left(C_{h1}^{-1} + C_{h2}^{-1} \right)^{-1} + C_1 \quad (15)$$

Applying constraints on the middle of the lower connecting rod and calculating the compliance of the two connecting rods.

As the constraints are applied on the middle of the lower connecting rod, the compliance of the lower connecting rod will inevitably affect the flexibility characteristics of the four-bar parallelogram. Then, the lower rod can also be regarded as two cantilevers with length d (as shown in Figure 5) which connect serially with two compliant

Figure 4. The local coordinate system of the upper connecting rod

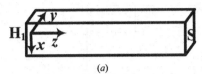

(a)

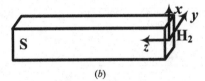

(b)

limbs and the upper rod after connecting in parallel with each other at middle point D. According to Equation (2), the compliance matrices of the lower cantilevers expressed in reference frame D can be respectively derived as:

$$C_{k1} = J_{k1} C_{prismd} J_{k1}^T \text{ and } C_{k2} = J_{k2} C_{prismd} J_{k2}^T \tag{16}$$

where

$$J_{k1} = \begin{bmatrix} I & -R_y\left(\dfrac{\pi}{2}\right) S\left(r_{k1}\right) \\ 0 & I \end{bmatrix}$$

and

$$J_{k2} = \begin{bmatrix} I & -R_y\left(-\dfrac{\pi}{2}\right) S\left(r_{k2}\right) \\ 0 & I \end{bmatrix}$$

are transformation matrices of the two lower cantilevers, $r_{k1} = r_{k2} = [0, 0, -d]^T$ are vectors pointing from the tips K_i to the reference point D.

Similarly, according to Equation (5), the compliance matrix of the four-bar parallelogram yields:

$$C_3 = \left(C_{k1}^{-1} + C_{k2}^{-1}\right)^{-1} + C_1 + C_2 \tag{17}$$

Case Study and Simulation

In order to verify the established models, an improved Delta compliant mechanism is chosen to conduct the stiffness analysis. The material of the four-bar parallelogram is selected as spring steel 65Mn with the Young's Modulus E 210GPa, the modulus of transverse elasticity G 80GPa, and the Poisson's ratio 0.27. To explore the influence of the compliance of the connecting rod with different boundary conditions, the simulation is carried out by changing the architectural parameters (half length of the connecting rod d, thickness b and width a/h of the connecting rod, length l of the compliant limb, the cutting radius r and the minimum hinge thickness t of the flexure hinge) of the four-bar parallelogram, and the computational results are shown in Figure 6, Figure 7, Figure 8, and Figure 9.

For convenience, the stiffness values of the planar four-bar parallelogram in three cases are renamed in the following contents as (1) Case 1: neglecting the compliance of the connecting rods; (2) Case 2: restricting completely the lower connecting rod and calculating the compliance of the upper connecting rod; (3) Case 3: applying the constraints on the middle of the lower connecting rod and calculating the compliance of the two connecting rods.

Figure 5. The local coordinate system of the lower connecting rod

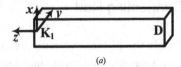

(a)

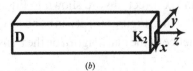

(b)

Figure 6. Stiffness versus the half length d of the connecting rod

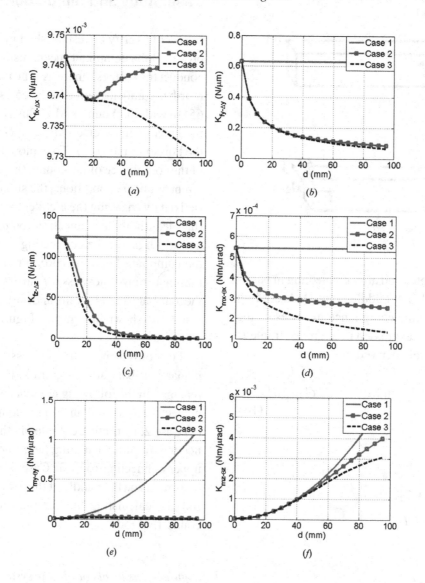

The Influence of the Half Length d of the Connecting Rod

In order to obtain the stiffness variation of the parallelogram with the changing of the half length of the connecting rod, the other geometry parameters of the parallelogram are set as r=3mm, t=0.4mm, a=h=6.4mm, b=6mm, l=40mm, and the stiffness curves in six directions are shown in Figure 6.

As shown in Figure 6(a), it can be observed that the influence of the half length d on the translation

stiffness in x direction ($K_{fx\text{-}\Delta x}$) of the prismatic pair is negligible small when the compliance of the connecting rod was neglected. Once the compliance of the rods being considered, both of the stiffness $K_{fx\text{-}\Delta x}$ decreases with the increasing of the length d under different boundary conditions at the beginning. However, the direction of the stiffness change in Case 2 will reverse when the length ratio of the rod and the compliant link ($2d/l$) reaches to 0.9. Namely, if the lower connecting rod is restricted completely and only the compliance of the upper

Figure 7. Stiffness difference values versus the section parameters of the connecting rod

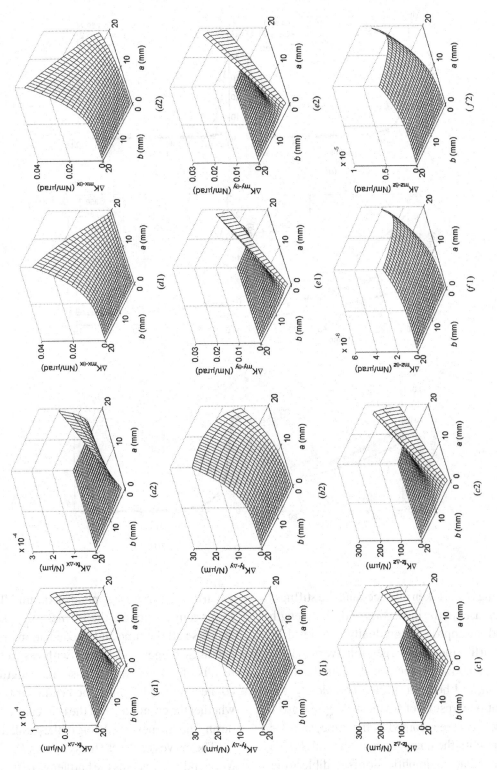

Figure 8. Stiffness versus the length l of the compliant link

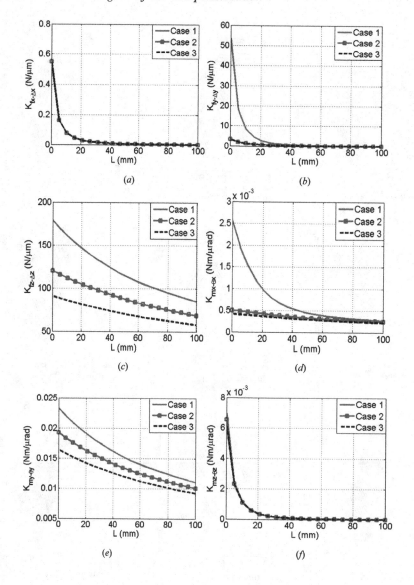

connecting rod is taken into account, the stiffness $K_{fx-\Delta x}$ increases as the increasing of the half length of the rod and approaches gradually to the result of Case 1. Furthermore, the largest relative deviation is less than 0.06%. In other words, with the lower connecting rod being restricted completely, the computational results of stiffness $K_{fx-\Delta x}$ in Case 1 and Case 2 can be regarded as the same, and it is irrelevant with the compliance of the rod. It also illuminates that the simplification is credible in the stiffness modeling of the parallelogram. However,

when the constraint is applied on the middle of the lower connecting rod, the stiffness $K_{fx-\Delta x}$ keeps to decrease all the time and the deviation gradually becomes bigger and bigger with respect to the results of Case 1. In this case, the length of the connecting rod becomes the key factor to decide whether the compliance of the rods can be ignored during the stiffness modeling of the parallelogram.

The curves of the stiffness factors $K_{fy-\Delta y}$, $K_{fz-\Delta z}$, $K_{mx-\theta x}$ and $K_{my-\theta y}$ versus the half length d of the connecting rod in three cases are illustrated in Figure

Figure 9. Stiffness difference values versus the geometry parameters of the flexure hinges

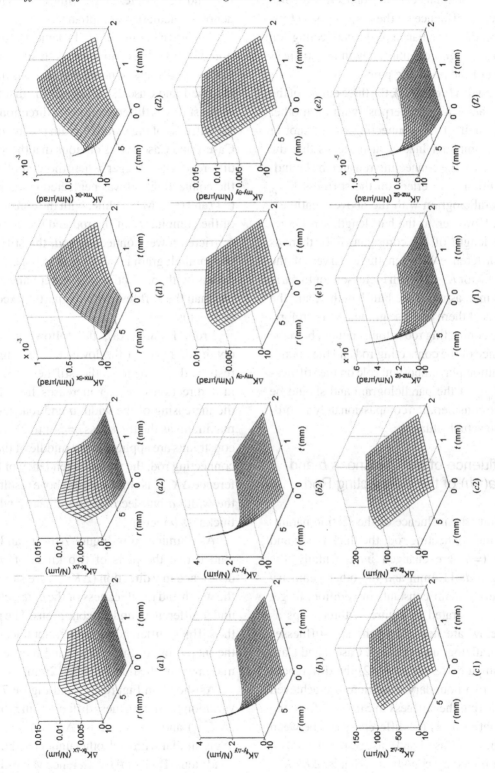

6(b-e). The compliance of the connecting rod has a significant influence on these stiffness factors. Therefore, the compliance of the connecting rod should not be ignored intuitively in the stiffness modeling of the prismatic pair.

As shown in Figure 6(f), the three curves of the stiffness factor $K_{mz-\theta z}$ superpose with each other when the half of the connecting rod is shorter than the compliant link. It also means that the compliance of the connecting rod and the boundary condition have effect on the stiffness $K_{mz-\theta z}$ of the parallelogram with the above mentioned premise. However, as the half length d is bigger than the length of the compliant link, there is visible difference between three curves of the stiffness factor $K_{mz-\theta z}$, which increases nonlinearly with the increasing of the half length d, and the deviations of them increase gradually. Therefore, when the connecting rod is long enough, both the compliance of the connecting rod and the boundary condition play important role on the stiffness factor $K_{mz-\theta z}$ of the parallelogram and should be taken into consideration compassionately according to the actual situation.

The Influence of the Thickness *b* and Width *a*(=*h*) of the Connecting Rod

To examine the influence of the section parameters of the connecting rod, the thickness b and width a (=h) are changed incrementally, the half length d=11.4mm and the other geometry parameters remain constant. In addition, to get the plots illustrated in Figure 7, three steps as listed below are needed: 1) the six stiffnesses of the parallelogram in Case 1, Case 2 and Case 3 are computed respectively; 2) the differences ΔK with two boundary conditions are achieved by subtracting the stiffness in Station 2 from the one in Station 1; 3) the differences ΔK between the stiffness of Case 1 and Case 3 are solved with different boundary conditions. That is, $\Delta K = K_{nc} -$

K_{cc}, and nc denotes neglecting compliance, cc denotes calculating compliance.

As shown in Figure 7, (*1) depict the distribution of the stiffness differences in six directions when the stiffness of parallelogram was modeled in Case 1 and Case 2, (*2) depict the distribution of the stiffness differences in six directions when the stiffness of the parallelogram was modeled in Case 1 and Case 3. It is obvious that the stiffness differences are bigger when the compliance of the connecting rod was neglected by comparing (*1) and (*2). Meanwhile, with the consideration of the compliance of the rod and the completely restricted lower connecting rod, the stiffness of the parallelogram is more close to the calculated results with neglecting the compliance of the rod than the stiffness with the partly fixed lower connecting rod.

From Figure 7(a), the followings can be obtained: 1) when the lower connecting rod is restricted completely, the stiffness difference in x direction ($\Delta K_{fx-\Delta x}$) increases linearly with the increasing of the width a and reaches to the maximum as the thickness b=3mm; 2) when the constraints are applied on the middle of the lower connecting rod, the variation tendency of the difference $\Delta K_{fx-\Delta x}$ is similar to the above results about the width a but decreases nonlinearly when the thickness b increases.

As illuminated in Figure 7(b), it can be concluded that the plots of two kinds of stiffness difference in y direction ($\Delta K_{fy-\Delta y}$) are same about the width and the thickness of the connecting rod under different boundary conditions. The plots of the stiffness difference decreases nonlinearly with the increasing of the width a and reaches to the maximum as the thickness b=12mm.

As shown in Figure 7(c) and Figure 7(e), the variations of the stiffness difference in z direction ($\Delta K_{fz-\Delta z}$) and about y axis ($\Delta K_{my-\theta y}$) are respectively similar with each other under two boundary conditions. Both of the difference values in these two directions are linearly incremental when the width a increase and reach the maximum as the thickness $b=2.3$mm.

As shown in Figure 7(d) and Figure 7(f), with the two boundary conditions, both of the stiffness difference, the stiffness around x direction ($\Delta K_{mx\text{-}\theta x}$) and about z axis ($\Delta K_{mz\text{-}\theta z}$), increase nonlinearly when the width a of the connecting rod increases gradually; When the width b increases, the variations of $\Delta K_{mx\text{-}\theta x}$ and $\Delta K_{mz\text{-}\theta z}$ can be described as the former increases but the latter decreases and both of them is nonlinear.

The Influence of the Length *l* of the Compliant Link

The cross-section parameters of the compliant link and the connecting rod are set to be identical, the other geometry parameters of the parallelogram are assigned to be constants, and the curves of the stiffness factors in six directions versus the length l of the compliant link are illustrated in Figure 8.

From Figure 8, we can obtain the following conclusions: all of the stiffness in six directions decrease with the increasing of the length l of the compliant link, and the three curves will be coincident gradually when the compliant link is long enough. It also obtained that the length of the compliant link is a key factor to influent the static characteristics of the parallelogram. In addition, to achieve satisfying stiffness, the length l must be discussed in detail and carefully, then, it can be decided whether the compliance of the connecting rod should be calculated during the stiffness modeling. In general, when analyzing flexibility characteristics of the parallelogram and developing the stiffness model of the spatial flexure-based stage, the compliance of the upper and lower connecting rods can be neglected no matter how the constraints is applied when the length ratio of the compliant link and the connecting rod is greater than or equal to 1.5.

The Influence of the Cutting Radius *r* and the Minimum Hinge Thickness *t* of the Flexure Hinge

It is indubitable that the flexure hinges play an important role in the four-bar parallelogram, and its geometry parameters are essential factors to determine the flexibility of the parallelogram. The six stiffness differences ΔK of the parallelogram are computed in terms of the three steps and the plots versus the cutting radius r and the minimum hinge thickness t of the flexure hinges are illustrated in Figure 9. As the stiffness of the parallelogram was analyzed in Case 1 and Case 2, respectively, the distribution plots of the stiffness difference in six directions are depicted one by one in Figure 9(*1). And the stiffness of the parallelogram was analyzed in Case 1 and Case 3, respectively, the distribution plots of the six stiffness difference are depicted individually in Figure 9(*2).

It can be seen from Figure 9 that the stiffness of the parallelogram are bigger if the compliance is neglected for the positive difference values in Figure 9 (*1), compared with the stiffness when the compliance of the connecting rod is calculated.

As shown in Figure 9(a), it is noted that the stiffness differences in x direction ($\Delta K_{fx\text{-}\Delta x}$) are similar to each other and increase nonlinearly with two defined boundary conditions when the cutting radius of the flexure hinge increases; the similar variation of the stiffness difference also takes place with the change of the minimum hinge thickness t and achieves the maximum at $t=0.5$mm.

In Figure 9(b) and Figure 9(f), the variation trends of the stiffness differences in y direction (about z axis) with the cutting radius r and the minimum hinge thickness t are respectively approximate in two defined boundary conditions, it can be described as that the stiffness difference increases nonlinearly when increasing r and remaining t unchanged, while decreases nonlinearly with

Figure 10. Finite element model of the planar four-bar parallelogram

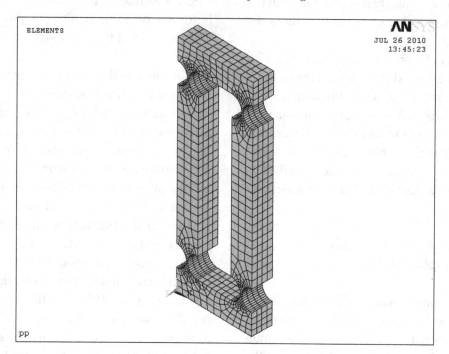

increasing *t* and *r* remaining constant. Besides, as shown in Figure 9(b), when the connecting rod was restricted completely, the stiffness differences in *y* direction are equivalent with those when the middle of the rod was restricted partly. However, for the stiffness about *z* direction, the difference is bigger under the condition of complete constraints.

Figure 9(c) and Figure 9(e) provide the influences of the cutting radius and the minimum hinge thickness on the stiffness differences in *z* direction ($\Delta K_{fz-\Delta z}$) and about *y* axis ($\Delta K_{my-\theta y}$) with different constraints. The distributing plots are similar with each other. Four of them all vary incrementally and linearly with the increasing of the cutting radius and reach to the maximum as *t*=0.5mm. In addition, stiffness differences $\Delta K_{fz-\Delta z}$ and $\Delta K_{my-\theta y}$ both are smal in the condition of the lower connecting rod being restricted completely than those when the constraints are applied partly in the middle of the rod.

The variation plots of the stiffness difference about *x* axis ($\Delta K_{mx-\theta x}$) are shown in Figure 9(d). When two constraints are applied on the lower connecting rod, respectively, the plots are similar with each other while the minimum hinge thickness *t* varies, that is, all of them decrease at the beginning and then increase as the increasing of *t*. Otherwise, the minimum of $\Delta K_{mx-\theta x}$ increase slightly rather than remaining unchanged once the cutting radius *r* varies incrementally. In addition, as shown in Figure 9(d1), on the condition of the lower connecting rod was restricted completely, $\Delta K_{mx-\theta x}$ decreases with the increasing of *r*, and *t* is less than 1mm, and then it increases when *t*>1mm. But in Figure 9(d2), it is clear that, as the constraints are applied on the middle of the lower connecting rod, the $\Delta K_{mx-\theta x}$ increases from beginning to end during the increasing of *r*.

Finite Element Simulation

To verify the stiffness modeling mentioned above, the numerical analysis and finite element analysis (FEA) are conducted using software package Matlab and ANSYS, respectively. The material of the stage is spring steel 65Mn with density

Table 1. The lower connecting rod is restricted completely

Stiffness	FEA	Calculating compliance	Neglecting compliance
$K_{fx-\Delta x}$	0.0105	0.0097	0.0098
$K_{fy-\Delta y}$	0.25	0.27	**0.63**
$K_{fz-\Delta z}$	72	92.73	**124.04**
$K_{mx-\theta x}$	5.7×10^{-4}	3.59×10^{-4}	5.43×10^{-4}
$K_{my-\theta y}$	0.018	0.014	0.016
$K_{mz-\theta z}$	8.61×10^{-5}	8.57×10^{-5}	8.57×10^{-5}

where the units are N/μm and Nm/μrad.

$\rho = 8.01 \times 10^3 \text{kg/m}^3$, Poisson ratio 0.27, and elastic modulus 210GPa. The geometric parameters are chosen: r=3mm, b=6mm, h=6mm, d=11.4mm, l=40mm, a=6.4mm. The parameters of the flexure hinges are r=3mm, b=6.4mm, t=0.4mm. The meshed three-dimensional finite element model of the planar four-bar parallelogram is shown in Figure 10, the element type is chosen as the 20-node solid element SOLID 95. The mesh of the flexure hinges is refined for more exact simulation results, for the torsion deformation of the flexure hinges is the largest in whole mechanism. The material and the architectural parameters of the parallelogram are same as the above mentioned. The lower connecting rod and its midpoint were fixed and a force/moment is exerted on the middle of the upper connecting rod. The six stiffness factors are calculated by the FEA approach and analytical method with the results shown in Table 1 and Table 2.

Comparing the results in Table 1 and Table 2, it can be observed that the simulation and theoretic results of the stiffness factors $K_{fy-\Delta y}$, $K_{fz-\Delta z}$ and $K_{my-\theta y}$ of the compliant four-bar parallelogram are evidently different, and the influence of the boundary conditions and the compliance of the connecting rods on the stiffness of the planar four-bar parallelogram is significant As for most of the stiffness, the results when the compliance of the connecting rod being calculated are closer to the

FEA results. Thus, the established mathematics models on the stiffness of the parallelogram are exact and the analysis results are credibility.

CONCLUSION

Aiming at the compliant characteristics of the prismatic pair in a novel precision positioning stage, the compliance matrix and matrix transformation method is utilized to establish the mathematics model with different premise. The comparisons are carried out under the conditions that the compliance of the connecting rods is considered and neglected, the lower connecting rod are restricted partly and fully. Furthermore, the influences of the geometric parameters on the stiffness factors were represented graphically. Finally, simulation experiments were conducted using finite element software ANSYS. The comparison of FEA and theoretical results indicates that the neglect of the compliance of the connecting rods and the simplifying of the mathematics model will result in low calculation accuracy, and further the constraint conditions should be taken into account during the stiffness modeling.

Table 2. The middle of the lower connecting rod is restricted

Stiffness	FEA	Calculating compliance	Neglecting compliance
$K_{fx-\Delta x}$	0.0148	0.0097	0.00975
$K_{fy-\Delta y}$	0.16	0.27	**0.63**
$K_{fz-\Delta z}$	60.46	74.044	**124.04**
$K_{mx-\theta x}$	5.16×10^{-4}	3.15×10^{-4}	5.43×10^{-4}
$K_{my-\theta y}$	0.0029	0.0074	0.016
$K_{mz-\theta z}$	**6.79×10^{-5}**	8.52×10^{-5}	8.58×10^{-5}

where the units is N/μm and Nm/μrad.

ACKNOWLEDGMENT

This research is supported by National Natural Science Foundation of China (Grant No. 50705064), and Natural Science Foundation of Tianjin (Grant No. 08JCYBJC01400).

REFERENCES

Dong, W., Sun, L. N., & Du, Z. J. (2008). Stiffness research on a high-precision, large-workspace parallel mechanism with compliant joints. *Precision Engineering*, *32*, 222–231. doi:10.1016/j. precisioneng.2007.08.002.

Koseki, Y., Tanikawa, T., & Koyachi, N. (2000). Kinematic analysis of translational 3-DOF micro parallel mechanism using matrix method. In *Proceedings of the IROS*, Kagawa, Japan (pp. 786-792).

Lobontin, N., & Garcia, E. (2003). Analytical model of displacement amplification and stiffness optimization for a class of flexure-based compliant mechanisms. *Computers & Structures*, *81*(32), 2797–2810. doi:10.1016/j.compstruc.2003.07.003.

Paros, J. M., & Weisbord, L. (1965). How to design flexure hinges. *Machine Design*, *37*, 151–157.

Pham, H. H., & Chen, I. M. (2005). Stiffness modeling of flexure parallel mechanism. *Precision Engineering*, *29*, 467–478. doi:10.1016/j. precisioneng.2004.12.006.

Ryu, J. W., & Gweon, D. G. (1997a). Error analysis of a flexure hinge mechanism induced by machining imperfection. *Precision Engineering*, *21*(2), 83–89. doi:10.1016/S0141-6359(97)00059-7.

Ryu, J. W., & Gweon, D. G. (1997b). Optimal design of a flexure hinge based *XYθ* wafer stage. *Precision Engineering*, *21*(1), 18–28. doi:10.1016/ S0141-6359(97)00064-0.

Smith, S. T., Badami, V. G., & Dale, J. S. (1997). Elliptical flexure hinges. *The Review of Scientific Instruments*, *68*(3), 1474–1483. doi:10.1063/1.1147635.

Tian, Y., Shirinzadeh, B., & Zhang, D. (2008). Development and dynamic modeling of a flexure-based Scott-Russell mechanism for nanomanipulation. *Mechanical Systems and Signal Processing*, *23*(3), 957–978. doi:10.1016/j. ymssp.2008.06.007.

Tian, Y., Shirinzadeh, B., & Zhang, D. (2008). Stiffness estimation of the flexure-based five-bar micro-manipulator. In *Proceedings of the 10th Intl. Conf. on Control, Automation, Robotics and Vision*, Hanoi, Vietnam (pp. 599-604).

Tian, Y., Shirinzadeh, B., & Zhang, D. (2009). A flexure-based mechanism and control methodology for ultra-precision turning operation. *Precision Engineering*, *33*(2), 160–166. doi:10.1016/j. precisioneng.2008.05.001.

Tian, Y., Shirinzadeh, B., & Zhang, D. (2009). Design and forward kinematics of the compliant micro-manipulator with lever mechanisms. *Precision Engineering*, *33*(4), 466–475. doi:10.1016/j. precisioneng.2009.01.003.

Tian, Y., Shirinzadeh, B., & Zhang, D. (2010). Three flexure hinges for compliant mechanism designs based on dimensionless graph analysis. *Precision Engineering*, *34*(1), 92–100. doi:10.1016/j. precisioneng.2009.03.004.

Tsai, L., & Stamper, R. (1996). A parallel manipulator with only translational degrees of freedom. In *Proceedings of the 1996 ASME Design Engineering Technical Conference (MECH)* (p. 1152).

Xu, Q. S., & Li, Y. M. (2006). Stiffness modeling for an orthogonal 3-PUU compliant parallel micromanipulator. In *Proceeding of the 2006 IEEE International Conference on Mechatronics and Automation*, Luoyang, China (pp. 124-129).

Yang, Q. Z., Yin, X. Q., & Ma, L. Z. (2005). Establishing stiffness of prismatic pair in fully compliant parallel micro-robot using energy method. *Journal of Jiangsu University, 26*(1), 12–15.

Yi, B. J., Na, H. Y., & Chung, G. B. (2002). Design and experiment of a 3 DOF parallel micro-mechanism utilizing flexure hinges. In *Proceedings of the IEEE International Conference on Robotics and Automation 2002* (Vol. 2, pp. 1167-1172).

Yu, J. J., Bi, S. S., & Zong, G. H. (2002). Analysis for the static stiffness of a 3-DOF parallel compliant micromanipulator. *Chinese journal of mechanical engineering, 38*(4), 7-10.

This work was previously published in the International Journal of Intelligent Mechatronics and Robotics, Volume 1, Issue 1, edited by Bijan Shirinzadeh, pp. 61-78, copyright 2011 by IGI Publishing (an imprint of IGI Global).

Chapter 5
Computation of the Output Torque, Power and Work of the Driving Motor for a Redundant Parallel Manipulator

Yongjie Zhao
Shantou University, China

ABSTRACT

Inverse dynamic analysis of the 8-PSS redundant parallel manipulator is carried out in the exhaustive decoupled way. The required output of the torque, the power and the work of the driving motor are achieved. The whole actuating torque is divided into four terms which are caused by the acceleration, the velocity, the gravity, and the external force. It is also decoupled into the components contributed by the moving platform, the strut, the slider, the lead screw, the motor rotor-coupler, and the external force. The required powers contributed by the component of torque caused by the acceleration term, the velocity term, the gravity term, the external force term, and the powers contributed by the moving platform, the strut, the slider, the lead screw, and the motor rotor-coupler are computed respectively. For a prescribed trajectory, the required output work generated by the ith driving motor is obtained by the presented numerical integration method. Simulation for the computation of the driving motor's output torque, power and work is illustrated.

INTRODUCTION

There are mainly two different types of redundancy for the parallel manipulators: a) kinematic redundancy and b) actuation redundancy. A parallel manipulator is said to be kinematically redundant manipulator when its mobility of the mechanism is greater than the required degrees of freedom of the moving platform. On the other hand, a parallel manipulator is called redundantly actuated manipulator when the number of actuators is greater than the mobility of the mechanism. It is believed that redundancy can improve the ability and performance of parallel manipulator (Kim, 2001; Merlet, 1996; Nokleby, 2005; Wang, 2004; Cheng, 2003; Müller, 2005; Mohamed, 2005;

DOI: 10.4018/978-1-4666-3634-7.ch005

Ebrahimi, 2007; Zhao, 2009). There are some advantages for the redundant parallel manipulator such as avoiding kinematic singularities, increasing workspace, improving dexterity, enlarging load capability and so on. It is shown that the redundantly actuated parallel manipulator has a better dynamic characteristic than its non-redundant counterpart considering the presented dynamic index (Zhao, 2009). Redundant actuation in a parallel manipulator can be implemented by the following approaches. The first one is to actuate some of the passive joints within the branches of parallel manipulator. The second one is to add some additional branches beyond the minimum necessary to actuate the parallel manipulator. The last one can be the hybrid of the above two approaches.

The dynamics of the manipulator are of importance in the areas of simulation, control and dynamic optimum design. Many works have been done on the dynamics of parallel manipulators. Several approaches, including the Newton-Euler formulation (Carvalho, 2001; Dasgupta, 1998), the Lagrangian formulation (Lee, 1988; Miller, 1992), the Kane formulation (Ben-Horin, 1998; Liu, 2000) and the virtual work principle (Li, 2005; Sokolov, 2007; Wang, 1998; Tsai, 2000; Zhu, 2005; Staicu, 2009) have been applied to the dynamics analysis of parallel manipulators. In fact, the inverse dynamics of parallel manipulators involve almost all of the mechanics principles. Along with these mechanics principles, many mathematic methods such as screw theory (Gallardo, 2003), Lie algebra (Muller, 2003), natural orthogonal complement (Khan, 2005), motor algebra (Sugimoto, 1987), group theory (Geike, 2003), symbolic programming (Geike, 2003; McPhee, 2002), geometric approach (Selig, 1999), parallel computational algorithms (Gosselin, 1996) and system identification (Wiens, 2002) have also been adopted to the dynamics of parallel manipulators. However, the results computed by different methods have been shown to be equivalent.

Though much attention has been paid to the dynamics of the parallel manipulator, little work has been done on the dynamics of the redundant parallel manipulator (Cheng, 2003), especially in the exhaustive decoupled way. Furthermore, maybe no paper has dealt with the required output power and work of the driving motor when the redundant parallel manipulator moves on a prescribed trajectory. It is one of the motivations for this work. By taking the 8-PSS redundant parallel manipulator as an object of study, this paper presents the inverse dynamic analysis in the exhaustive decoupled way. The actuating torques caused by the following term: acceleration, velocity, gravity, external force, moving platform, strut, slider, lead screw and motor rotor-coupler are computed respectively. The required output powers generated by the motor corresponding to the above terms of torque are achieved. For a prescribed trajectory, the computation of the required output work generated by the *i*th driving motor is implemented by the numerical integration. The paper is organized as follows: the description and the dynamic model of the redundant parallel manipulator are presented in section two. The rigid dynamic model is decoupled in the exhaustive decoupled way. The computation of the required output power and work of the driving motor are given in section three. Investigations of the system dynamic characteristics through simulation and conclusions are presented in section four and section five, respectively.

SYSTEM DESCRIPTION AND DYNAMIC MODEL

System Description

As shown in Figure 1 and Figure 2, the 8-PSS redundant parallel manipulator consists of a moving platform and eight sliders. In each kinematic chain, the platform and the slider are connected via spherical ball bearing joints by a strut of fixed length. Each slider is driven by a DC motor via a linear ball screw. So the 8-PSS is an out actuated redundant parallel manipulator. The lead screws of B_1, B_2, B_3 and B_4 are vertical to the ground. The lead screws of B_5, B_6 and the lead screws of B_7, B_8 are parallel with the ground. They are orthogonal to each other.

For the purpose of analysis, the following coordinate systems shown in Figure 1 and Figure 2 are defined: the coordinate system $O - xyz$ is attached to the fixed base and another moving coordinate frame $O' - uvw$ is located at the center of mass of the moving platform. The orientation of the moving frame with respect to the fixed frame is described by the matrix which is a function of three rotation angles ϕ_x, ϕ_y and ϕ_z about the fixed x, y and z axis. So the angular velocity of the moving platform is given by (Tsai, 2000)

$$\omega = \begin{bmatrix} \dot{\phi}_x & \dot{\phi}_y & \dot{\phi}_z \end{bmatrix}^T \tag{1}$$

The orientation of each kinematic strut with respect to the fixed base can be described by two Euler angles. As shown in Figure 3, the local coordinate system of the ith strut can be thought of as a rotation of ϕ_i about the z axis resulting in a $C_i - x'_i y'_i z'_i$ system followed by another rotation of φ_i about the rotated y'_i axis. So the rotation matrix of the ith strut can be written as

$$^0\mathbf{R}_i = \text{Rot}(z, \phi_i)\,\text{Rot}(y'_i, \varphi_i) \tag{2}$$

The unit vector along the strut in the coordinate system $O - xyz$ is

$$\mathbf{w}_i = {}^0\mathbf{R}_i\,{}^i\mathbf{w}_i = {}^0\mathbf{R}_i \begin{bmatrix} 0 \\ 0 \\ 1 \end{bmatrix} = \begin{bmatrix} c\phi_i\,s\varphi_i \\ s\phi_i\,s\varphi_i \\ c\varphi_i \end{bmatrix} \tag{3}$$

So the Euler angles ϕ_i and φ_i can be computed as follows

Figure 1. Diagram of an 8-PSS redundant parallel manipulator

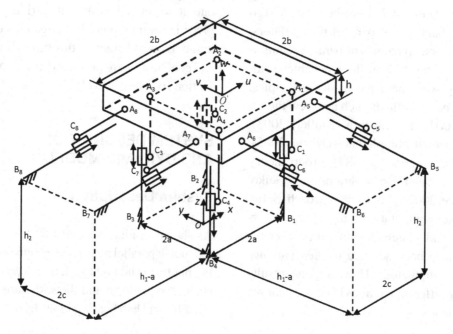

Figure 2. Vector diagram of a PSS kinematic chain

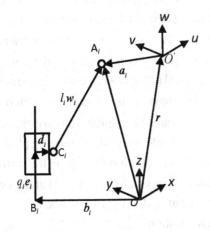

$$
\begin{cases}
c\varphi_i = \mathbf{w}_{iz} \\
s\varphi_i = \sqrt{\mathbf{w}_{ix}^2 + \mathbf{w}_{iy}^2}, \ (0 \le \varphi_i < \mathrm{\AA}) \\
s\phi_i = \mathbf{w}_{iy} / s\varphi_i \\
c\phi_i = \mathbf{w}_{ix} / s\varphi_i \\
\text{if } \varphi_i = 0, \text{ then } \phi_i = 0
\end{cases}
\tag{4}
$$

Dynamic Model

The rigid dynamic model of the 8-PSS redundant parallel manipulator had been formulated by means of the principle of virtual work and the concept of the link Jacobian matrices (Zhao, 2009) while considering the rotation inertia of the lead screw and the rotation inertia of motor rotor and coupler. There are eight unknown quantities in the six linear consistent equations for the 8-PSS redundant parallel manipulator. The most common strategy to solve this kind of problem with infinite solutions is by minimizing the Euclidean norm of the actuating forces. The rigid dynamic model of the two parallel manipulators can be expressed as

$$
\begin{aligned}
\tau &= -A^{-T} \left(J^T \right)^+ \left(Q_p + \sum_{i=1}^{8} J_{iv\omega}^T {}^iQ_i + J^T f_q \right) + I_{LCM} \ddot{\theta} \\
&= -A^{-T} \left(J^T \right)^+ \left(Q_p + \sum_{i=1}^{8} J_{iv\omega}^T {}^iQ_i + J^T f_q \right) + I_{LCM} A\ddot{q}
\end{aligned}
\tag{5}
$$

Box 1.

$$
J = diag\left(\frac{1}{w_1^T e_1}, \frac{1}{w_2^T e_2}, \cdots, \frac{1}{w_i^T e_i}, \cdots, \frac{1}{w_n^T e_n} \right) \begin{bmatrix} w_1, w_2, \cdots, w_i, \cdots, w_n \\ a_1 \times w_1, a_2 \times w_2, \cdots, a_i \times w_i, \cdots, a_n \times w_n \end{bmatrix}^T
\tag{6a}
$$
$$
i = 1, 2, \cdots, n; \ n = 6 \ or \ 8
$$

$$
J_{iv\omega} = \begin{bmatrix} \left[{}^iR_o \quad -S({}^ia_i) {}^iR_o \right] + \dfrac{l_i}{2} S({}^iw_i) J_{i\omega} \\ \dfrac{1}{l_i} \left\{ \left[S({}^iw_i) {}^iR_o \quad -S({}^iw_i)S({}^ia_i) {}^iR_o \right] - ({}^iw_i \times {}^ie_i) \left[\dfrac{w_i^T}{w_i^T e_i} \quad \dfrac{(a_i \times w_i)^T}{w_i^T e_i} \right] \right\} \end{bmatrix} = \begin{bmatrix} J_{iv} \\ J_{i\omega} \end{bmatrix}
\tag{6b}
$$

$$
S({}^iw_i) = \begin{bmatrix} 0 & -{}^iw_{iz} & {}^iw_{iy} \\ {}^iw_{iz} & 0 & -{}^iw_{ix} \\ -{}^iw_{iy} & {}^iw_{ix} & 0 \end{bmatrix}
\tag{6c}
$$

$$
S({}^ia_i) = \begin{bmatrix} 0 & -{}^ia_{iz} & {}^ia_{iy} \\ {}^ia_{iz} & 0 & -{}^ia_{ix} \\ -{}^ia_{iy} & {}^ia_{ix} & 0 \end{bmatrix}
\tag{6d}
$$

Figure 3. The local coordinate system of the ith strut

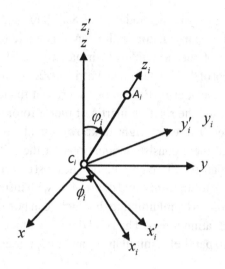

The physical significance of Equation (5) is that among the possible actuating torque vectors, the optimum solution is with minimum norm and least quadratic sum. τ is the actuating torques, $\mathbf{J}^+$ is the Moore-Penrose inverse of the Jacobian matrix and when $\mathbf{J} \in \mathbf{R}_m^{m \times n}$, $\mathbf{J}^+ = \mathbf{J}^T(\mathbf{J}\mathbf{J}^T)^{-1}$, $\mathbf{J}_{iv\omega}$ is the *link Jacobian matrix* which maps the velocity of the moving platform into the velocity of the ith strut in the $C_i - x_i y_i z_i$ coordinate system, and can be expressed as that shown in Box 1.

In Equation (5), $\mathbf{Q}_P$ is the resultant of applied and inertia forces exerted at the center of mass of the moving platform

$$Q_P = \begin{bmatrix} f_P \\ n_P \end{bmatrix} = \begin{bmatrix} f_e + m_p g - m_p \dot{v} \\ n_e - {}^oI_p \dot{\omega} - \omega \times ({}^oI_p \omega) \end{bmatrix} \quad (7)$$

where $\mathbf{f}_e$ and $\mathbf{n}_e$ are the external force and moment exerted at the center of mass of the moving platform, ${}^o\mathbf{I}_p = {}^o\mathbf{R}_{o'} {}^{o'}\mathbf{I}_p {}^{o'}\mathbf{R}_o$ is the inertia matrix of the moving platform taken about the center of mass expressed in the $O - xyz$ coordinate system and m_p is its mass. $\mathbf{g}$ is the gravity acceleration. ω, $\dot{v}$ and $\dot{\omega}$ are the angular velocity, the linear and angular acceleration of the moving platform, respectively. In Equation (5), ${}^i\mathbf{Q}_i$ is the resultant of applied and inertia forces exerted at the center of the ith strut and expressed in the $C_i - x_i y_i z_i$ coordinate system

$$^iQ_i = \begin{bmatrix} {}^if_i \\ {}^in_i \end{bmatrix} = \begin{bmatrix} m_i\,{}^iR_o g - m_i\,{}^i\dot{v}_i \\ -{}^iI_i\,{}^i\dot{\omega}_i - {}^i\omega_i \times ({}^iI_i\,{}^i\omega_i) \end{bmatrix} \quad (8)$$

where ${}^i\mathbf{I}_i$ is the inertia matrix of the ith cylindrical strut about its center of mass expressed in the $C_i - x_i y_i z_i$ coordinate system and m_i is its mass. ${}^i\dot{v}_i$, ${}^i\omega_i$ and ${}^i\dot{\omega}_i$ are the linear acceleration, the angular velocity and acceleration of the ith strut expressed in the $C_i - x_i y_i z_i$ coordinate system respectively, and can be given by that shown in Box 2.

In Equation (5), $\mathbf{f}_q$ is the resultant of applied and inertia forces exerted at the center of the slider expressed in the $O - xyz$ coordinate system

$$\mathbf{f}_q = \begin{bmatrix} f_{q1} & f_{q2} & f_{q3} & f_{q4} & f_{q5} & f_{q6} & f_{q7} & f_{q8} \end{bmatrix}^T \quad (14)$$

Table 1. The parameters of the base platform (unit: m)

	1	2	3	4	5	6	7	8
x_{Bi}	0.400000	0.400000	-0.400000	-0.400000	0.400000	-0.400000	-2.000000	-2.000000
y_{Bi}	-0.400000	0.400000	0.400000	-0.400000	-2.000000	-2.000000	-0.400000	0.400000
z_{Bi}	0.000000	0.000000	0.000000	0.000000	1.500000	1.500000	1.500000	1.500000

Table 2. The parameters of the moving platform which are measured in the coordinate frame $O' - uvw$ *(unit: m)*

	1	2	3	4	5	6	7	8
x_{Ai}	0.400000	0.400000	-0.400000	-0.400000	0.400000	-0.400000	-0.681000	-0.681000
y_{Ai}	-0.400000	0.400000	0.400000	-0.400000	-0.681000	-0.681000	-0.400000	0.400000
z_{Ai}	-0.166000	-0.166000	-0.166000	-0.166000	-0.037500	-0.037500	-0.037500	-0.037500

and

$$f_{qi} = (m_{qi}g - m_{qi}\ddot{q}_i)^T e_i \qquad (15)$$

where m_{qi} and $\ddot{q}_i$ are the mass and the acceleration of the slider. $\ddot{q}_i$ is the acceleration of the slider (see Box 3).

In Equation (5), $\mathbf{I}_{LCM}$ is the diagonal inertia matrix (see Box 4).

I_{Li}, I_{Ci}, and I_{Mi} are the rotary inertia of the lead screw, coupler and motor rotor, respectively.

In Equation (5), $\ddot{\theta}$ is the diagonal matrix of the angular acceleration of the screw-coupler-rotor

$$\ddot{\theta} = \begin{bmatrix} \ddot{\theta}_1 & \ddot{\theta}_2 & \ddot{\theta}_3 & \ddot{\theta}_4 & \ddot{\theta}_5 & \ddot{\theta}_6 & \ddot{\theta}_7 & \ddot{\theta}_8 \end{bmatrix}^T$$
$$= diag\left(\frac{2\pi}{p_1} \quad \frac{2\pi}{p_2} \quad \frac{2\pi}{p_3} \quad \frac{2\pi}{p_4} \quad \frac{2\pi}{p_5} \quad \frac{2\pi}{p_6} \quad \frac{2\pi}{p_7} \quad \frac{2\pi}{p_8} \right) \ddot{q}$$
$$= A\ddot{q} \qquad (19)$$

Box 2.

$$^i\omega_i = J_{i\omega} \begin{bmatrix} v \\ \omega \end{bmatrix} \qquad (9)$$

$$^i\dot{\omega}_i = J_{i\omega} \begin{bmatrix} \dot{v} \\ \dot{\omega} \end{bmatrix} + \frac{1}{l_i}(\Delta_1 + \Delta_2) \qquad (10)$$

$$\Delta_1 = -\frac{(^iw_i \times {}^ie_i)}{w_i^T e_i}((w_i^T \omega)(a_i^T \omega) - (w_i^T a_i)(\omega^T \omega) + l_i|\omega_i \times w_i|^2) \qquad (11)$$

$$\Delta_2 = ({}^i\omega_i^T {}^ia_i)({}^iw_i \times {}^i\omega_i) - ({}^i\omega^T {}^i\omega)({}^iw_i \times {}^ia_i) \qquad (12)$$

$$^i\dot{v}_i = J_{iv} \begin{bmatrix} \dot{v} \\ \dot{\omega} \end{bmatrix} + S({}^i\omega_i)S({}^i\omega_i){}^ia_i + \frac{1}{2}S({}^iw_i)(\Delta_1 + \Delta_2) - \frac{l_i}{2}S({}^i\omega_i)S({}^i\omega_i){}^iw_i \qquad (13)$$

Box 3.

$$\ddot{q}_i = \frac{1}{w_i^T e_i}(w_i^T \dot{v} + (a_i \times w_i)^T \dot{\omega} + w_i^T(\omega \times (\omega \times a_i)) - w_i^T(\omega_i \times (\omega_i \times l_i w_i))) \qquad (16)$$

Box 4.

$$\mathbf{I}_{LCM} = \mathrm{diag}\begin{pmatrix} I_{LCM1} & I_{LCM2} & I_{LCM3} & I_{LCM4} & I_{LCM5} & I_{LCM6} & I_{LCM7} & I_{LCM8} \end{pmatrix} \qquad (17)$$

$$I_{LCMi} = I_{Li} + I_{Ci} + I_{Mi} \qquad (18)$$

Table 3. The length of the strut C_iA_i (unit: m)

l_i	1	2	3	4	5	6	7	8
	1.000000	1.000000	1.000000	1.000000	1.000000	1.000000	1.000000	1.000000

Table 4. The mass parameters of the manipulator (unit: kg)

	1	2	3	4	5	6	7	8
m_i	20	20	20	20	20	20	20	20
m_{qi}	50	50	50	50	50	50	50	50

where $p_i = 0.05\mathrm{m}^{-1}$ is the lead of the linear ball screw. Other than the speed reduction caused by the pitch of the lead screws there is no speed reducer for the redundant parallel manipulator, otherwise the reduction ration should be included in Equation (19).

Substituting Equation (7), Equation (8) and Equation (14) into Equation (5) yields that seen in Box 5, where $\mathbf{0} = \begin{bmatrix} 0 & 0 & 0 \end{bmatrix}^T$. Simplifying the inverse dynamics model of the redundant parallel manipulator yields

$$\tau = M(X)\ddot{X} + V(X,\dot{X}) + g(X) - A^{-T}J^{-T}\begin{bmatrix} f_e \\ n_e \end{bmatrix}$$

$$= \tau_a + \tau_v + \tau_g + \tau_e \qquad (21)$$

where $\mathbf{X}$ is the pose and position of the moving platform and given by that seen in Box 6. Equation

25 is the generalized inertia matrix of the manipulator which maps the acceleration of the moving platform into the actuating torques. $\mathbf{E}_3$ denotes the 3×3 unit matrix. τ_a, τ_v, τ_g and τ_e are the acceleration term, the velocity term, the gravity term and the external force term of the inverse dynamic equations respectively.

From Equation (20), it is shown that the actuating torques are also determined by the inertia parameters of the moving platform, the strut, the slider, the lead screw, the motor rotor-coupler and the external forces. So Equation (20) can be written as

$$\tau = \tau_{mov} + \tau_{strut} + \tau_{slider}$$
$$+ \tau_{lead\ screw} + \tau_{motor\ rotor-coupler} + \tau_e \qquad (26)$$

Table 5. The required output work generated by the ith driving motor and their sum for the prescribed trajectory (unit: J)

	1	2	3	4	5	6	7	8	sum
W_i	99.697	1544.119	2819.088	401.408	166.805	753.256	6495.371	2962.475	15242.219

Box 5.

$$
\begin{aligned}
\tau = {} & -A^{-T}\left(J^T\right)^+ \begin{bmatrix} f_e \\ n_e \end{bmatrix} - A^{-T}\left(J^T\right)^+ \left\{ \begin{bmatrix} m_p g \\ 0 \end{bmatrix} + \sum_{i=1}^{8} J_{iv\omega}^T \begin{bmatrix} m_i\,{}^i R_o g \\ 0 \end{bmatrix} \right. \\
& + J^T \left[(m_{q1}g)^T e_1 \quad (m_{q2}g)^T e_2 \quad (m_{q3}g)^T e_3 \quad (m_{q4}g)^T e_4 \right. \\
& \left. (m_{q5}g)^T e_5 \quad (m_{q6}g)^T e_6 \quad (m_{q7}g)^T e_7 \quad (m_{q8}g)^T e_8 \right]^T \right\} \\
& + A^{-T}\left(J^T\right)^+ \left\{ \begin{bmatrix} m_p \dot{v} \\ {}^o I_p \dot{\omega} \end{bmatrix} + \sum_{i=1}^{8} J_{iv\omega}^T \begin{bmatrix} m_i\,{}^i \dot{v}_i \\ {}^i I_i\,{}^i \dot{\omega}_i \end{bmatrix} \right. \\
& + J^T \left[m_{q1}\ddot{q}_1 \quad m_{q2}\ddot{q}_2 \quad m_{q3}\ddot{q}_3 \quad m_{q4}\ddot{q}_4 \quad m_{q5}\ddot{q}_5 \quad m_{q6}\ddot{q}_6 \quad m_{q7}\ddot{q}_7 \quad m_{q8}\ddot{q}_8 \right]^T \right\} + I_{LCM} A \ddot{q} \\
& + A^{-T}\left(J^T\right)^+ \left\{ \begin{bmatrix} 0 \\ \omega \times ({}^o I_p \omega) \end{bmatrix} + \sum_{i=1}^{8} J_{iv\omega}^T \begin{bmatrix} 0 \\ {}^i\omega_i \times ({}^i I_i\,{}^i\omega_i) \end{bmatrix} \right\}
\end{aligned}
\tag{20}
$$

Box 6.

$$
X = \begin{bmatrix} x & y & z & \phi_x & \phi_y & \phi_z \end{bmatrix}^T
\tag{22}
$$

$$
\dot{X} = \begin{bmatrix} v \\ \omega \end{bmatrix}
\tag{23}
$$

$$
\ddot{X} = \begin{bmatrix} \dot{v} \\ \dot{\omega} \end{bmatrix}
\tag{24}
$$

$$
\begin{aligned}
M(X) = {} & A^{-T}\left(J^T\right)^+ \left(\begin{bmatrix} m_p E_3 & \\ & {}^o I_p \end{bmatrix} + \sum_{i=1}^{8} J_{iv}^T m_i J_{iv} + \sum_{i=1}^{8} J_{i\omega}^T\,{}^i I_i J_{i\omega} \right) \\
& + A^{-T} diag\left(m_{q1} \quad m_{q2} \quad m_{q3} \quad m_{q4} \quad m_{q5} \quad m_{q6} \quad m_{q7} \quad m_{q8} \right) J + A I_{LCM} J
\end{aligned}
\tag{25}
$$

COMPUTATIONS OF THE REQUIRED OUTPUT POWER AND WORK OF THE DRIVING MOTOR

The instantaneous required power of the ith actuated joint is

$$P_i = \tau_i \dot{q}_i \qquad (27)$$

where τ_i and $\dot{q}_i$ are the input torques and the angular velocity of the ith actuated joint. Substituting Equation (21) and Equation (26) into Equation (27) yields the equations presented in Box 7, where P_{ia}, P_{iv}, P_{ig} and P_{ie} are the required power of the ith actuated joint caused by the acceleration term, the velocity term, the gravity term and the external force term of the actuating torques, respectively. P_{imov}, P_{istrut}, $P_{islider}$, $P_{ilead\,screw}$ and $P_{imotor\,rotor-coupler}$ are the respective power term caused by the inertia of the moving platform, the strut, the slider, the lead screw and the motor rotor-coupler.

For the actuated joint, the sign of the required power is determined by the signs of the actuated joint's angular velocity and the torque which is exerted at the joint by the motor. However, the power generated by the driving motor should be regarded as positive since energy consumption is always necessary. So the required output work generated by the ith driving motor for a specific motion period can be achieved

$$W_i = \int_{t_s}^{t_e} |P_i| \, dt \qquad (30)$$

where t_s and t_e are the start time and the end time of the motion.

The required output work generated by the ith driving motor can be implemented by the following numerical integration technique

$$W_i = \frac{\Delta t}{2}\left[|P_i(t_s)| + |P_i(t_e)| + 2\sum_{k=1}^{num-1} |P_i(t_k)| \right] \qquad (31)$$

where

$$\Delta t = \frac{t_e - t_s}{num} \qquad (32)$$

num is the number of the intervals. In general

$$num = 2^n \qquad (33)$$

where n is a positive integral. The required approximation error of the computation of the output work should satisfy

$$\left| W_i(num = 2^{n_1}) - W_i(num = 2^{n_1-1}) \right| \le e \qquad (34)$$

where e is the desired approximation error. n_1 is the positive integral adopted by the numerical integration.

Box 7.

$$P_i = (\tau_{ia} + \tau_{iv} + \tau_{ig} + \tau_{ie})\dot{q}_i = P_{ia} + P_{iv} + P_{ig} + P_{ie} \qquad (28)$$

and

$$P_i = (\tau_{imov} + \tau_{istrut} + \tau_{islider} + \tau_{ilead\,screw} + \tau_{imotor\,rotor-coupler} + \tau_{ie})\dot{q}_i$$
$$= P_{imov} + P_{istrut} + P_{islider} + P_{ilead\,screw} + P_{imotor\,rotor-coupler} + P_{ie} \qquad (29)$$

SIMULATION

In this section, a numerical example for the inverse dynamics analysis of the 8-PSS redundant parallel manipulator in the exhaustive way is presented. The simulation is implemented by using the MATLAB software. The parameters of the manipulator used for the simulation are given in Table 1 through Table 4.

The mass of the moving platform is $m_p = 200\text{kg}$. The inertia parameters used in the simulation are chosen as

$$
{}^{o'}\mathbf{I}_p = \begin{bmatrix} 17.333 & 0 & 0 \\ 0 & 17.333 & 0 \\ 0 & 0 & 33.333 \end{bmatrix} \text{kg} \cdot \text{m}^2,
$$

$$
{}^{i}\mathbf{I}_i = \begin{bmatrix} 1.279 & 0 & 0 \\ 0 & 1.279 & 0 \\ 0 & 0 & 0.005 \end{bmatrix} \text{kg} \cdot \text{m}^2,
$$

$$
I_{Li} = 10.5 \times 10^{-4} \text{kg} \cdot \text{m}^2,
$$

$$
I_{Ci} + I_{Mi} = 248 \times 10^{-4} \text{kg} \cdot \text{m}^2.
$$

Another parameter used in the simulation is given as $d_i = 0.244\text{m}$.

The motion of the moving platform used in the numerical simulation is expressed as

$$
\begin{cases}
x = -0.1 + \dfrac{a_{max}T^2}{2\pi}\left(\tau - \dfrac{1}{2\pi}\sin\left(2\pi\tau\right)\right) \\[2mm]
y = -0.1 + \dfrac{a_{max}T^2}{2\pi}\left(\tau - \dfrac{1}{2\pi}\sin\left(2\pi\tau\right)\right) \\[2mm]
z = 1.644 + \dfrac{a_{max}T^2}{2\pi}\left(\tau - \dfrac{1}{2\pi}\sin\left(2\pi\tau\right)\right) \\[2mm]
\phi_x = -0.1 + \dfrac{a_{max}T^2}{2\pi}\left(\tau - \dfrac{1}{2\pi}\sin\left(2\pi\tau\right)\right) \\[2mm]
\phi_y = -0.1 + \dfrac{a_{max}T^2}{2\pi}\left(\tau - \dfrac{1}{2\pi}\sin\left(2\pi\tau\right)\right) \\[2mm]
\phi_z = -0.1 + \dfrac{a_{max}T^2}{2\pi}\left(\tau - \dfrac{1}{2\pi}\sin\left(2\pi\tau\right)\right)
\end{cases} \quad (35)
$$

where $a_{max} = 9.8 m/s^2$, $\tau = \dfrac{t}{T}$, $T = \sqrt{\dfrac{2\pi S}{a_{max}}}s$

and $S = 0.2\text{m(rad)}$.

The position, velocity and acceleration of the sliders are shown in Figure 4.

The external force and moment exerted at the center of the moving platform are not considered since the redundant parallel manipulator is usually used for the motion simulator while not for the machine tool. According to Equation (20) and Equation (21), the whole actuating torques and the torques caused by the acceleration term, the velocity term and the gravity term can be computed respectively. The results shown in the Figure 5 indicate that the torques caused by the acceleration terms should be much bigger than the other two terms in the simulation.

From Equation (20) and Equation (26), the torques caused by moving platform, strut, slider, lead screw, motor rotor and coupler are computed respectively. They are shown in the Figure 6.

According to Equation (28), the required powers and the powers caused by the acceleration term, the velocity term and the gravity term of torques can be achieved respectively. The results of the simulation shown in Figure 7 indicate that the required powers are primarily dominated by the component caused by the acceleration term of torque.

Figure 4. Variations of slider position (a), velocity (b) and acceleration (c) vs time

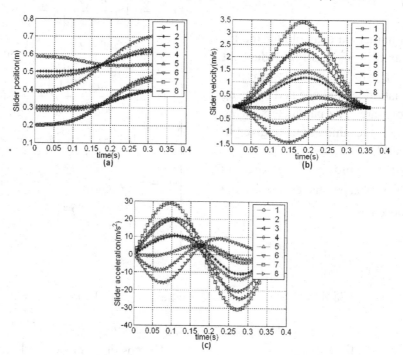

Figure 5. Variations of whole actuating torques (a), torques caused by acceleration term (b), torques caused by the velocity term (c) and torques caused by the gravity term (d) vs time

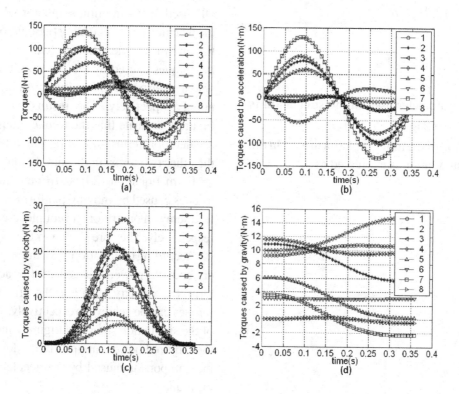

Figure 6. Variations of torques caused by the moving platform (a), torques caused by the strut (b), torques caused by the slider (c), torques caused by the lead screw (d) and torques caused by the motor rotor and coupler (e) vs time

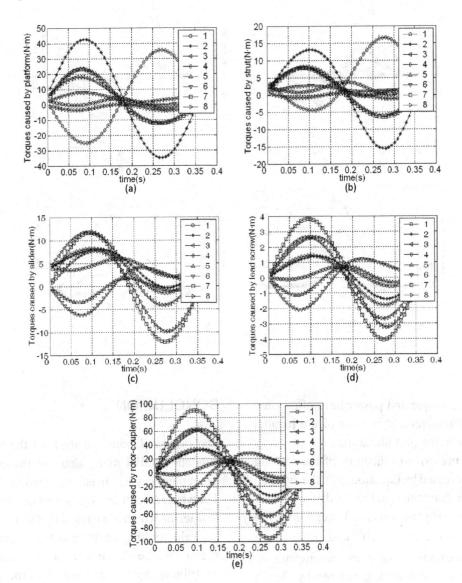

According to Equation (29), the required powers caused by the moving platform, the strut, the slider, the lead screw, the coupler and the motor rotor are computed respectively. The results of the simulation shown in Figure 8 indicate that the required powers caused by the rotation inertia of the motor rotor-coupler are much bigger than the others.

The results shown in the Figure 6 and Figure 8 turn out that the required torques and powers caused by the rotation inertia of motor-coupler-screw should be included for the exact dynamic model of the out actuated parallel manipulator used for the design of control law or the estimation of servomotor parameters. However, the effects of the rotation inertia of motor-coupler-screw on

Figure 7. Variations of whole actuating powers (a), powers caused by acceleration term of torque (b), powers caused by the velocity term of torque (c) and powers caused by the gravity term of torque (d) vs time

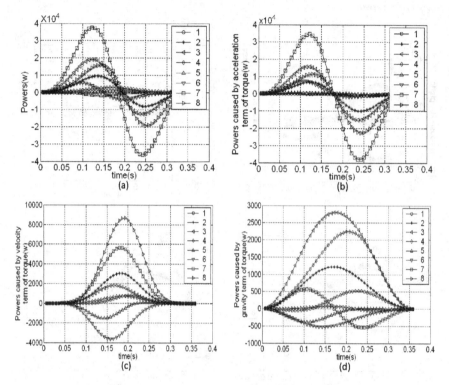

the actuating torque and power have often been omitted in the inverse dynamics of the parallel manipulator in the past literatures.

For the prescribed trajectory of the moving platform expressed by Equation (25), the required output work generated by the ith driving motor can be achieved by the numerical integration. The approximation error $e = 10^{-3}$ and $n_1 = 12$ are adopted for the numerical integration computation. The required output work generated by the ith driving motor and their sum for the prescribed motion is shown in Table 5. For the prescribed trajectory, the required output work generated by the seventh driving motor is biggest while the required output work generated by the first driving motor is smallest.

CONCLUSION

The inverse dynamic analysis of the 8-PSS redundant parallel manipulator in the exhaustive decoupled way has been presented in this paper. The whole actuating torques has been divided into four terms which are caused by the acceleration, the velocity, the gravity and the external force. It can also be decoupled into the components contributed by the moving platform, the strut, the slider, the lead screw, the motor rotor-coupler and the external force. The power contributed by the components of torque caused by the acceleration term, the velocity term, the gravity term, the external force term and the powers contributed by the moving platform, the strut, the slider, the lead screw, the motor rotor-coupler have been achieved respectively. For a prescribed trajectory, the required output work generated by the ith driving motor is obtained by the numeri-

Figure 8. Variations of powers caused by the moving platform (a), powers caused by the strut (b), powers caused by the slider (c), powers caused by the lead screw (d) and powers caused by the motor rotor and coupler (e) vs time

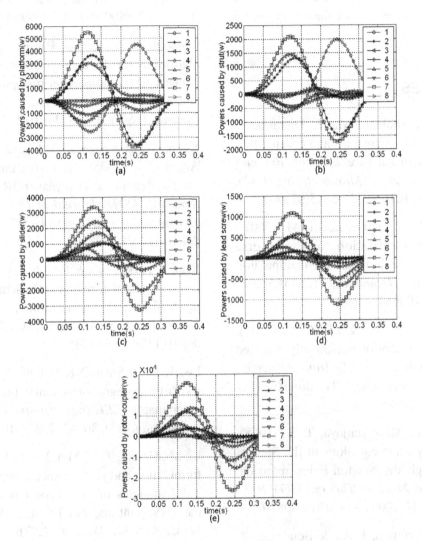

cal integration technique in the simulation. The simulation results show that torques and powers caused by the rotation inertia of the motor rotor-coupler are bigger than those of caused by the other component inertia of the manipulator. So the rotation inertia of the motor rotor-coupler should be included for the dynamic analysis of the out actuated parallel manipulator. The torque caused by the acceleration term is much bigger than the other two terms in the simulation. The required output power generated by the *i*th driving motor is primarily dominated by the component caused by the acceleration term of torque. The required output work generated by the seventh driving motor is biggest while the required output work generated by the first driving motor is smallest.

ACKNOWLEDGMENT

This research is jointly sponsored by the National Natural Science Foundation of China (Grant No.50905102), the China Postdoctoral Science Foundation (Grant No. 200801199) and the Natural Science Foundation of Guangdong Province (Grant No. 10151503101000033 and 8351503101000001). I would also like to thank the anonymous reviewers for their very useful comments.

REFERENCES

Ben-Horin, R., Shoham, M., & Djerassi, S. (1998). Kinematics, dynamics and construction of a planarly actuated parallel robot. *Robotics and Computer-integrated Manufacturing*, *14*(2), 163–172. doi:10.1016/S0736-5845(97)00035-5.

Carvalho, J. C. M., & Ceccarelli, M. (2001). A closed-form formulation for the inverse dynamics of a Cassino parallel manipulator. *Multibody System Dynamics*, *5*(2), 185–210. doi:10.1023/A:1009845926734.

Cheng, H., Yiu, Y. K., & Li, Z. X. (2003). Dynamics and control of redundantly actuated parallel manipulators. *IEEE Transactions on Mechatronics*, *8*(4), 483–491. doi:10.1109/TMECH.2003.820006.

Dasgupta, B., & Mruthyunjaya, T. S. (1998). Closed-form dynamic equations of the Stewart platform through the Newton-Euler approach. *Mechanism and Machine Theory*, *33*(7), 993–1012. doi:10.1016/S0094-114X(97)00087-6.

Ebrahimi, I., Carretero, J. A., & Boudreau, R. (2007). 3-PRRR redundant planar parallel manipulator: Inverse displacement, workspace and singularity analyses. *Mechanism and Machine Theory*, *42*(8), 1007–1016. doi:10.1016/j.mechmachtheory.2006.07.006.

Gallardo, J., Rico, J. M., & Frisoli, A. (2003). Dynamics of parallel manipulators by means of screw theory. *Mechanism and Machine Theory*, *38*(11), 1113–1131. doi:10.1016/S0094-114X(03)00054-5.

Geike, T., & McPhee, J. (2003). Inverse dynamic analysis of parallel manipulators with full mobility. *Mechanism and Machine Theory*, *38*(6), 549–562. doi:10.1016/S0094-114X(03)00008-9.

Gosselin, C. M. (1996). Parallel computational algorithms for the kinematics and dynamics of planar and spatial parallel manipulators. *ASME Journal of Dynamic Systems. Measurement and Control*, *118*(1), 22–28. doi:10.1115/1.2801147.

Khan, W. A., Krovi, V. A., Saha, S. K., & Angeles, J. (2005). Recursive kinematics and inverse dynamics for a planar 3R parallel manipulator. *ASME Journal of Dynamic Systems. Measurement and Control*, *127*(4), 529–536. doi:10.1115/1.2098890.

Kim, J., Park, F. C., & Ryu, S. J. (2001). Design and analysis of a redundantly actuated parallel mechanism for rapid machining. *IEEE Transactions on Robotics and Automation*, *17*(4), 423–434. doi:10.1109/70.954755.

Lee, K. M., & Shan, D. K. (1988). Dynamic analysis of a three-degrees-freedom in-parallel actuated manipulator. *IEEE Transactions on Robotics and Automation*, *4*(3), 361–367. doi:10.1109/56.797.

Li, M., Huang, T., & Mei, J. P. (2005). Dynamic formulation and performance comparison of the 3-dof modules of two reconfigurable PKMs – the TriVariant and the Tricept. *ASME Journal of Mechanical Design*, *127*(6), 1129–1136. doi:10.1115/1.1992511.

Liu, M. J., Li, C. X., & Li, C. N. (2000). Dynamics analysis of the Gough-Stewart platform manipulator. *IEEE Transactions on Robotics and Automation*, *16*(1), 94–98. doi:10.1109/70.833196.

McPhee, J., Shi, P., & Piedboeuf, J. C. (2002). Dynamics of multibody systems using virtual work and symbolic programming. *Mathematical and Computer Modelling of Dynamical Systems*, *8*(3), 137–155. doi:10.1076/mcmd.8.2.137.8591.

Merlet, J. P. (1996). Redundant parallel manipulators. *Laboratory Robotics and Automation, 8*(1), 17–24. doi:10.1002/(SICI)1098-2728(1996)8:1<17::AID-LRA3>3.0.CO;2-#.

Miller, K., & Clavel, R. (1992). The Lagrange-based model of Delta-4 robot dynamics. *Robotersysteme, 8*(4), 49–54.

Mohamed, M. G., & Gosselin, C. M. (2005). Design and analysis of kinematically redundant parallel manipulators with configurable platforms. *IEEE Transactions on Robotics, 21*(3), 277–287. doi:10.1109/TRO.2004.837234.

Müller, A. (2005). Internal preload control of redundantly actuated parallel manipulators-its application to backlash avoiding control. *IEEE Transactions on Robotics, 21*(4), 668–677. doi:10.1109/TRO.2004.842341.

Muller, A., & Maißer, P. (2003). A Lie-group formulation of kinematics and dynamics of constrained MBS and its application to analytical mechanics. *Multibody System Dynamics, 9*(4), 311–352. doi:10.1023/A:1023321630764.

Nokleby, S. B., Fisher, R., & Podhorodeski, R. P. (2005). Force capabilities of redundantly-actuated parallel manipulators. *Mechanism and Machine Theory, 40*(5), 578–599. doi:10.1016/j.mechmachtheory.2004.10.005.

Selig, J. M., & McAree, P. R. (1999). Constrained robot dynamics II: Parallel machines. *Journal of Robotic Systems, 16*(9), 487–498. doi:10.1002/(SICI)1097-4563(199909)16:9<487::AID-ROB2>3.0.CO;2-R.

Sokolov, A., & Xirouchakis, P. (2007). Dynamics analysis of a 3-dof parallel manipulator with R-P-S joint structure. *Mechanism and Machine Theory, 42*(5), 541–557. doi:10.1016/j.mechmachtheory.2006.05.004.

Staicu, S. (2009). Power requirement comparison in the 3-RPR planar parallel robot dynamics. *Mechanism and Machine Theory, 44*(5), 1045–1057. doi:10.1016/j.mechmachtheory.2008.05.009.

Sugimoto, K. (1987). Kinematics and dynamic analysis of parallel manipulator by means of motor algebra. *ASME Journal of Mechanisms, Transmissions, and Automation in Design, 109*(1), 3–7. doi:10.1115/1.3258783.

Tsai, L. W. (2000). Solving the inverse dynamics of a Stewart-Gough manipulator by the principle of virtual work. *ASME Journal of Mechanical Design, 122*(1), 3–9. doi:10.1115/1.533540.

Wang, J., & Gosselin, C. M. (1998). A new approach for the dynamic analysis of parallel manipulators. *Multibody System Dynamics, 2*(3), 317–334. doi:10.1023/A:1009740326195.

Wang, J., & Gosselin, C. M. (2004). Kinematic analysis and design of kinematically redundant parallel mechanisms. *ASME Journal of Mechanical Design, 126*(1), 109–118. doi:10.1115/1.1641189.

Wiens, G. J., Shamblin, S. A., & Oh, Y. H. (2002). Characterization of PKM dynamics in terms of system identification. *Journal of Multi-body Dynamics. Part K, 216*(1), 59–72.

Zhao, Y. J., & Gao, F. (2009). Dynamic performance comparison of the 8PSS redundant parallel manipulator and its non-redundant counterpart—the 6PSS parallel manipulator. *Mechanism and Machine Theory, 44*(5), 991–1008. doi:10.1016/j.mechmachtheory.2008.05.015.

Zhu, Z. Q., Li, J. S., Gan, Z. X., & Zhang, H. (2005). Kinematic and dynamic modeling for real-time control of Tau parallel robot. *Mechanism and Machine Theory, 40*(9), 1051–1067. doi:10.1016/j.mechmachtheory.2004.12.024.

This work was previously published in the International Journal of Intelligent Mechatronics and Robotics, Volume 1, Issue 2, edited by Bijan Shirinzadeh, pp. 1-17, copyright 2011 by IGI Publishing (an imprint of IGI Global).

Chapter 6
Random Weighting Estimation of One-Sided Confidence Intervals in Discrete Distributions

Yalin Jiao
Northwestern Polytechnical University, China

Shesheng Gao
Northwestern Polytechnical University, China

Yongmin Zhong
RMIT University, Australia

Bijan Shirinzadeh
Monash University, Australia

ABSTRACT

This paper presents a new random weighting method for estimation of one-sided confidence intervals in discrete distributions. It establishes random weighting estimations for the Wald and Score intervals. Based on this, a theorem of coverage probability is rigorously proved by using the Edgeworth expansion for random weighting estimation of the Wald interval. Experimental results demonstrate that the proposed random weighting method can effectively estimate one-sided confidence intervals, and the estimation accuracy is much higher than that of the bootstrap method.

INTRODUCTION

Random weighting method is an emerging computing concept in statistics (Gao & Zhong, 2010). It has received great attention in the recent years, and has been used to solve different problems (Gao & Zhong, 2010; Gao et al., 2008; Barvinok & Samorodnitsky, 2007; Fang & Zhao, 2006; Xue & Zhu, 2005; Gao, Zhang & Yang, 2004; Gao, Zhang

& Zhou, 2003). However, there has been little research to investigate random weighting estimation of one-sided confidence intervals. Currently, Bootstrap is a commonly used method for estimation of confidence intervals and distribution errors (Karlis & Patilea, 2008; Mandel & Betensky, 2008; Mantalosa & Zografos, 2008; Chakraborti & Li, 2007; Dogan, 2007; Haukoos & Lewis, 2005). This method can determine confidence intervals for an

DOI: 10.4018/978-1-4666-3634-7.ch006

unknown parameter of the underlying distribution function. Although both the random weighting method and Bootstrap method conduct statistical computations and analyses in a parallel manner, the random weighting method has advantages in comparison with the Bootstrap method (Gao & Zhong, 2010; Gao et al., 2008; Gao, Zhang & Yang, 2004; Gao, Zhang & Zhou, 2003; Zheng, 1987). The random weighting method is simple in computation and suitable for large samples. It does not need to know the distribution function. It can also be used to calculate a statistic with a probability density function, since the resultant statistical distribution provides a probability density function. Therefore, the random weighting method provides a promising solution for confidence interval estimation.

One-sided confidence intervals such as the commonly used Wald and Score intervals play an important role in many applications (Montgomery, 2001; Duncan, 1996). The one-sided interval estimation problem significantly differs from the two-sided problem, although there are some common features. In particular, despite the good performance of the Score interval in the two-sided case, the one-sided Score interval does not perform well for three important distributions, i.e., the binomial, negative binomial and Poisson distributions, in terms of coverage probability and expected length. Studies have shown that both the one-sided Wald and Score intervals suffer a pronounced systematic bias in the coverage, although the severity and direction are different (Kott & Liu, 2009; Cai, 2005).

Focusing on the problems of one-sided confidence intervals in the binomial, negative binomial and Poisson distributions, this paper adopts the random weighting method to estimation of one-sided confidence intervals in discrete distributions. Random weighting estimations are constructed for the Wald and Score intervals. Based on this, the coverage probability for the random weighting estimations is also studied. A theorem of coverage probability is rigorously proved for

random weighting estimation of the Wald interval. Experiments and comparison analysis have been conducted to comprehensively evaluate the performance of the proposed methodology for estimation of one-sided confidence intervals.

RANDOM WEIGHTING METHOD

Assume that $X_1, X_2, \cdots, X_n$ is a sample of independent and identically distributed random variables with common distribution function F. Let $x_1, x_2, \cdots, x_n$ be the corresponding observed realizations of $X_1, X_2, \cdots, X_n$. Further, we shall denote $\tilde{X}_n = (X_1, X_2, \cdots, X_n)$ and $\tilde{x}_n = (x_1, x_2, \cdots, x_n)$. Then, the random weighting process can be described as follows:

- Construct the sample (empirical) distribution function F_n from $\tilde{x}$, i.e.

$$F_n = \frac{1}{n}\sum_{i=1}^{n} X_i. \tag{1}$$

- The random weighting estimation of F_n is

$$H_n(x) = \sum_{i=1}^{n} V_i I_{(X_i \leq x)} \tag{2}$$

where $I_{(X_i \leq x)}$ is the characteristic function, and random vector $(V_1, \cdots, V_n)$ obeys Dirichlet distribution $D(1, \cdots, 1)$, that is, $\sum_{i=1}^{n} V_i = 1$ and the joint density function of $(V_1, V_2, \cdots, V_n)$ is

$$f(V_1, V_2, \cdots, V_n) = \Gamma(n),$$

where

$$(V_1, V_2, \cdots, V_{n-1}) \in D_{n-1}$$

and

$$D_{n-1} = \left\{ \begin{array}{l} (V_1, V_2, \cdots, V_{n-1}) : V_i \geq \\ 0, i = 1, \cdots, n-1, \sum_{i=1}^{n-1} V_i \leq 1 \end{array} \right\}.$$

RANDOM WEIGHTING ESTIMATION OF ONE-SIDED CONFIDENCE INTERVALS

Assume that $X_1, X_2, \cdots, X_n$ is a sample of independent and identically distributed random variables with common distribution function $F(x)$ which consists of three most important discrete distributions, i.e. binomial distribution *Bin(1, p)*, negative binomial distribution *N-bin(l, p)*, and Poisson distribution $Poi(\lambda)$. In the following, we will construct the random weighting estimations of one-sided confidence intervals for mean μ.

The binomial, negative binomial, and Poisson distributions are members of the discrete natural exponential family with a quadratic variance function (Cai, 2005; Brown, 1986; Morris, 1982). A common feature of these distributions is that variance σ^2 is a quadratic function of mean μ, i.e.

$$\sigma^2 \equiv V(\mu) = \mu + b\mu^2 \tag{3}$$

where $\mu = p$ and $b = -1$ for distribution *Bin(l, p)*, $\mu = \lambda$ and $b = 0$ for distribution $Poi(\lambda)$, and $\mu = p / (1 - p)$ and $b = 1$ for distribution *N-bin(l, p)*.

Throughout the paper, we shall denote by k the $100(1 - \alpha)th$ percentile of the standard normal distribution, and let

$$X = \sum_{i=1}^{n} X_i, \quad \hat{\mu} \equiv \bar{X} = \frac{1}{n} \sum_{i=1}^{n} X_i$$

and

$$\hat{\mu}^* = H(x) = \sum_{i=1}^{n} V_i X_i.$$

Wald and Score Intervals

Based on the normal approximation, the Wald interval can be constructed as (Wald & Wolfowitz, 1939)

$$W_n = \frac{\sqrt{n}(\hat{\mu} - \mu)}{V^{1/2}(\hat{\mu})} \to N(0,1) \tag{4}$$

Accordingly, the random weighting estimation of the Wald interval can be described as

$$W_n^* = \frac{\sqrt{n}(\hat{\mu}^* - \mu)}{V^{1/2}(\hat{\mu}^*)} \to N(0,1) \tag{5}$$

The $100(1 - \alpha)\%$ upper limit of the Wald interval is

$$CI_w^u = [0, \hat{u} + \kappa V^{1/2}(\hat{\mu})n^{-1/2}] = \\ [0, \hat{u} + \kappa(\hat{u} + b(\hat{u})^2)^{1/2}n^{-1/2}] \tag{6}$$

Accordingly, the $100(1 - \alpha)\%$ upper limit for the random weighting estimation of the Wald interval may be written as

$$\hat{C}I_w^u = [0, \hat{u}^* + \kappa V^{1/2}(\hat{\mu}^*)n^{-1/2}] = \\ [0, \hat{u}^* + \kappa(\hat{u}^* + b(\hat{u}^*)^2)^{1/2}n^{-1/2}] \tag{7}$$

The $100(1 - \alpha)\%$ lower limit of the Wald interval is defined as

$$CI_w^l = [\hat{u} - \kappa V^{1/2}(\hat{\mu})n^{-1/2}, 1] \tag{8}$$

Therefore, the $100(1 - \alpha)\%$ lower limit for random weighting estimation of the Wald interval can be represented as

$$\hat{C}I_w^l = [\hat{u}^* - \kappa V^{1/2}(\hat{\mu}^*)n^{-1/2}, 1] \tag{9}$$

$$\hat{CI}_w^l = [\hat{u} - \kappa V^{1/2}(\hat{\mu})n^{-1/2}, \infty] \qquad (10)$$

and

$$\hat{CI}_w^l = [\hat{u}^* - \kappa V^{1/2}(\hat{\mu}^*)n^{-1/2}, \infty] \qquad (11)$$

for the binomial, Poisson and negative binomial cases, respectively.

Using the normal approximation, the Score interval can be constructed as (Wilson, 1927)

$$Z_n = \frac{\sqrt{n}(\hat{\mu} - \mu)}{V^{1/2}(\hat{\mu})} \to N(0,1) \qquad (12)$$

Accordingly, the random weighting estimation of the Score interval can be described as

$$Z^*_n = \frac{\sqrt{n}(\hat{\mu}^* - \mu)}{V^{1/2}(\hat{\mu}^*)} \to N(0,1) \qquad (13)$$

Coverage Probability of One-Sided Confidence Intervals

In the following, for the sake of concise description, we shall focus on analysis of the upper limit intervals. The lower limit intervals can be analyzed in the similar way.

Consider the coverage of a general upper limit interval

$$CI^* = [0, \frac{X + s_1}{n - bs_2} + k\left\{V(\hat{u}^*) + (V(\hat{u}^*) + r_2)n^{-1}\right\}^{1/2} n^{-1/2}] \qquad (14)$$

The random weighting estimation of CI^* may be written as

$$\hat{CI}^* = [0, \frac{X + s_1}{n - bs_2} + k\left\{V(\hat{u}^*) + (V(\hat{u}^*) + r_2)n^{-1}\right\}^{1/2} n^{-1/2}] \qquad (15)$$

where s_1, s_2, r_1 and r_2 are constants. It is not difficult to see that confidence interval $\hat{CI}_w^u$ is a special case of $\hat{CI}^*$.

By solving a quadratic equation, we have

$$P(\mu \in CI^*) = P(\frac{n^{1/2}(\hat{\mu} - \mu)}{\sigma} \geq z_W) \qquad (16)$$

and

$$P(\mu \in \hat{CI}^*) = P(\frac{n^{1/2}(\hat{\mu}^* - \mu)}{\sigma} \geq z_W) \qquad (17)$$

where

$$z_W = (\frac{B - \sqrt{B^2 - 4AC}}{2A} - \mu)\sigma^{-1}n^{1/2} \qquad (18)$$

where

$$A = n - bk^2(1 + r_1 n^{-1})(1 - bs_2 n^{-1})^2$$

$$B = 2n\mu - 2(s_1 + bs_2\mu) + k^2(1 + r_1 n^{-1})(1 - bs_2 n^{-1})$$

$$C = n(\mu - (s_1 + bs_2\mu)n^{-1})^2 - r_2 k^2 n^{-1}(1 - bs_2 n^{-1})^2 \qquad (19)$$

COVERAGE PROBABILITY ANALYSIS

Theorem: *Assuming $n\mu + n^{1/2}\sigma z_W$ is not an integer, the coverage probability of Wald interval $\hat{CI}_w^u$ is that displayed in Box 1.*

The following lemma is used to prove the theorem.

Lemma: (Morris, 1982) *Assume that $X_1, X_2, \cdots, X_n$ is a sample of independent and identically distributed random variables with common*

Box 1.

$$P_W = P(\mu \in \hat{C}I_W^u) = (1-\alpha) - \frac{1}{6}(2k^2+1)(1+2b\mu)\sigma^{-1}\phi(k)n^{1/2} + \Theta_1(\mu, z_W)\sigma^{-1}\phi(k)n^{-1/2}$$

$$+\{-\frac{b}{36}(8k^5 - 11k^3 + 3k) - \frac{1}{36\sigma^2}(2k^5 - k^3 + 3k)\}\phi(k)n^{-1}$$

$$+\left\{\frac{1}{6}(2k^3+3)(1+2b\mu)\Theta_1(\mu, z_W) + \Theta_2(\mu, z_W)\right\} \times \sigma^{-2}k\phi(k)n^{-1} + O(n^{-3/2})$$

(20)

distribution function $F(x)$ which is represented by Bin(1, p), Poi(λ) or N-bin(1, p). Denote $Z_n = n^{1/2}(\hat{\mu} - \mu)/\sigma$ and $F_n(z) = P(Z_n \leq z)$. Then, the two-term Edgeworth expansion of $F_n(z)$ is that displayed in Box 2.

If z depends on n, i.e.

$$z = z_0 + c_1 n^{-\frac{1}{2}} + c_2 n^{-1} + O(n^{-\frac{3}{2}}) \quad (24)$$

where z_0, c_1 and c_2 are constants, then Equation 25 in Box 3, where

$$\tilde{p}_1(z) = c_1 + \frac{1}{6}(1 - z_0^2)(1 + 2b\mu)\sigma^{-1} \quad (26)$$

$$\tilde{p}_2(z) = c_2 - \frac{1}{2}z_0 c_1^2 + \frac{1}{6}(z_0^3 - 3z_0)(1+2b\mu)\sigma^{-1} + p_2(z_0) \quad (27)$$

$$\tilde{p}_3(z) = c_1 - \frac{1}{6}(z_0^3 - 3)(1+2b\mu)\sigma^{-1} \quad (28)$$

We now start to prove the theorem. Expand Equation 29 in Box 4.

Denoting Equation 30 in Box 5, we have

$$z_W = -k + c_1^* n^{-\frac{1}{2}} + c_2^* n^{-1} + O(n^{-\frac{3}{2}}) \quad (31)$$

From (25), the coefficient of the non-oscillatory term $O(n^{-\frac{1}{2}})$ in the Edgeworth expansion of coverage probability $P(\mu \in \hat{C}I^*) = 1 - F_n(z_W)$ is

$$P_1^*(z_W) = -\tilde{P}_1(z_W) = -c_1^* + \frac{1}{6}(k^2-1)(1+2b\mu)\sigma^{-1}$$

$$= \{(s_1 - (\frac{1}{3}k^2 + \frac{1}{6})) + (s_2 - 2(\frac{1}{3}k^2 + \frac{1}{6}))b\mu\}\sigma^{-1} \quad (32)$$

Choose

$$s_1 = \frac{1}{6}(2k^2 + 1)$$

$$s_2 = \frac{1}{3}(2k^2 + 1) \quad (33)$$

such that $P_1^*(z_W)$ becomes zero for all μ.

By (33),

$$c_2^* = -\frac{1}{2}kr_1 + \frac{1}{3}(k^3 + 2k)b - \frac{1}{2}k\sigma^{-2}r_2 + \frac{1}{24}(k^3+2)\sigma^{-2} \quad (34)$$

From (27), the coefficient of the non-oscillatory term $O(n^{-1})$ in the Edgeworth expansion of $P(\mu \in \hat{C}I^*)$ is Equation 35 in Box 6.

Choose

Box 2.

$$F_n(z) = \phi(z) + p_1(z)\phi(z)n^{-\frac{1}{2}} - \Theta_1(\mu, z)\sigma^{-1}\phi(z)n^{-\frac{1}{2}} + p_2(z)\phi(z)n^{-1}$$
$$+ \left\{\Theta_1(\mu, z)\sigma p_3(z) + \Theta_2(\mu, z)\right\}\sigma^{-2}z\phi(z)n^{-1} + O(n^{-\frac{3}{2}})$$

(21)

where

$$p_1(z) = \frac{1}{6}(1 - z^2)(1 + 2b\mu)\sigma^{-1}$$

$$p_2(z) = -\frac{1}{36}(2z^5 - 11z^3 + 3z)b - \frac{1}{72}(z^5 - 7z^3 + 6z)\sigma^{-2}$$

(22)

$$p_3(z) = -\frac{1}{6}(z^2 - 3)(1 + 2b\mu)\sigma^{-1}$$

and Θ_1 *and* Θ_2 are described as

$$\Theta_1(\mu, z) = g(\mu, z) - \frac{1}{2}$$

$$\Theta_2(\mu, z) = -\frac{1}{2}g^2(\mu, z) + \frac{1}{2}g(\mu, z) - \frac{1}{12}$$

$$g(\mu, z) = g(\mu, z, n) = n\mu + n^{-\frac{1}{2}}\sigma z - (n\mu + n^{-\frac{1}{2}}\sigma z)_-$$

(23)

where "(x)_" denotes the largest integer less than or equal to x.

Box 3.

$$F_n(z) = \phi(z_0) + \tilde{p}_1(z)\phi(z_0)n^{-\frac{1}{2}} - \Theta_1(\mu, z)\sigma^{-1}\phi(z)n^{-\frac{1}{2}} + \tilde{p}_2(z)\phi(z_0)n^{-1}$$
$$+ \left\{\Theta_1(\mu, z)\sigma\tilde{p}_3(z_0) + \Theta_2(\mu, z)\right\}\sigma^{-2}z_0\phi(z_0)n^{-1} + O(n^{-\frac{3}{2}})$$

(25)

Box 4.

$$z_W = -k - (s_1 + bs_2\mu - \frac{1}{2}k^2(1 - 2b\mu))\sigma^{-1}n^{-\frac{1}{2}} - \{(\frac{1}{2}r_1 + bk^2 - bs_2)k$$
$$- \frac{1}{2}k(s_1 + bs_2\mu)(1 + 2b\mu)\sigma^{-2} + (\frac{1}{2}r_2 + \frac{1}{8}k^3)\sigma^{-2}\}n^{-1} + O(n^{-\frac{3}{2}})$$

(29)

Box 5.

$$c_1^* = -(s_1 + bs_2\mu - \frac{1}{2}k^2(1 + 2b\mu))\sigma^{-1}$$

$$c_2^* = -\{(\frac{1}{2}r_1 + bk^2 - bs_2)k - \frac{1}{2}k(s_1 + bs_2\mu)(1 + 2b\mu)\sigma^{-2} + (\frac{1}{2}r_2 + \frac{1}{8}k^3)\sigma^{-2}\}$$

(30)

Box 6.

$$P_2^*(z_W) = -\tilde{P}_2(z_W) = -c_2^* - \frac{1}{36}(k^3 - 7k)b + \frac{1}{72}(k^3 - k)\sigma^{-2}$$

$$= \frac{1}{2}k\left\{r_1 - \frac{1}{18}(13k^2 + 17)b\right\} + \frac{1}{2}k\sigma^{-2}\left\{r_2 - \frac{1}{36}(2k^{2+7})\right\}$$

(35)

$$r_1 = \frac{1}{18}(13k^2 + 17)b$$

$$r_2 = \frac{1}{36}(2k^2 + 7) \tag{36}$$

such that $P_2^*(z_W)$ becomes zero. As a result, the non-oscillatory term $O(n^{-1})$ is disappeared in (25). Therefore, we have

$$s_1 = s_2 = r_1 = r_2 = 0 \tag{37}$$

for z_W which is defined by (18).

Substituting (37) into (29), Equation 38 in Box 7.

Since the addition of (20) and (25) is equal to one,

$$P(\mu \in \hat{C}I_W^u) = 1 - F_n(z_W) \tag{39}$$

Accordingly, (20) follows from (25). The proof of the theorem is completed.

EXPERIMENTAL ANALYSIS AND DISCUSSIONS

Experiments have been conducted to comprehensively evaluate the performance of the proposed random weighting method for estimation of one-sided confidence intervals in discrete distributions. Comparison analysis with the commonly used confidence interval estimation methods is also discussed in this section.

Trials have been conducted to evaluate the capability of the proposed random weighting method for estimating the upper and lower limits

Box 7.

$$z_W = -k + \frac{1}{2}k^2(1 + 2b\mu)\sigma^{-1}n^{-\frac{1}{2}} - (b + \frac{1}{8}\sigma^{-2})k^3n^{-1} + O(n^{-\frac{3}{2}})$$

(38)

Table 1. One-sided confidence interval estimation for parameter λ of the Poisson distribution

λ	Estimation Methods	Upper Limits of Confidence Intervals	Lower Limits of Confidence Intervals
$\lambda_1 = 4.15$	Bootstrap estimation	$(0, \ 458)$	$(3.73, \ +\infty)$
	Random weighting estimation	$(0, \ 356)$	$(3.84, \ +\infty)$
$\lambda_2 = 4.25$	Bootstrap estimation	$(0, \ 467)$	$(3.75, \ +\infty)$
	Random weighting estimation	$(0, \ 428)$	$(3.86, \ +\infty)$
$\lambda_3 = 4.35$	Bootstrap estimation	$(0, \ 477)$	$(3.76, \ +\infty)$
	Random weighting estimation	$(0, \ 436)$	$(3.94, \ +\infty)$
$\lambda_4 = 4.45$	Bootstrap estimation	$(0, \ 486)$	$(3.93, \ +\infty)$
	Random weighting estimation	$(0, \ 458)$	$(4.05, \ +\infty)$

Table 2. One-sided confidence interval estimation for parameter p of the binomial distribution

p	Estimation Methods	Upper Limits of Confidence Intervals	Lower Limits of Confidence Intervals
$p_1 = 0.45$	Bootstrap estimation	$(0, \ 0.491)$	$(0.420, \ 1)$
	Random weighting estimation	$(0, \ 0.489)$	$(0.423, \ 1)$
$p_2 = 0.55$	Bootstrap estimation	$(0, \ 0.589)$	$(0.519, \ 1)$
	Random weighting estimation	$(0, \ 0.587)$	$(0.520, \ 1)$
$p_3 = 0.65$	Bootstrap estimation	$(0, \ 0.690)$	$(0.621, \ 1)$
	Random weighting estimation	$(0, \ 0.688)$	$(0.623, \ 1)$
$p_4 = 0.75$	Bootstrap estimation	$(0, \ 0.792)$	$(0.719, \ 1)$
	Random weighting estimation	$(0, \ 0.790)$	$(0.721, \ 1)$

for the one-sided confidence intervals of parameter λ of the Poisson distribution. For the purpose of comparison analysis, the upper and lower limits for the one-sided confidence intervals of parameter λ have been calculated by both the proposed method and the Bootstrap method, respectively, and the results are shown in Table 1. From Table 1, it can be seen that the one-sided confidence intervals of parameter λ obtained by the proposed random weighting method have smaller lengths than those obtained by the Bootstrap method. This demonstrates that the proposed random weighting method has higher accuracy for estimating one-sided confidence intervals in the Poisson distribution than the Bootstrap method.

Experiments have also been conducted to evaluate the performance of the proposed random weighting method for estimating the upper and lower limits for the one-sided confidence intervals of parameter p of the binomial distribution. For the comparison purpose, the upper and lower limits for the one-sided confidence intervals of parameter p have been estimated by both the proposed method and the Bootstrap method, respectively. Table 2 shows the estimated results obtained by the both methods. It can be seen that the one-sided confidence intervals of parameter p estimated by the proposed random weighting method have smaller lengths than those by the Bootstrap method. This demonstrates that the proposed random weighting method has higher estimation accuracy than the Bootstrap method for estimating one-sided confidence intervals in the binomial distribution.

From the above experiments, it is evident that the proposed random weighting method has higher accuracy for estimation of one-sided confidence intervals in discrete distributions than the Bootstrap method.

CONCLUSION

This paper adopts the random weighting method to estimation of one-sided confidence intervals in discrete distributions. The contribution of the paper is that random weighting estimations are established for the Wald and Score intervals, and a theorem of coverage probability is rigorously proved by using the Edgeworth expansion for random weighting estimation of the Wald interval. Future research work will focus on applications of the established theory in engineering practices.

ACKNOWLEDGMENT

The work of this paper is supported by the Natural Science Foundation of Shaanxi Province (Project Number: SJ08F04).

REFERENCES

Barvinok, A., & Samorodnitsky, A. (2007). Random weighting, asymptotic counting, and inverse is operimetry. *Israel Journal of Mathematics, 158*(1), 159–191. doi:10.1007/s11856-007-0008-8.

Brown, L. D. (1986). *Fundamentals of statistical exponential families with applications in statistical decision theory*. Hayward, CA: Institute of Mathematical Statistics.

Cai, T. T. (2005). One-sided confidence intervals in discrete distributions. *Journal of Statistical Planning and Inference, 131*(1), 63–88. doi:10.1016/j.jspi.2004.01.005.

Chakraborti, S., & Li, J. (2007). Confidence interval estimation of a normal percentile. *The American Statistician, 61*(4), 331–336. doi:10.1198/000313007X244457.

Dogan, G. (2007). Bootstrapping for confidence interval estimation and hypothesis testing for parameters of system dynamics models. *System Dynamics Review, 23*(4), 415–436. doi:10.1002/sdr.362.

Duncan, A. J. (1986). *Quality control and industrial statistics* (5th ed.). Homewood, IL: R. D. Irwin.

Fang, Y., & Zhao, L. (2006). Approximation to the distribution of LAD estimators for censored regression by random weighting method. *Journal of Statistical Planning and Inference, 136*(4), 1302–1316. doi:10.1016/j.jspi.2004.09.010.

Gao, S., Feng, Z., Zhong, Y., & Shirinzadeh, B. (2008). Random weighting estimation of parameters in generalized Gaussian distribution. *Information Sciences, 178*(9), 2275–2281. doi:10.1016/j.ins.2007.12.011.

Gao, S., Zhang, J., & Zhou, T. (2003). Large numbers law for sample mean of random weighting estimation. *Information Sciences, 155*(1-2), 151–156. doi:10.1016/S0020-0255(03)00158-0.

Gao, S., Zhang, Z., & Yang, B. (2004). The random weighting estimation of quantile process. *Information Sciences, 164*(1-4), 139–146. doi:10.1016/j.ins.2003.10.002.

Gao, S., & Zhong, Y. (2010). Random weighting estimation of kernel density. *Journal of Statistical Planning and Inference, 140*(9), 2403–2407. doi:10.1016/j.jspi.2010.02.009.

Haukoos, J. S., & Lewis, R. J. (2005). Advanced statistics: Bootstrapping confidence intervals for statistics with "difficult" distributions. *Academic Emergency Medicine, 12*(4), 360–365. doi:10.1111/j.1553-2712.2005.tb01958.x.

Karlis, D., & Patilea, V. (2008). Bootstrap confidence intervals in mixtures of discrete distributions. *Journal of Statistical Planning and Inference, 138*(8), 2313–2329. doi:10.1016/j.jspi.2007.10.026.

Kott, P. S., & Liu, Y. K. (2009). One-sided coverage intervals for a proportion estimated from a stratified simple random sample. *International Statistical Review, 77*(2), 251–265. doi:10.1111/j.1751-5823.2009.00081.x.

Mandel, M., & Betensky, R. A. (2008). Simultaneous confidence intervals based on the percentile bootstrap approach. *Computational Statistics & Data Analysis, 52*(4), 2158–2165. doi:10.1016/j.csda.2007.07.005.

Mantalosa, P., & Zografos, K. (2008). Interval estimation for a binomial proportion: A bootstrap approach. *Journal of Statistical Computation and Simulation, 78*(2), 1251–1265. doi:10.1080/00949650701749356.

Montgomery, D. C. (2001). *Introduction to statistical quality control* (4th ed.). New York, NY: John Wiley & Sons.

Morris, C. N. (1982). Natural exponential families with quadratic variance functions. *Annals of Statistics, 10*(1), 65–80. doi:10.1214/aos/1176345690.

Wald, A., & Wolfowitz, J. (1939). Confidence limits for continuous distribution functions. *Annals of Mathematical Statistics, 10*(2), 105–118. doi:10.1214/aoms/1177732209.

Wilson, E. B. (1927). Probable inference, the law of succession, and statistical inference. *Journal of the American Statistical Association, 22*(158), 209–212. doi:10.2307/2276774.

Xue, L., & Zhu, L. (2005). L1-norm estimation and random weighting method in a semiparametric model. *Acta Mathematicae Applicatae Sinica, 21*(2), 295–30. doi:10.1007/s10255-005-0237-8.

Zheng, Z. (1987). Random weighting method. *Acta Mathematicae Applicatae Sinica, 10*(2), 247–253.

This work was previously published in the International Journal of Intelligent Mechatronics and Robotics, Volume 1, Issue 2, edited by Bijan Shirinzadeh, pp. 18-26, copyright 2011 by IGI Publishing (an imprint of IGI Global).

Chapter 7
Dynamic Modeling, Simulation and Velocity Control of Rocker–Bogie Rover for Space Exploration

Pushpendra Kumar
Indian Institute of Technology Roorkee, India

Pushparaj Mani Pathak
Indian Institute of Technology Roorkee, India

ABSTRACT

Wheeled mobile rovers are being used in various missions for planetary surface exploration. In this paper a six-wheeled rover with rocker-bogie structure has been analyzed for planar case. The detailed kinematic model of the rover was built and the dynamic model was derived based on bond graph. The simulation studies were performed for obstacle climbing capability of the rover. It was observed from the study that rover can pass through plane surface, inclined surface, and inclined ditch without any control on the actuators of the rover. However, it fails to cross a vertical ditch so a velocity controller was designed. It consists of a proportional integral (PI) controller and reduced model of the rover. It is found from simulation and animation studies that with the proposed velocity controller the rover is able to cross the vertical ditch.

1. INTRODUCTION

Many planetary exploration rovers have been developed since 1960s for planetary exploration. Based on the features of the mobile mechanism in robots, planetary exploration rovers have various types such as wheeled, legged and tracked.

Wheeled mobile mechanisms have excellent features, such as high speed on a relatively flat terrain and easy control; so many researchers have designed their exploration rovers with wheeled structure. Wheeled structure planetary rovers have four-wheel, six-wheel and eight-wheel, etc. Among these rovers, the six-wheeled mobile rover

DOI: 10.4018/978-1-4666-3634-7.ch007

Figure 1. Schematic diagram of rocker bogie rover

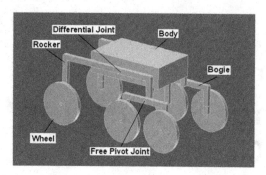

with rocker bogie mechanism has superior adaptability and obstacle climbing capability. These rovers have been used in Mars Rovers (Lindemann, Bickler, Harrington, Ortiz, & Voorhees, 2006; Lindemann, 2005).

The research in exploration rovers has been devoted to analyze the performance of exploration rovers based on different types of suspension systems. Bai-chao, Rong-ben, Lu, Li-sheng, and Lie (2007) proposed a new type of suspension system formed by a positive and a negative quadrilateral levers mechanism. Thueer, Siegwart, and Backes, Planetary Vehicle Suspension Options (2008) studied the locomotion performance of different suspension types in order to find the rover that matches best for any mission requirement. Their performance was evaluated on hard and loose soil. Thueer, Krebs, Siegwart, and Lamon (2007) presented the performance optimization tool based

on a static approach to compare and optimize the rover chassis in quick and efficient way. Thueer and Siegwart (2007) analyzed different rovers based on kinematic model and the optimal velocities of all wheels were used for characterization of the suspension of different rovers. Hacot, Dubowsky, and Bidaud (1998) presented the mechanics of rover and method for solving the inverse kinematics of the rocker-bogie rover. The quasi-static force analysis has been described for the rover. Li, Gao, and Deng (2008) built the mobility performance indexes based on work conditions and established an optimized mathematical model of suspension parameters of rocker-bogie rover. Mann and Shiller (2005) described a measure of stability of rocker-bogie rover for a range of acceptable velocities and accelerations that satisfy a set of dynamic constraints. Yongming, Xiaoliu, and Wencheng (2009) described the quasi-static mathematical model of rocker-bogie rover. Obstacle climbing capabilities of the rover were analyzed. Most of the rover models reported in literature is based on kinematic relations. Rover dynamic relations have not been derived.

In this paper the kinematic model of the rocker-bogie rover is developed. This kinematic model is used in deriving the dynamic model using bond graph technique (Mukherjee, Karmakar, & Samantaray, 2006; Karnopp, Margolis, & Rosenberg, 2000). The dynamic model includes the model of the motor and ground reaction. The

Figure 2. Planar model of rocker bogie rover

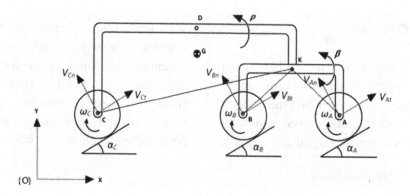

simulation studies are performed to the obstacle climbing capability of the rover. It has been observed from the study that rover can pass through plane surface, inclined surface, and inclined ditch without any control on the motors of the rover. However it failed to cross a vertical ditch. So a velocity controller has been designed. It consists of a proportional integral controller and reduced model of the rover. It has been found from simulation and animation studies that with the proposed velocity controller the rover is able to cross the vertical ditch.

2. KINEMATICS FORMULATION

In this section the rover is briefly presented and the kinematic modeling is derived. Figure 1 shows the schematic diagram of rocker bogie rover. The rover has different elements, which are assumed to be rigid. The rover consists of several joints. By adjusting its joints the rover is capable of locomotion over various uneven terrains. The rocker bogie structure has six independently driven wheels which are mounted on an articulated passive suspension system.

Figure 3. Assignment of frames on planar model of rocker bogie rover

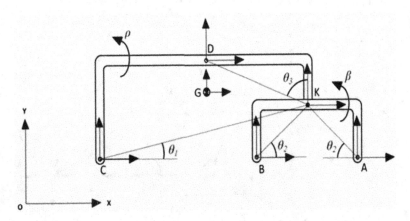

Figure 4. Word bond graph of rocker bogie rover

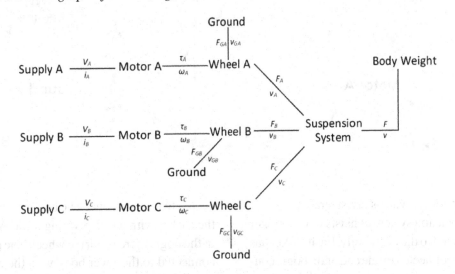

Figure 5. Bond graph model of rocker bogie rover

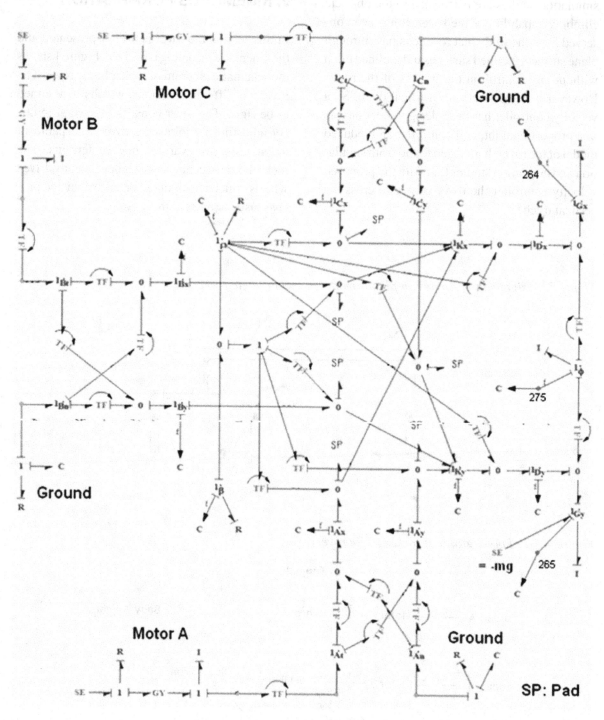

The four corner wheels are steerable.

The suspension system consists of two rocker arms connected to the rover body. Each rocker has a rear wheel connected to one end and a bogie con-

nected to the other end. The bogie is connected to the rocker with a free pivoting joint. At each end of the bogie there is a drive wheel. The rockers are connected to the rover body with the differential

Figure 6. Parameter description of rocker bogie rover

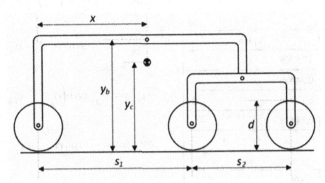

Figure 7. Rover's trajectory on 300 slopes: (a) front wheel trajectory, (b) middle wheel trajectory, (c) rear wheel trajectory, (d) body CM trajectory

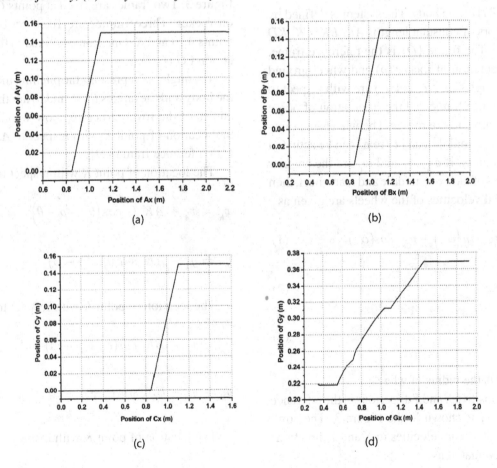

joint. In our analysis rigid body kinematics has been used to model the rover.

The planar model of the rover is shown in Figure 2. In this figure *A*, *B* and *C* are the points on the centre of wheel axles and the point *K* represents the free pivot joint between rocker

and bogie. Point *D* is the differential joint between rocker and body and point *G* is centre of

Table 1. Parameter values used for simulation of rocker bogie rover

Parameter	Value
Main body mass	17.5 Kg
Main body position (x)	0.344 m
Main body position (y_b)	0.229 m
Body's CM position (x)	0.344 m
Body's CM position (y_c)	0.22 m
Wheel diameter (d)	0.2 m
Wheel distance (s_1)	0.389 m
Wheel distance (s_2)	0.259 m

mass (CM) of the body. The system is defined by the vectors of constant lengths *AK, BK, CK, KD* and *DG*. The frame {O} is the reference frame. The orientation of rocker (ρ) and wheel-ground contact angles (α_A, α_B and α_C) are with respect to the reference frame and the orientation of bogie (β) is relative to the rocker. The wheels at points *A* (front), *B*(middle), and *C*(rear) can be actuated by the motors. If ω_A, ω_B, and ω_C are the angular velocities of the front, middle and rear wheel, then tangential velocities of the wheels are given as

$$v_{Cy} = v_{Ct}\ sin\left(\alpha_C\right) + v_{Cn}\ cos(\alpha_C) \tag{1}$$

$$v_{Bt} = r\omega_B \tag{2}$$

$$v_{Ct} = r\omega_C \tag{3}$$

where *r* is the radius of wheel.

The positive *X* and *Y* directions of the reference frame {O} is shown in the Figure 3. The components of linear velocities in *X* and *Y* directions can be evaluated as

$$v_{Ax} = v_{At}\ cos\left(\alpha_A\right) - v_{An}\ sin(\alpha_A) \tag{4}$$

$$v_{Ay} = v_{At}\ sin\left(\alpha_A\right) + v_{An}\ cos(\alpha_A) \tag{5}$$

$$v_{Bx} = v_{Bt}\ cos\left(\alpha_B\right) - v_{Bn}\ sin(\alpha_B) \tag{6}$$

$$v_{By} = v_{Bt}\ sin\left(\alpha_B\right) + v_{Bn}\ cos(\alpha_B) \tag{7}$$

$$v_{Cx} = v_{Ct}\ cos\left(\alpha_C\right) - v_{Cn}sin(\alpha) \tag{8}$$

$$v_{Cy} = v_{Ct}\ sin\left(\alpha_C\right) + v_{Cn}\ cos(\alpha_C) \tag{9}$$

Here *n* subscript refers to normal velocity component. For further analysis first of all we assign frames at different points as shown in Figure 3. Two frames are fixed at points *C* and *D* on rocker. Three frames are fixed on bogie at points *A*, *B* and *K*. One frame is fixed to the body at point *G* (*CM*).

The angle 'φ' represents the pitching motion of the body and the angles θ_1, θ_2 and θ_3 are the fixed angles of the rover structure. Now we calculate the location of point *K* with respect to *A*, *B* and *C* in reference frame {O}.

The location of point *K* with respect to *A* is:

$$x_K = x_A + AK\ (-cos\left(\theta_2 - \rho - \theta\right)) \tag{10}$$

$$y_K = y_A + AK\ sin(\theta_2 - \rho - \beta) \tag{11}$$

The location of point *K* with respect to *B* is:

$$x_K = x_B + BK\ cos(\theta_2 + \rho + \beta) \tag{12}$$

$$y_K = y_B + BK\ sin(\theta_2 + v + \beta) \tag{13}$$

The location of point *K* with respect to *C* is:

$$x_K = x_C + CK\ cos(\theta_1 + \rho) \tag{14}$$

$$y_K = y_C + CK\ sin(\theta_1 + \rho) \tag{15}$$

Figure 8. (a) CM velocity of rover in x direction on 30^0 slope (b) CM velocity of rover in y direction on 30^0 slope (c) and (d) animation frames of rover climbing 30^0 slopes

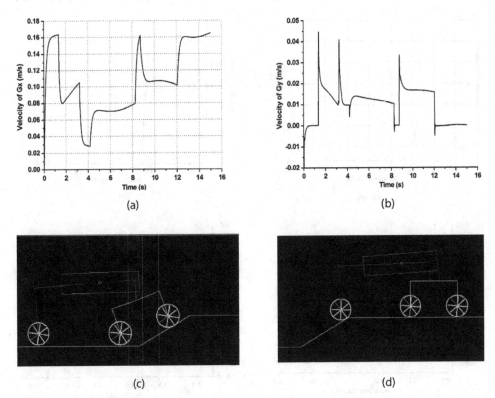

(a)

(b)

(c)

(d)

By differentiating above equations we can find the velocity relationships as follow:

$$v_{Kx} = v_{Ax} - \left(\dot{\rho} + \dot{\beta}\right) AK \sin\left(\theta_2 - \rho - \beta\right) \quad (16)$$

$$v_{Ky} = v_{Ay} - \left(\dot{\rho} + \dot{\beta}\right) AK \cos\left(\theta_2 - \rho - \beta\right) \quad (17)$$

$$v_{Kx} = v_{Bx} - \left(\dot{\rho} + \dot{\beta}\right) BK \sin\left(\theta_2 + \rho + \beta\right) \quad (18)$$

$$v_{Ky} = v_{By} + \left(\dot{\rho} + \dot{\beta}\right) BK \cos\left(\theta_2 + \rho + \beta\right) \quad (19)$$

$$v_{Ky} = v_{Cx} - \left(\dot{\rho}\right) CK \sin\left(\theta_1 + \rho\right) \quad (20)$$

$$v_{Ky} = v_{Cy} - \left(\dot{\rho}\right) CK \cos\left(\theta_1 + \rho\right) \quad (21)$$

Also the location of point D and G are given by:

$$x_D = x_K - KD \; \sin\left(\theta_3 + \rho\right) \quad (22)$$

$$y_D = y_K + KD \; \cos\left(\theta_3 + \rho\right) \quad (23)$$

$$x_G = x_D + DG \; \sin\left(\varphi\right) \quad (24)$$

$$y_G = y_D - DG \; \cos\left(\varphi\right) \quad (25)$$

Differentiating above expressions the velocity relationship can be derived as

$$v_{Dx} = v_{Kx} - \left(\dot{\rho}\right) KD \cos\left(\theta_3 + \rho\right) \quad (26)$$

Figure 9. Rover's trajectory on a ditch: (a) front wheel trajectory, (b) middle wheel trajectory, (c) rear wheel trajectory, (d) body CM trajectory

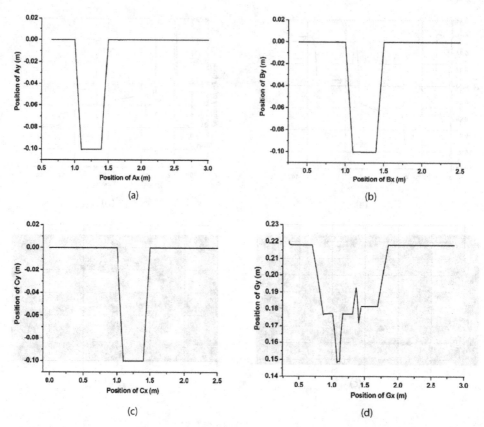

(a) (b)

(c) (d)

$$v_{Dy} = v_{Ky} - \left(\dot{\rho}\right) KD \sin\left(\theta_3 + \rho\right) \qquad (27)$$

$$v_{Gx} = v_{Dx} + \left(\dot{\varphi}\right) DG \cos\left(\varphi\right) \qquad (28)$$

$$v_{Gy} = v_{Dy} + \left(\dot{\varphi}\right) DG \sin\left(\varphi\right) \qquad (29)$$

In this section kinematic model of the rover has been developed considering the planar case. Now we can use this kinematic model to derive various transformer moduli which can be used to draw bond graph model. From this bond graph model dynamic equations of rocker bogie rover are derived.

3. DYNAMIC MODEL OF ROVER BASED ON BOND GRAPH

In bond graph models the dynamics is automatically represented correctly since the bond graph junction structure is power conservative. In drawing bond graph model following assumptions are made.

1. Rocker and Bogie are assumed to be mass less as compared to the body.
2. The mass of the body is assumed to be acted on *CM* at point *G*.
3. Each wheel is independently driven by the motors.

Figure 10. (a) CM velocity of rover in x direction on 45⁰ ditch (b) CM velocity of rover in y direction on 45⁰ ditch (c) and (d) animation frames of rover passing by 45⁰ ditches

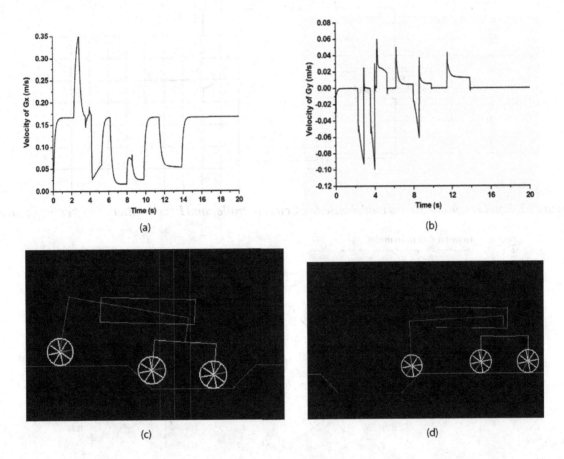

Figure 4, shows the word bond graph of rocker bogie rover system. The voltage supply is given as input to the motors and the motor output is given to the wheels. The ground also influences the wheel motion. The effect of wheel motion is transferred to the body via suspension system and the weight of the body acts on the body *CM*.

Figure 5 represents the bond graph model of rocker bogie rover. In this Figure three 1 junctions $(1_{\dot{A}t}, 1_{\dot{B}t},$ and $1_{\dot{C}t})$ corresponds to tangential velocity of wheel A, B and C, three 1 junctions $(1_{\dot{A}n}, 1_{\dot{B}n},$ and $1_{\dot{C}n})$ corresponds to normal velocity of wheel A, B and C. Also in this figure six 1 junctions

$(1_{\dot{A}x}, 1_{\dot{B}x}, 1_{\dot{C}x}, 1_{\dot{D}x}, 1_{\dot{K}x},$ and $1_{\dot{G}x})$ corresponds for velocity in *X* direction, six 1 junctions $(1_{\dot{A}y}, 1_{\dot{B}y}, 1_{\dot{C}y}, 1_{\dot{D}y}, 1_{\dot{K}y},$ and $1_{\dot{G}y})$ for velocity in *Y* direction of wheel A, B, C, point D, K and G. Also three 1 junctions $(1_{\dot{\rho}}, 1_{\dot{\beta}},$ and $1_{\dot{\varphi}})$ represents angular velocity of rocker, bogie and body. The mass and the rotary inertia of the body are modeled using *I* elements.

The three wheels are independently driven by the motors. In Figure 5, the angular velocities of three motors are represented by 1 junctions (,, and). The angular velocity is multiplied by radius of wheel to get the tangential velocity represented

Figure 11. Rover velocity control

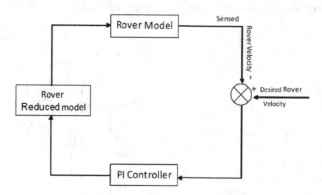

Figure 12. Bond graph of reduced model based velocity controller and PI controller of rocker bogie rover

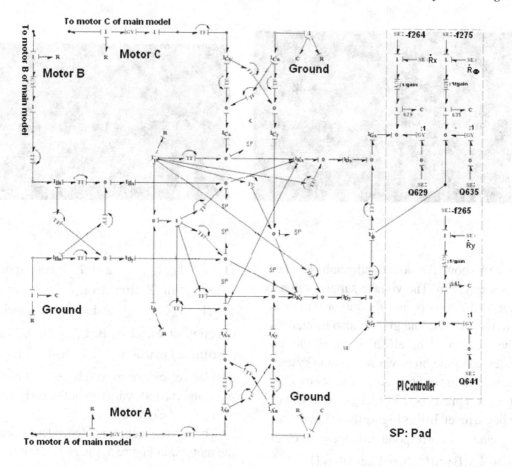

by junctions (,, and). The flow in normal direction of the rover comes from ground. It is represented by junctions (,, and). The ground has been mod-

eled as spring damper system using *C* and *R* elements as shown in Figure 5. In Figure 5 pads are used to avoid differential causality.

4. SIMULATION AND ANIMATION RESULTS

The parameters descriptions of rocker bogie rover are shown in Figure 6. The numerical values of parameters used in simulation are shown in Table 1. The parameter values are same as that used for mars exploration rover (MER). The simulation has been carried out for terrain at 30^0 slope, ditch inclined at 45^0 from vertical, and step.

4.1. Rover on Terrain at 30^0 Slope

To carry out the simulation study the rear wheel motor has been supplied with a voltage of 4.5V.

Here it is assumed that the rover is moving over a 30^0 slope. Figure 7 shows the trajectory of the wheels and body *CM*. Figure 7 (a) shows the front wheel trajectory, Figure 7(b) shows the middle wheel trajectory, Figure 7(c) shows the rear wheel trajectory and Figure 7(d) shows the body *CM* trajectory. From these Figures one can conclude that the rover is able to move on an inclined terrain.

Figure 8 (a) shows the *CM* velocity of rover in *x* direction on 30^0 slopes, and Figure 8(b) shows *CM* velocity of rover in *y* direction on 30^0 slopes. From Figure 8 (a) and (b), it is seen that after transients the rover moves at speed 16 cm/s, then at time 1.36 sec the front wheel touches the slope and begins to climb up, and the rover velocity

Figure 13. Rover's trajectory on a step with velocity control: (a) front wheel trajectory

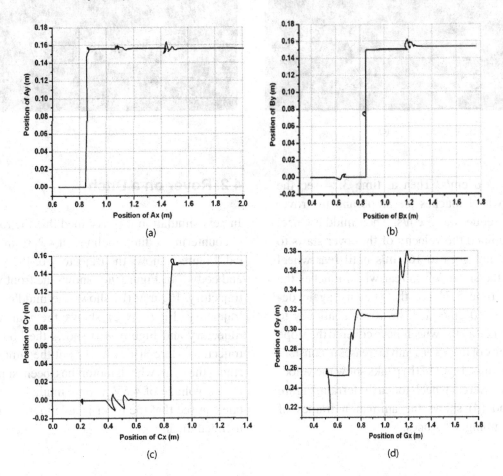

(a)

(b)

(c)

(d)

Figure 14. Rover moving on a step with velocity control (a) CM velocity of rover in x direction on a step (b) CM velocity of rover in y direction on step (c) and (d) animation frames of rover passing by a step

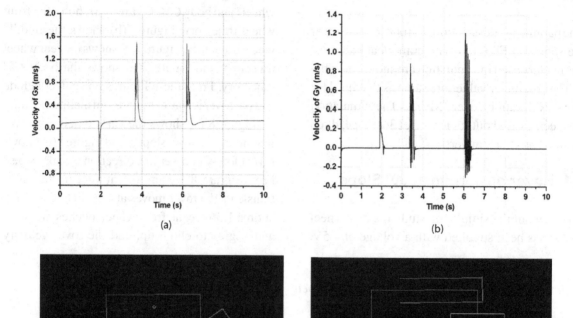

(a)

(b)

(c)

(d)

reduces to 8 cm/s. Then at time 3.24 sec the middle wheel touches the slope and the rover velocity reduces to 3 cm/s. Then middle wheel climbs up and the velocity of the rover starts to increase and goes to 16 cm/s until rear wheel touches the slope. When rear wheel touches the slope at time 8.74 sec the velocity again decreases to 10 cm/s and the rover continues to climb. When all the wheels have covered the slope the velocity of the rover again increases to constant value 16 cm/s. Figure 8(b) peaks show the velocity when different wheels touch the terrain. Figure 8 (c) and (d) shows the animation frames of rover climbing 30^0 slopes.

4.2. Rover on a Ditch

In next simulation it is assumed that the rover is encountering a ditch inclined at 45^0 from vertical. Figure 9 shows the trajectory of the wheels and body *CM*. Figure 9(a) shows the front wheel trajectory, Figure9 (b) shows the middle wheel trajectory, Figure 9(c) shows the rear wheel trajectory and Figure 9(d) shows the body *CM* trajectory. Here also to carry out the simulation study the rear wheel motor has been supplied with a voltage of 4.5V. From simulation results one can see that the rover is able to move on an inclined ditch.

Figure 10 (a) shows the *CM* velocity of rover in *x* direction on 45^0 ditches; Figure 10 (b) shows *CM* velocity of rover in *y* direction on 45^0 ditches. From Figure 10(a) and (b) it is seen that the rover moves at speed of 16 cm/s, then at time 2.24 sec the front wheel touches the ditch and rover velocity increases to 35 cm/s. The middle wheel touches the ditch at time 3.54 sec. When middle wheel comes down, rover velocity becomes 17 cm/s. Then at time 4.16 sec front wheel touches the up of the ditch and begins to climb up, and then the rover velocity decreases and reaches to 3 cm/s. When front wheel is over the ditch, rover velocity again starts to increase and becomes 16 cm/s. Then at time 6.17 sec the middle wheel touches the up of the ditch and rover velocity decreases to 2 cm/s. The rear wheel touches the ditch at 7.93 sec and rover velocity again increases to 8 cm/s. When rear wheel comes down to the ditch rover velocity again decreases and becomes 2 cm/s. Then at time 9.8 sec middle wheel is over the ditch and rover velocity again reaches to 16 cm/s. The rear wheel touches the up of ditch at time 11.46 sec and rover velocity again starts to decrease and reaches to 5 cm/s. Then at time 13.82 sec rear wheel is over the ditch and the rover moves with a constant speed of 16 cm/s. Similarly the rover behavior can be explained by Figure 10(b). Here downward peaks in the graph correspond to rover touching the ditch whereas upward peaks correspond to rover touching the slope of the ditch. The animation frame of the rover is shown in Figure 10 (c) and (d).

If it is assumed that the rover is moving over a step like obstacle and the rear wheel motor has been supplied with a voltage of 4.5V, then from simulation study it is observed that the rover is not able to cross the step as the rear wheel velocity reduces to zero when it encounters the step. So a reduced model based velocity controller is designed which removes this problem.

5. A REDUCED MODEL BASED CONTROLLER FOR VELOCITY CONTROL

As stated in previous section the rover is not able to cross the step like obstacle so a control strategy is required to apply on the system. Here we incorporate a reduced model-based proportional-integral (PI) controller to control the velocity of the rover. The block diagram of rover control scheme is shown in Figure 11.

The difference of the rover velocity sensed by the sensor and the desired rover velocity is taken as an error signal. This error signal is given to the reduced model of rover and a PI controller is incorporated in the reduced model. The output of the reduced model is fed back to three motors in main model. Figure 12 shows the bond graph of reduced model based velocity controller with PI control. In this reduced model the rover translational and rotational inertia is not considered.

5.1. Rover on a Step

To validate the controller it is assumed that the rover is moving on a step like obstacle. Figure 13 shows the rover's wheel and body *CM* trajectory on a step with velocity control. Figure 13(a) shows the front wheel trajectory, Figure 13 (b) shows the middle wheel trajectory, Figure 13(c) shows the rear wheel trajectory and Figure 13(d) shows the body *CM* trajectory of rover. From these Figures we can see that the rover is able to cross the step.

Figure 14 (a) shows the *CM* velocity of rover in *x* direction on a step. From Figure 14 (a), it is seen that the rover moves with a uniform desired velocity of 15 cm/s in *x* direction. There are three peaks in the graph. These peaks occur when front wheel, middle wheel and rear wheel touch the step respectively. Figure 14(b) shows *CM* velocity of rover in *y* direction on a step. In Figure 14(b) different peaks show the rover velocity when front wheel, middle wheel and rear wheel touch the step respectively. Figure 14 (c) and (d) shows the animation frames of rover moving on a step like obstacle. Thus we can conclude that with the proposed combined reduced model based controller and PI controller rover is able to cross the step.

CONCLUSION

A detailed kinematic model of the rocker-bogie rover is presented. The dynamics of the system is also presented based on bond graph model of the rover under reasonable assumptions. The motor and ground dynamics are added to the system. The performance of rover is analyzed for obstacle climbing capability. It is seen that without a controller rover is not able to cross a step. So a combined reduced model based and PI controller is proposed. Simulation and animation of the rover are done over various terrains to validate the modeling and control strategy. In future work two dimensional kinematics model of rover can be extended to three dimensional models. Using the three dimensional kinematic model dynamic model of rover can be derived using bond graph and proposed control strategy can be implemented in the derived model.

REFERENCES

Bai-chao, C., Rong-ben, W., Lu, Y., Li-sheng, J., & Lie, G. (2007). Design and simulation research on a new type of suspension for lunar rover. In *Proceedings of the IEEE International Symposium on Computational Intelligence in Robotics and Automation*, Jacksonville, FL (pp. 173-177).

Hacot, H., Dubowsky, S., & Bidaud, P. (1998). Analysis and simulation of a rocker-bogie exploration rover. In *Proceedings of the Twelfth Symposium on Theory and Practice of Robots and Manipulators*, Paris, France.

Karnopp, D., Margolis, D., & Rosenberg, R. (2000). *System dynamics- modeling and simulation of mechatronics systems*. New York, NY: John Wiley & Sons.

Li, S., Gao, H., & Deng, Z. (2008). Mobility performance evaluation of lunar rover and optimization of rocker-bogie suspension parameters. In *Proceedings of the International Symposium on Systems and Control in Aerospace and Astronautics*, Shenzhen, China.

Lindemann, R. A. (2005). Dynamic testing and simulation of the mars exploration rover. In *Proceedings of the ASME International Design Engineering Technical Conference and Computers and Information in Engineering Conference* (pp. 24-28).

Lindemann, R. A., Bickler, D. B., Harrington, B. D., Ortiz, G. M., & Voorhees, C. J. (2006). Mars exploration rover mobility development. *IEEE Robotics & Automation Magazine*, 19–26. doi:10.1109/MRA.2006.1638012.

Mann, M. P., & Shiller, Z. (2005). Dynamic stability of a rocker bogie vehicle: Longitudinal motion. In *Proceedings of the IEEE International Conference on Robotics and Automation*, Barcelona, Spain (pp. 861-866).

Mukherjee, A., Karmakar, R., & Samantaray, A. K. (2006). *Bond graph in modeling, simulation and fault identification*. Boca Raton, FL: CRC Press.

Thueer, T., Krebs, A., Siegwart, R., & Lamon, P. (2007). Performance comparison of rough-terrain robots- simulation and hardware. *Journal of Field Robotics*, 251–271. doi:10.1002/rob.20185.

Thueer, T., & Siegwart, R. (2007). Characterization and comparison of rover locomotion performance based on kinematic aspects. In *Proceedings of the International Conference on Field and Service Robotics*, Chamonix, France (pp. 189-198).

Thueer, T., Siegwart, R., & Backes, P. G. (2008). Planetary vehicle suspension options. In *Proceedings of the IEEE Aerospace Conference* (pp. 1-13).

Yongming, W., Xiaoliu, Y., & Wencheng, T. (2009). Analysis of obstacle-climbing capability of planetary exploration rover with rocker-bogie structure. In *Proceedings of the International Conference on Information Technology and Computer Science*, Kiev, Ukraine (pp. 329-332).

This work was previously published in the International Journal of Intelligent Mechatronics and Robotics, Volume 1, Issue 2, edited by Bijan Shirinzadeh, pp. 27-41, copyright 2011 by IGI Publishing (an imprint of IGI Global).

Chapter 8
Processing of 3D Unstructured Measurement Data for Reverse Engineering

Yongmin Zhong
RMIT University, Australia

ABSTRACT

One of the most difficult problems in reverse engineering is the processing of unstructured data. NURBS (Non-uniform Rational B-splines) surfaces are a popular tool for surface modeling. However, they cannot be directly created from unstructured data, as they are defined on a four-sided domain with explicit parametric directions. Therefore, in reverse engineering, it is necessary to process unstructured data into structured data which enables the creation of NURBS surfaces. This paper presents a methodology to processing unstructured data into the structured data for creating NURBS surfaces. A projection based method is established for constructing 3D triangulation from unstructured data. An optimization method is also established to optimize the 3D triangulation to ensure that the resulted NURBS surfaces have a better form. A triangular surface interpolation method is established for constructing triangular surfaces from the triangulation. This method creates five-degree triangular surfaces with C^1 continuity. A series of segment data are obtained by cutting the triangular surfaces with a series of parallel planes. Finally, the structured data is obtained by deleting repetitive data points in each segment data. Results demonstrate the efficacy of the proposed methodology.

INTRODUCTION

Reverse engineering has become an effective method to create a 3D computer model of a physical object by dimensional measurement and surface modelling (Varady et al., 1997; Beccari et al., 2010). It has many applications in different fields, such as medical imaging, entertainment, cultural heritage, web commerce, collaborative design and obviously engineering. The physical object can be measured using 3D scanning technologies such as coordinate measuring machines or computed tomography scanners, which provide outputs in the form of a large set of vertices in a

DOI: 10.4018/978-1-4666-3634-7.ch008

3D coordinate system. For the objects with good structured shapes, unstructured data can be avoided if an appropriate scanning direction is selected during the measurement process. However, for the objects with complex shapes, unstructured data cannot be avoided no matter what scanning directions are used. The unstructured data do not have explicit parametric directions, especially in comparison with the structured four-sided data which has explicit parametric directions (Figure 1). NURBS (Non-uniform Rational B-splines) surfaces are the popular surfaces modelling tools and have been extensively used in reverse engineering and product design activities (Ma & Kruth, 1998; Yin et al., 2003; Jiang & Wang, 2006; Pal & Ballav, 2007; Pal, 2008). As they are defined on a structured four-sided domain with explicit parametric directions, they cannot be created from unstructured data directly. Therefore, in reverse engineering, it is necessary to process unstructured data into the structured data for creating NURBS surfaces, thus enabling surface operations such as joining, filleting and blending for creating complex-shaped objects. In this paper, a methodology is presented for processing unstructured data into the structured data for 3D object reconstruction. A projection based method is established for triangulation of 3D unstructured data. This method simplifies the complex 3D triangulation problem by converting it into the 2D triangulation problem. The 3D triangulation is directly obtained from the 2D triangulation obtained by an improved 2D triangulation algorithm. An optimization method is established for optimization of the 3D triangulation to ensure that the surfaces constructed from the spatial triangulation have the better form. A triangular surface interpolation method is established to construct five-degree triangular surfaces with C^1 continuity by interpolating each triangle. Subsequently, a series of segment data are obtained by cutting the triangular interpolation surfaces with a series of parallel planes. Structured data are finally generated by deleting repetitive points in each segment

data, and smoothing and uniforming each segment data for NURBS surface modelling.

Reverse engineering is the process of measuring an object and then generating the object's computer model which captures the object's physical features (Lin et al., 1997; Meng et al., 1996; Tuohy et al., 1997; Varady et al., 1997; Wang & Wang, 1997). The processing of unstructured data, which is the one of key issues in reverse engineering (Ma & He, 1998; Seiler et al., 1996; Yau & Chen, 1997; Jiang & Wang, 2006, Pal & Ballav, 2007; Pal, 2008; Beccari et al., 2010, Tang et al., 2010). However, because of the complexity of unstructured data, the existing studies suffer from different problems. Focusing on engineering applications, this paper presents an approach for processing unstructured data into structured data for creating NURBS surfaces.

METHOD DESCRIPTION

The procedure for processing unstructured data into structured data is illustrated in Figure 2. The first step is to construct 3D triangulation over 3D unstructured data. A projection based method is established for this purpose. The 3D unstructured data is projected onto a reference plane to generate the 2D unstructured data with the same topological structure as the 3D unstructured data.

Figure 1. Structured data in the form of structured four-sided grid

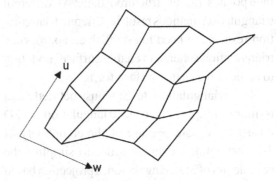

A 2D triangulation is then constructed from the 2D unstructured data by using an improved 2D triangulation algorithm to triangulate the 2D unstructured data. Subsequently, the 3D triangulation is obtained by projecting the 2D triangulation to 3D unstructured data.

The second step is to optimize the 3D triangulation to ensure that the subsequent surface modelling will generate a better-form surface. An optimization method is established to adjust the triangles in the triangulation according to the physical object surface to make them in the same convexities or concavities as the physical object surface.

The third step is to construct the five-degree triangular surfaces with C^1 continuity by interpolating each triangle in the spatial triangulation. Subsequently, a series of segment data can be obtained by cutting the triangular surfaces with a series of parallel planes. Finally, the structured data is obtained by deleting repetitive data points in each segment data, and then smoothing and uniforming each segment data. It can be seen that the key step of processing unstructured data into structured data lies in constructing the triangular surfaces that interpolate the unstructured data.

TRANGULATION

The triangulation of 3D unstructured data is a difficult problem for creating interpolation surfaces from 3D unstructured data. The triangulation directly influences the form of the surface that interpolates the unstructured data. As different triangulation methods result in different triangulation results, we need to establish an appropriate triangulation scheme to enable surface modeling to achieve the better surface form.

The triangulation to 3D unstructured data is more complex than the triangulation to 2D unstructured data because of the complexity of 3D unstructured data. In order to simplify the complexity of 3D triangulation, a projection based

method is established in this paper by transforming 3D triangulation into 2D triangulation. First, 3D unstructured data is projected onto a reference plane, without changing the topological structure of 3D unstructured data. Subsequently, triangulation is conducted over 2D unstructured data by an improved 2D triangulation algorithm. The 3D triangulation can be easily obtained from the 2D triangulation since the connecting relationships between unstructured data points in 2D are the same as in 3D.

There have been considerable research efforts reported on 2D triangulation over 2D unstructured data. These studies can be divided into the following two types. One is the direct triangulation to randomly multi-connected planar domains (Fang

Figure 2. Procedure for processing unstructured data into the structured data

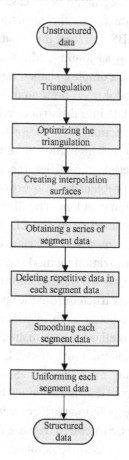

& Piegl, 1993; Joe & Simpson, 1986; Sapidis & Perucchio, 1991). This method constructs the Delaunay triangulation using the circumscribed circles of triangles. However, it needs a large amount of the calculation and testing for the circumscribed circles of triangles. The other is the indirect triangulation to random multi-connected planar domains (De Floriani, 1985; Sloan, 1987; Subramanian, 1994). This method constructs an appropriate triangulation grid at first, and then refines the triangular grid. It simplifies the complexity of triangulation. However, the efficiency of triangulation depends on the initial triangulation grid.

In this paper, a direct triangulation algorithm is developed to efficiently triangulate 2D unstructured data. This algorithm is an improvement of the triangulation algorithm reported by Karamete et al. (1997). In the Karamete's algorithm, the triangulation process is controlled by the two edges in the current triangle. Therefore, it is not necessary to check which edge will be used to construct the next Delaunay triangle. Furthermore, since there are at most two triangles that can be generated from the two edges of the current triangle, the efficiency of triangulation is improved. However, the detection for the bound of a boundary data point is very complex in the Karamete's algorithm. In order to solve this problem, we adopt the concept of boundary edges (Du, 1996) into the Karamete's algorithm to simplify the detection for the bound of a boundary data point. If

Figure 3. Two triangulation examples created by the improved triangulation algorithm

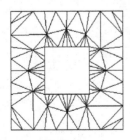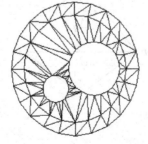

there is not any boundary edge that connects to a boundary point, the boundary point is bounded and is deleted from the data point list. Figure 3 gives two triangulation examples created by the proposed improved triangulation algorithm. The 3D triangulation can be easily obtained from the 2D triangulation since the connecting relationships between unstructured data points are the same in both 2D and 3D.

OPTIMIZATION OF 3D TRIANGULATION

The form of the triangular surfaces that interpolate 3D unstructured data depends on the spatial triangulation. In fact, the spatial triangulation is a linear interpolation to the given 3D unstructured data $\{\vec{p}_i\}_{i=1}^{i=N}$. Therefore, it has a significant influence on the triangular interpolation surfaces. If the form of a 3D triangulation is bumpy, the interpolation surfaces constructed from the spatial triangulation is also bumpy. Therefore, we have to optimize the 3D triangulation obtained through the project based method to ensure that the interpolation surfaces created from the 3D triangulation has a better form. A surface has a better form means that:

- The surface has as fewer bumps as possible, i.e., the surface has fewer changes between convex and concave domains;
- The surface varies evenly, i.e. the surface has as fewer drastic variations as possible.

According to the above two rules, the objective of the optimization of the 3D triangulation is to adjust the triangles in the 3D triangulation to make them have the same convexity or concavity as that of the physical object surface. We check the convexity or concavity between any two adjacent triangles in the 3D triangulation along the common edge between the two triangles, and change the convexity or concavity between the two triangles according to the convexity or concavity of the physical object surface. For example, as shown in Figure 4(a), suppose that we have a triangulation $\{\Delta_i\}_{i=1}^{i=N}$ and the physical surface is convex. For any two triangles Δ_1 and Δ_2 sharing a common edge in the triangulation $\{\Delta_i\}_{i=1}^{i=N}$, the vertexes of Δ_1 are labeled as $\vec{p}_1$, $\vec{p}_3$ and $\vec{p}_2$, the vertexes of Δ_2 are labeled as $\vec{p}_1$, $\vec{p}_3$ and $\vec{p}_4$, and the intersection between Δ_1 and Δ_2 is the edge $\vec{E} = \overrightarrow{p_1 p_3}$. The unit normal vectors of Δ_1 and Δ_2 are $\vec{n}_1$ and $\vec{n}_2$ respectively.

$$\vec{n}_1 = \frac{(\vec{p}_1 - \vec{p}_2) \times (\vec{p}_3 - \vec{p}_2)}{\left|(\vec{p}_1 - \vec{p}_2) \times (\vec{p}_3 - \vec{p}_2)\right|} \qquad (1)$$

$$\vec{n}_2 = \frac{(\vec{p}_3 - \vec{p}_4) \times (\vec{p}_1 - \vec{p}_4)}{\left|(\vec{p}_3 - \vec{p}_4) \times (\vec{p}_1 - \vec{p}_4)\right|} \qquad (2)$$

Therefore, we can define

$$I = \begin{vmatrix} \vec{n}_{1x} & \vec{n}_{2x} & \vec{E}_x \\ \vec{n}_{1y} & \vec{n}_{2y} & \vec{E}_y \\ \vec{n}_{1z} & \vec{n}_{2z} & \vec{E}_z \end{vmatrix} \qquad (3)$$

If

$I > 0$, Δ_1 and Δ_2 is convex at $\vec{E}$.

$I < 0$, Δ_1 and Δ_2 is concave at $\vec{E}$.

$I = 0$, Δ_1 and Δ_2 is co-planar at $\vec{E}$.

For the case that Δ_1 and Δ_2 is concave at $\vec{E}$, by changing the common edge $\overrightarrow{p_1 p_3}$ into $\overrightarrow{p_2 p_4}$, two new triangles can be obtained. As shown in Figure 4(b), the vertexes of Δ_1 are $\vec{p}_1$, $\vec{p}_2$ and $\vec{p}_4$, the vertexes of Δ_2 are $\vec{p}_2$, $\vec{p}_3$ and $\vec{p}_4$, the new common edge is $\overrightarrow{p_2 p_4}$, and Δ_1 and Δ_2 is convex at the new common edge.

Construction of Five-Degree Interpolation Surfaces

After optimization of the 3D triangulation, we can construct the interpolation surfaces from the optimized 3D triangulation. A local triangular interpolation method is established in this paper for construction of local five-degree C^1 polynomial interpolation surfaces. This method has advantages such as the simplicity, fast computational speed, and C^1 continuity between triangular patches. It also provides the convenience for obtaining the segment data.

A five-degree surface is defined on a triangle, i.e.

Figure 4. Transforming two concave triangles (a) into two convex triangles (b)

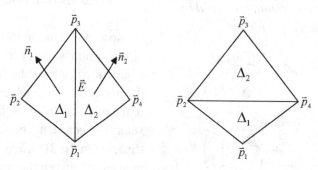

$$F(x,y) = \sum_{i+j \leq 5} a_{ij} x^i y^j \qquad (4)$$

where a_{ij} are the 21 coefficients to be determined. In order to enable the interpolation surface to satisfy C^1 continuity [7], the five-degree C^1 interpolation surface is constructed as follows (see Figure 5):

- Given the C^0, C^1, C^2 values at each vertex on triangle ΔT, that is, there are six interpolation conditions at each vertex of triangle ΔT.
- Given the directional derivatives along the normal vectors at the centers of each edge of triangle ΔT.

With the above 21 interpolation conditions at the vertexes and edge centers of ΔT, the coefficients a_{ij} can be solely determined. Therefore, the five-degree interpolation surfaces $F(x,y)$ with C^1 continuity can be obtained.

Calculating the First-Order and Second-Order Partial Derivatives at the Vertexes of Triangle ΔT

If the first-order and second-order partial derivatives at a vertex are given by the user or are defined by the adjacent surfaces, there is no need to calculate them. Otherwise, these values must be calculated. The first-order and second-order partial derivatives at vertex $\vec{p}_i$ are calculated from the five points nearest to $\vec{p}_i$. With the six points including point $\vec{p}_i$, we can solely define a quadric surface

$$S(x,y) = \sum_{i+j \leq 2} b_{ij} x^i y^j \qquad (5)$$

The first-order and second-order partial derivatives of the quadric surface $S(x,y)$ at $\vec{p}_i$ are

calculated as the first-order and second-order partial derivatives at $\vec{p}_i$.

Calculating the Directional Derivatives along the Normal Vectors at Each Edge Centers of Triangle ΔT

After calculating the first-order partial derivatives at each vertex of triangle ΔT, we can calculate the directional derivatives along any directions at each vertex. For example, if the unit normal vector along edge AC is $\vec{n}_1$, we can easily calculate the directional derivatives along $\vec{n}_1$ at vertexes A and C. By taking the average of these two directional derivatives, the directional derivatives along $\vec{n}_1$ at the center of the edge AC are obtained (Figure 5).

GENERATION OF STRUCTURED DATA

After constructing the C^1 triangular interpolation surfaces on the spatial triangulation, we can use a series of parallel planes to cut the triangular interpolation surfaces along a given direction to get a series of parallel segment data. The series of segment data are the intersections between the triangular interpolation surfaces and the series of

Figure 5. Five-degree C^1 surface interpolation

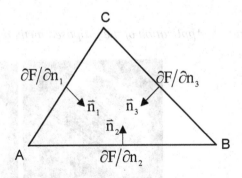

parallel planes. If the distance between any two adjacent parallel planes is small, the generated segment data are distributed densely. Otherwise, they are loosely distributed. This can help us to distribute the series of segment data according to the features of the physical object shape. The more dense series of segment data can be generated if the object measure data drastically vary. In contrast, the more loose series of segment data can be generated if the object measure data vary evenly. In this way, the time for generating the series of segment data from the triangular interpolation surfaces can be saved. Furthermore, each segment

data needs to be further processed to obtain the better structured data. The processing operations include deleting the repetitive data in each segment data, and smoothing and uniforming each segment data. After performing these processing operations, the structured data is obtained and can be used to generate NURBS surfaces.

IMPLEMENTATION AND RESULTS

The proposed approach for processing unstructured data into structured data has been imple-

Figure 6. Processing unstructured data into structured data

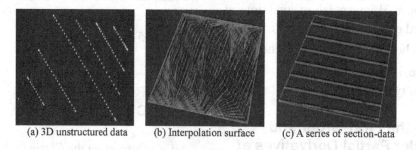

(a) 3D unstructured data   (b) Interpolation surface   (c) A series of section-data

Figure 7. Application of the proposed method for reverse engineering of an aircraft component

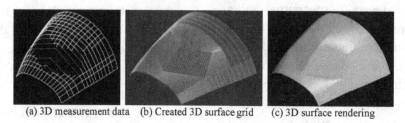

(a) 3D measurement data   (b) Created 3D surface grid   (c) 3D surface rendering

Figure 8. Application of the proposed method for reverse engineering of a mould

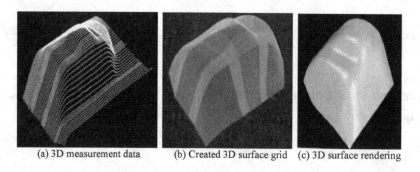

(a) 3D measurement data   (b) Created 3D surface grid   (c) 3D surface rendering

mented in a reverse engineering prototype system. Experiments have been conducted to evaluate the performance of the proposed methodology for processing unstructured data into structured data for surface modelling. Figure 6 shows a typical example on processing unstructured data into structured data by the proposed methodology. Figure 6(a) is the unstructured data measured from the left front window of an aircraft. Figure 6(b) is the five-degree C^1 triangular surfaces that interpolate the unstructured data. Figure 6(c) shows a series of segment data generated by cutting the triangular interpolation surfaces using a series of parallel planes. It can be seen that the proposed methodology is effective in processing unstructured data into structured data. Figure 7 illustrates an example on the application of the proposed methodology in reverse engineering of an aircraft component. Figure 7(a) is the measurement data measured from the framework of the front window of an aircraft. The measurement data include the unstructured measurement data shown in Figure 6(a). Figure 7(b) is the NURBS surface model constructed from the measure data in Figure 7(a). Figure 7(c) is the 3D graphics rendering of the surface model.

Figure 8 illustrates another example on the application of the proposed approach in reverse engineering of a mould. Figure 8(a) is the measurement data measured from the mould. The measurement data include the unstructured data. Figure 8(b) is the NURBS surface model constructed from the measure data in Figure 8(a). Figure 8(c) is the 3D graphics rendering of the surface model.

CONCLUSION

One of the difficult problems in reverse engineering is unstructured data processing. This paper presents a methodology to process unstructured data into structured data for surface modelling. A projection based triangulation method is estab-lished for 3D unstructured data by transforming the complex 3D triangulation problem to the simple 2D triangulation problem. A method for optimizing the 3D triangulation is established to ensure that the resulting interpolation surfaces have a better form. A local triangular interpolation approach is established for constructing five-degree C^1 interpolation surfaces from optimized 3D triangulation. A series of segment data are obtained by cutting the interpolation surfaces with a series of parallel planes. The structured data is finally obtained by further processing such as deleting the repetitive data in each segment data, and smoothing and uniforming each segment data. The proposed methodology has been applied in reverse engineering. The experimental results demonstrate that the proposed methodology provides an effective solution for processing 3D unstructured data in reverse engineering.

Future research will focus on the automation of the proposed methodology for unstructured data processing. Advanced artificial intelligence technologies such as neural networks, genetic algorithms, expert systems and knowledge engineering will be used to automate the data processing process of the proposed methodology.

REFERENCES

Beccari, C. V., Farella, E., Liverani, A., Morigia, S., & Rucci, M. (2010). A fast interactive reverse-engineering system. *Computer Aided Design*, *42*(10), 860–873. doi:10.1016/j.cad.2010.06.001.

De Floriani, L., Faicidieno, B., & Pienovi, C. (1985). Delaunay-based representation of surface defined over arbitrarily shaped domains. *Computer Vision Graphics and Image Processing*, *32*(1), 127–140. doi:10.1016/0734-189X(85)90005-2.

Du, C. (1996). An algorithm for automatic Delaunay triangulation of arbitrary planar domains. *Advances in Engineering Software*, *27*(1-2), 21–26. doi:10.1016/0965-9978(96)00004-X.

Fang, T. P., & Piegl, L. A. (1993). Algorithm for Delaunay triangulation and convex-hull computation using a sparse matrix. *Computer Aided Design, 24*(8), 425–536. doi:10.1016/0010-4485(92)90010-8.

Jiang, D., & Wang, L. C. (2006). An algorithm of NURBS surface fitting for reverse engineering. *International Journal of Advanced Manufacturing Technology, 31*(1-2), 92–97. doi:10.1007/s00170-005-0161-3.

Joe, B., & Simpson, R. B. (1986). Triangulation meshes for regions of complicated shape. *International Journal for Numerical Methods in Engineering, 23*(5), 751–778. doi:10.1002/nme.1620230503.

Karamete, B. K., Tokdemir, T., & Ger, M. (1997). Unstructured grid generation and a simple triangulation algorithm for arbitrary 2D geometries using object oriented programming. *International Journal for Numerical Methods in Engineering, 40*(2), 251–268. doi:10.1002/(SICI)1097-0207(19970130)40:2<251::AID-NME62>3.0.CO;2-U.

Lawson, C. L. (1977). Software for C^1 surface interpolation. In Rice, J. R. (Ed.), *Mathematical software III* (pp. 161–194). New York, NY: Academic Press.

Lin, C.-Y., Liou, C.-S., & Lai, J.-Y. (1997). A surface-lofting approach for smooth-surface reconstruction from 3D measurement data. *Computers in Industry, 34*(1), 73–85. doi:10.1016/S0166-3615(96)00082-6.

Ma, W., & He, P. (1998). B-spline surface local updating with unorganized points. *Computer Aided Design, 30*(11), 853–862. doi:10.1016/S0010-4485(98)00042-6.

Ma, W., & Kruth, J. P. (1998). NURBS curve and surface fitting for reverse engineering. *International Journal of Advanced Manufacturing Technology, 14*(12), 918–927. doi:10.1007/BF01179082.

Meng, C., & Chen, F. L. (1996). Curve and surface approximation from CMM measurement data. *Computers & Industrial Engineering, 30*(2), 211–225. doi:10.1016/0360-8352(95)00165-4.

Pal, P. (2008). A reconstruction method using geometric subdivision and NURBS interpolation. *International Journal of Advanced Manufacturing Technology, 38*(3-4), 296–308. doi:10.1007/s00170-007-1102-0.

Pal, P., & Ballav, R. (2007). Object shape reconstruction through NURBS surface interpolation. *International Journal of Production Research, 45*(2), 287–307. doi:10.1080/00207540600688481.

Sapidis, N., & Perucchio, R. (1991). Delaunay triangulation of arbitrarily shaped planar domains. *Computer Aided Geometric Design, 8*(6), 421–437. doi:10.1016/0167-8396(91)90028-A.

Seiler, A., Balendran, V., Sivayoganathan, K., & Sackfield, A. (1996). Reverse engineering from uni-directional CMM scan data. *International Journal of Advanced Manufacturing Technology, 11*(4), 276–284. doi:10.1007/BF01351285.

Sloan, S. W. (1987). A fast algorithm for constructing Delaunay triangulations in the plane. *Advances in Engineering Software, 9*(1), 34–55. doi:10.1016/0141-1195(87)90043-X.

Subramanian, G., Raveendra, V. V. S., & Kamath, M. G. (1994). Robust boundary triangulation and Delaunay triangulation of arbitrary planar domains. *International Journal for Numerical Methods in Engineering, 37*(10), 1779–1789. doi:10.1002/nme.1620371009.

Tang, P. B., Huber, D., Akinci, B., Lipman, R., & Lytle, A. (2010). Automatic reconstruction of as-built building information models from laser-scanned point clouds: A review of related techniques. *Automation in Construction, 19*(7), 829–843. doi:10.1016/j.autcon.2010.06.007.

Tuohy, S. T., Maekawa, T., Shen, G., & Patrika-lakis, N. M. (1997). Approximation of measured data with interval B-splines. *Computer Aided Design*, *29*(11), 791–799. doi:10.1016/S0010-4485(97)00025-0.

Varady, T., Martin, R. R., & Cox, J. (1997). An integrated reverse engineering approach to reconstructing free-form surfaces. *Computer Integrated Manufacturing Systems*, *10*(1), 49–60. doi:10.1016/S0951-5240(96)00019-5.

Varady, T., Martin, R. R., & Cox, J. (1997). Reverse engineering of geometric models-an introduction. *Computer Aided Design*, *29*(4), 255–268. doi:10.1016/S0010-4485(96)00054-1.

Wang, G., & Wang, C. (1997). Reconstruction of sculptured surface on reverse engineering. In *Proceedings of the ASME Design Engineering Technical Conference*, Sacramento, CA (pp14-17).

Yau, H. (1997). Reverse engineering of engine intake ports by digitization and surface approximation. *International Journal of Machine Tools & Manufacture*, *37*(6), 871–875. doi:10.1016/S0890-6955(95)00100-X.

Yau, H., & Chen, J. (1997). Reverse engineering of complex geometry using rational B-splines. *International Journal of Advanced Manufacturing Technology*, *13*(8), 548–555. doi:10.1007/BF01176298.

Yin, Z., Zhang, Y., & Jiang, S. (2003). A methodology of sculptured surface fitting from CMM measurement data. *International Journal of Production Research*, *41*(14), 3375–3384. doi:10.1080/00207540310000112815.

Yin, Z., Zhang, Y., & Jiang, S. (2003). Methodology of NURBS surface fitting based on off-line software compensation for errors of a CMM. *Precision Engineering*, *27*(3), 299–303. doi:10.1016/S0141-6359(03)00033-3.

This work was previously published in the International Journal of Intelligent Mechatronics and Robotics, Volume 1, Issue 2, edited by Bijan Shirinzadeh, pp. 42-51, copyright 2011 by IGI Publishing (an imprint of IGI Global).

Chapter 9
Modeling and Simulation of Digital Systems Using Bond Graphs

Majid Habibi
K. N. Toosi University of Technology, Iran

Alireza B. Novinzadeh
K. N. Toosi University of Technology, Iran

ABSTRACT

Bond graphs are suitable tools for modeling many types of dynamical systems and can model these systems consisting of mechanical, electrical, fluidic, and pneumatic sub-systems. The advantage of a bond graph is that it can model non-linear systems and combinational systems. In this paper, the authors utilize bond graphs for modeling mechatronics systems. Mechatronics systems consist of mechanics, electronics, and intelligent software. Many of these systems have digital sections that are constructed by logical circuits (hardware by transistors and now mostly by chips). The authors present a methodology to implement these mechatronics systems by bond graphs.

INTRODUCTION

Bond graph was introduced by the late Henry M. Paynter (1923-2002), professor at MIT & UT Austin, with the introduction of junctions in April 1959. In a period of about a decade, most of the underlying concepts were formed and were put together into a conceptual framework and corresponding notation (Paynter, 1961, 1992). In the sixties the notation, e.g. the half arrow to represent positive orientation and insightful node labeling, was further elaborated by researchers, in particular Dean C. Karnopp at UC Davis (Ca); and Roland C. Rosenberg (1968, 1974, 1990), at Michigan State University (Michigan) who also designed the first computer tool (ENPORT)

DOI: 10.4018/978-1-4666-3634-7.ch009

Figure 1. The voltage range that show 0 and 1

that supported simulation of bond graph models [Rosenberg, 1965, 1974]. In the early seventies Jan J. Van Dixhoorn (1972; Evans et al., 1974) at the University of Twente, NL, and Jean U. Thoma (1975) at the University of Waterloo (Ontario) were the first to introduce bond graphs in Canada and Europe, respectively.

These pioneers in the field and their students have been developing these ideas worldwide (Karnopp et al., 1979). Jan Van Dixhoorn realized that an early prototype of the block-diagram-based software TUTSIM could be used to input simple causal bond graphs, and about a decade later, resulted in a PC-based tool (Beukeboom et al., 1985). This laid the foundation for the development of truly port-based computer tool 20-sim at the University of Twente (Broenink & Breedveld, 1988) (www.20sim.com). He also initiated research in modeling more complex physical systems, in particular thermofluid systems (Breedveld, 1979). In the last two decades, bond graphs either have been a research topic or are used in research projects at many universities worldwide and have become part of engineering curricula at a steadily growing number of universities. In the last decade, their industrial use has become more and more important.

Logic circuits are also very important tools for implementing logic by electronic circuits. Initially, these circuits were used for implementing necessary logic in computers. In fact the existence of computers owes a lot to these circuits. After production of electronic chips the size of these circuits decreased considerably and thus increased their popularity. However, implementing

logic by transistors is still used. Nowadays these circuits exist in most of the electronic systems, and transistors and chips can be readily found on such electronic circuits and boards.

Mechatrornics systems utilize mechanics, electronics and software. These systems have very large usage, and their usage is increased daily. In electronic sub-systems of these systems logical circuits are used very largely. Because bond graph is a powerful tool for modeling electrical and mechanical systems, it is also very suitable for modeling mechatronics systems. The use of bond graph for modeling electrical and mechanical systems can be found in many areas (Gawthrop, 1991; Amerongen & Breedveld, 2003; Breedveld, 2004; Karnopp, 2006). Further, Chhabra and Emami developed an alternative modeling scheme for mechatronics systems by combining bond graph and block diagram and presented these findings (Chhabra & Emami, 2010). These models hitherto were not used for modeling systems that have

Table 1. Truth table of digital gates

AND			OR		
X	Y	OUT	X	Y	OUT
T	T	T	T	T	T
T	F	F	T	F	T
F	T	F	F	T	T
F	F	F	F	F	F
NAND			NOR		
X	Y	OUT	X	Y	OUT
T	T	F	T	T	F
T	F	T	T	F	F
F	T	T	F	T	F
F	F	T	F	F	T
XOR			XNOR		
X	Y	OUT	X	Y	OUT
T	T	T	T	T	F
T	F	F	T	F	T
F	T	F	F	T	T
F	F	T	F	F	F

Figure 2. The equivalent bond graph of gates

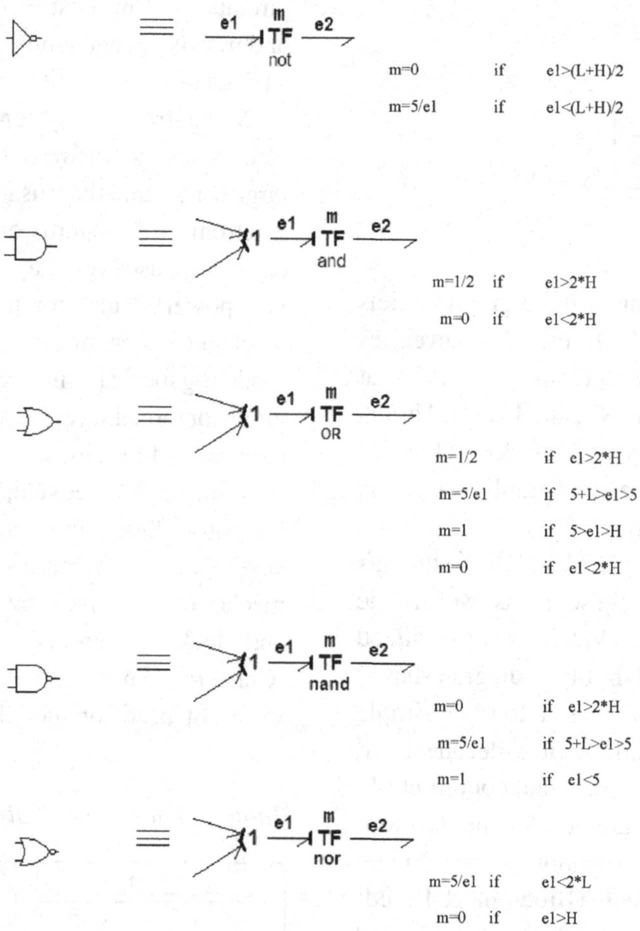

$$m=0 \quad \text{if} \quad e1>(L+H)/2$$
$$m=5/e1 \quad \text{if} \quad e1<(L+H)/2$$

$$m=1/2 \quad \text{if} \quad e1>2*H$$
$$m=0 \quad \text{if} \quad e1<2*H$$

$$m=1/2 \quad \text{if} \quad e1>2*H$$
$$m=5/e1 \quad \text{if} \quad 5+L>e1>5$$
$$m=1 \quad \text{if} \quad 5>e1>H$$
$$m=0 \quad \text{if} \quad e1<2*H$$

$$m=0 \quad \text{if} \quad e1>2*H$$
$$m=5/e1 \quad \text{if} \quad 5+L>e1>5$$
$$m=1 \quad \text{if} \quad e1<5$$

$$m=5/e1 \quad \text{if} \quad e1<2*L$$
$$m=0 \quad \text{if} \quad e1>H$$

logical circuits (that are implemented by chip or transistor). Here we present a way for modeling logical circuits by bond graph that will increase the domain of usage of bond graphs, and they can cover a larger portion of mechatronics systems.

STRACTURE OF DIGITAL GATES

Digital gates are based on the logic rules and the values of inputs, providing what the outputs should be. The logic that is used in these gates is binary logic or 0 and 1, also known as Boolean algebra. In the implementation of these gates each of 0 and 1 values are shown by a domain of voltage (Figure 1) (We use the range that is used in TTL ICs).

The gates that are used and their truth table are shown in Table 1.

As mentioned earlier, the electronic hardware for such mechatronics systems are constructed by transistors or integrated circuits (ICs).

MODELING GATES BY BOND GRAPH

Now we want to show these gates by equivalent bond graphs. We show the input of gates, that is a voltage domain, by an effort variable. The input efforts are added to each other and then by a conditional transformer the result value reaches to the desired value. Details are shown in Figure 2.

Figure 3. The equivalent bond graph of a gate with more than 2 inputs

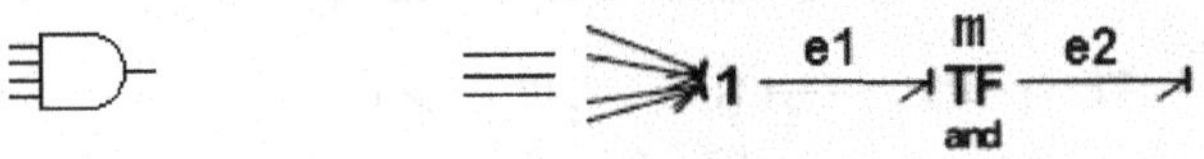

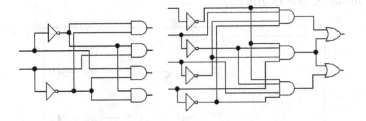

$$m = 1/4 \quad \text{if} \quad e1 > 4*H$$
$$m = 0 \quad \text{if} \quad e1 < 4*H$$

Figure 4. 2 in 4 decoder (top) and 4 in 2 encoder (bottom) Boolean circuits

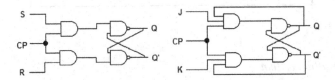

Figure 5. SR flip flop (left) and JK flip flop (right) Boolean circuits

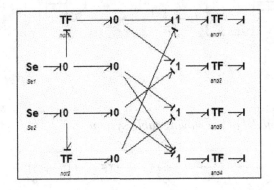

Figure 6. Bond graph model of a 2 in 4 decoder

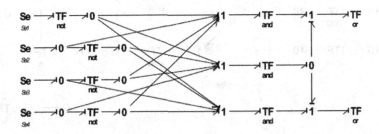

Figure 7. Bond graph model of a 4 in 2 encoder

Figure 8. Bond graph model of a RS flip flop

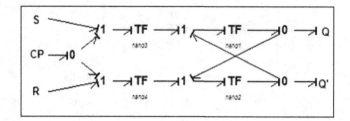

Figure 9. Bond graph model of a JK flip flop

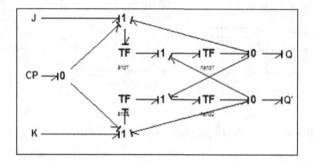

Figure 10. The moving table system

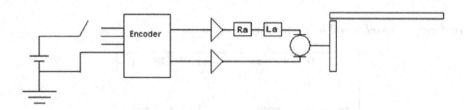

Figure 11. The bond graph model of the moving table system

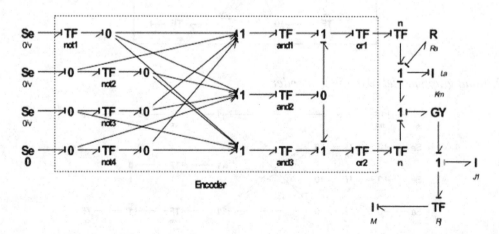

If we have more than two inputs, these gates are constructed by integration of 2-input gates, but we can also show them by one gate and one equivalent bond graph such as Figure 3.

MODELING OF EXAMPLES OF LOGICAL CIRCUITS

In this section we present some examples of modeling integrated combinational and sequential circuits by bond graph. Combinational circuits are circuits that do not any feedback from output

to input while in sequential circuits this feedback exists. For example, in combinational circuits we select decoder and encoder that have very popular usage in logical circuits. We model a 2 in 4 decoder and a 4 in 2 encoder (Figure 4) by bond graph that are shown in Figures 6 and 7, respectively.

In sequential circuits, we present flip flops that are the based on sequential circuits, as shown in Figure 5. In Figure 8 the model of RS flip flop, and in Figure 9 the model of JK flip flop are shown. For additional details for these circuits, the reader can refer to (Mano, 2006).

Box 1.

$$P_J = Jf_J = \frac{J}{R_j}f_M = \frac{J}{MR_j}P_M$$

$$\dot{P}_M = e_M = \frac{1}{R_j}(K_m f_{L_a} - \dot{P}_j) = \frac{1}{R_j}(\frac{K_m}{L_a}P_{L_a} - \frac{J}{MR_j}\dot{P}_M) \Rightarrow \dot{P}_M = \frac{MR_j^2 K_m}{MR_j^2 + J}P_{L_a}$$

$$\dot{P}_{L_a} = e_{L_a} = n_1 or_1(and_1(Se_1 not_1 + Se_2 + Se_3 not_3 + Se_4 not_4)$$

$$+and_2(Se_1 not_1 + Se_2 not_2 + Se_3 not_3 + Se_4)) - R_a f_{L_a} - K_m f_J$$

$$+n_2 or_2(and_2(Se_1 not_1 + Se_2 not_2 + Se_3 not_3 + Se_4)$$

$$+and_3(Se_2 not_2 + Se_1 not_1 + Se_3 + Se_4 not_4))$$

$$= -\frac{R_a}{L_a}P_{L_a} - \frac{K_m}{MR_j}P_M + and_2(n_1 or_1 + n_2 or_2)(Se_1 not_1 + Se_2 not_2 + Se_3 not_3 + Se_4)$$

$$+n_1 or_1 and_1(Se_1 not_1 + Se_2 + Se_3 not_3 + Se_4 not_4)$$

$$+n_2 or_2 and_3(Se_1 not_1 + Se_2 not_2 + Se_3 + Se_4 not_4)$$

therefore

$$\begin{bmatrix} \dot{P}_{L_a} \\ \dot{P}_M \end{bmatrix} = \begin{bmatrix} -\dfrac{R_a}{L_a} & -\dfrac{K_m}{MR_j} \\ \dfrac{MR_j^2 K_m}{MR_j^2 + J} & 0 \end{bmatrix} \begin{bmatrix} P_{L_a} \\ P_M \end{bmatrix} + \begin{bmatrix} 1 \\ 0 \end{bmatrix} U$$

where

$$U = and_2(n_1 or_1 + n_2 or_2)(Se_1 not_1 + Se_2 not_2 + Se_3 not_3 + Se_4)$$

$$+n_1 or_1 and_1(Se_1 not_1 + Se_2 + Se_3 not_3 + Se_4 not_4)$$

$$+n_2 or_2 and_3(Se_1 not_1 + Se_2 not_2 + Se_3 + Se_4 not_4)$$

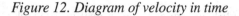

Figure 12. Diagram of velocity in time

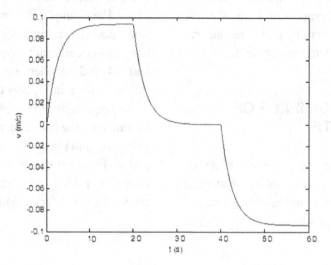

Figure 13. Diagram of state in time

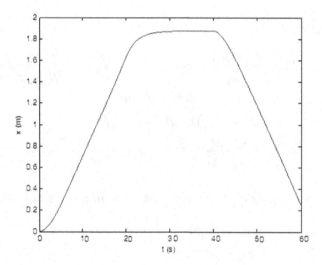

MODELING MECHATRONICS SYSTEMS COMPRISING LOGICAL CIRCUITS

In this section, we present an example of mechatronics system with mechanical and electronic subsystems using logic circuits incorporated in its structures, and then develop its bond graph using the method presented in the previous section. We also show that full modeling of this system without using this method is impossible.

The system we present and model in this section is a table that moves to the left and right by a rack and pinion that is rotated by an electrical motor (Figure 10). As shown, we use a three state switch that has states left, right and stop. The outputs of the switch are connected to a 4 in 2 encoder that has one of its inputs connected to the ground, and one of their outputs is zero, and the other has voltage (and in stop state both are zero), with two amplifiers are connected to the electrical motor resulting in a potential difference in motor.

As shows in the structure of this system, logical circuits are used in the case of an encoder, thus in order to perform the modeling of the system, we must model the encoder by the proposed method, and this model is shown in Figure 7. Hence, we can model the full of system by bond graph, as is shown in Figure 11.

We can now extract state equations from the bond graph (see Box 1).

Subsequently, we select the problem parameters as shown:

$$M = 40kg$$
$$J = .05kgm^2$$
$$R_j = .15m$$
$$L_a = 100mH$$
$$R_a = 1\Omega$$
$$K_m = 0.6$$
$$n_1 = n_2 = 5$$

The motion of the table may be presented, as shown in Figure 12. We use a scenario that our switch is in the right position for 20 seconds, then its position is changed to stop the table for 20 seconds, and finally it stays in left position for 20 seconds.

CONCLUSION

This paper presented an approach to model logic circuits using bond graph technique. Such logic circuits may be sued in the structures of many systems. The proposed bond graph technique can be used to fully model electronic and electromechanical systems that are consisted of circuits. Subsequently, the methodology allows extraction of full state equations from the bond graph. We showed that conditional gain of gates, that are dependent on the state they have, may be different; and that the inputs of circuits may be shown in the

resulting equations. Thus, by specifying numerical values to the parameters of the gates in different states, we can readily solve these equations.

REFERENCES

Beukeboom, J. J. A., van Dixhoorn, J. J., & Meerman, J. W. (1985). Simulation of mixed bond graphs and block diagrams on personal computer using TUTSIM. *Journal of the Franklin Institute, 319*(1-2), 257–267. doi:10.1016/0016-0032(85)90079-1.

Breedveld, P. C. (1979). *Irreversible thermodynamics and bond graphs: A synthesis with some practical examples*. Unpublished master's thesis, University of Twente, The Netherlands.

Breedveld, P. C. (2004). *Port-based modeling of mechatronic systems*. Amsterdam, The Netherlands: Elsevier.

Chhabra, R., & Emami, M. R. (2010). *Holistic system modeling in mechatronics*. Amsterdam, The Netherlands: Elsevier.

Gawthrop, P. J. (1991). *Bond graphs: A representation for mechatronic systems*. Amsterdam, The Netherlands: Elsevier.

Karnopp, D. C. (2006). *System dynamics: Modeling and simulation of mechatronics systems*. New York, NY: John Wiley & Sons.

Karnopp, D. C., Pomerantz, M. A., Rosenberg, R. C., & van Dixhoorn, J. J. (Eds.). (1979). Bond graph techniques for dynamic systems in engineering and biology. *Journal of the Franklin Institute, 308*(3).

Mano, M. M. (2006). *Digital design* (4th ed.). Upper Saddle River, NJ: Prentice Hall.

Paynter, H. M. (1961). *Analysis and design of engineering systems*. Cambridge, MA: MIT Press.

Paynter, H. M. (1992). An epistemic prehistory of bond graphs. In Breedveld, P. C., & Dauphin-Tanguy, G. (Eds.), *Bond graphs for engineers* (pp. 3–17). Amsterdam, The Netherlands: Elsevier.

Rosenberg, R. C. (1965). *Computer-aided teaching of dynamic system behavior*. Unpublished doctoral dissertation, MIT, Cambridge, MA.

Rosenberg, R. C. (1974). *A user's guide to EN-PORT-4*. New York, NY: John Wiley & Sons.

Rosenberg, R. C., & Karnopp, D. C. (1983). *Introduction to physical system dynamics*. New York, NY: McGraw-Hill.

Thoma, J. U. (1975). *Introduction to bond graphs and their applications*. Oxford, UK: Pergamon Press.

van Amerongen, J., & Breedveld, P. C. (2003). *Modelling of physical systems for the design and control of mechatronic systems*. Amsterdam, The Netherlands: Elsevier.

van Dixhoorn, J. J. (1972). Network graphs and bond graphs in engineering modeling. *Annals of System Research*, 2, 22–38.

van Dixhoorn, J. J., & Evans, F. J. (Eds.). (1974). *Physical structure in system theory: Network approaches to engineering and economics* (p. 305). London, UK: Academic Press.

This work was previously published in the International Journal of Intelligent Mechatronics and Robotics, Volume 1, Issue 2, edited by Bijan Shirinzadeh, pp. 52-61, copyright 2011 by IGI Publishing (an imprint of IGI Global).

Chapter 10
Modeling, Simulation and Motion Cues Visualization of a Six–DOF Motion Platform for Micro–Manipulations

Umar Asif
National University of Sciences & Technology- Islamabad, Pakistan

Javaid Iqbal
College of Electrical and Mechanical Engineering- Rawalpindi, Pakistan

ABSTRACT

This paper examines the problem of realizing a 6-DOF motion platform by proposing a closed loop kinematic architecture that benefits from an anthropological serial manipulator design. In contrast to standard motion platforms based on linear actuators, a mechanism with actuator design inspired from anthropological kinematic structure offers a relatively larger motion envelope and higher dexterity making it a viable motion platform for micromanipulations. The design consists of a motion plate connected through only revolute hinges for the passive joints, and three legs located at the base as the active elements. In this hybrid kinematic structure, each leg is connected to the top (motion) plate through three revolute hinges and to the bottom (fixed) plate through a single revolute joint forming a closed-loop kinematic chain. The paper describes the mathematical modeling of the proposed design and demonstrates its simulation model using SimMechanics and xPC Target for real-time simulations and visualization of the motion cues.

DOI: 10.4018/978-1-4666-3634-7.ch010

1. INTRODUCTION

In literature, research studies (Mehregany, Gabriel, & Trimmer, 1988; Jokiel, Benavides, Bieg, & Allen, 2001) investigate the initial use of Micro Electro-Mechanical Systems technology to design and fabricate parallel mechanisms. Furthermore, a parallel mechanism was built based on MEMS fabrication technology that provided translational motions in three axes as reported in Fan, Wu, Choquette, and Crawford (1997). Motion platforms with two and three degrees of freedom were presented in Jokiel, Benavides, Bieg, and Allen (2001) and Hollar, Bergbreiter, and Pister (2003).

1.1 Parallel Structures

In the design and development of manipulators, it has been realized through several research studies (Merlet, 2000; Gosselin & Angeles, 1991) that parallel structures possess intrinsic advantages over serial manipulators due to their high rigidity, large payload capacity, high velocity and notable precision. However, rigid joints with multi degrees of freedom used in conventional parallel robots affect the precision and accuracy of the system due to backlash, hysteresis and other manufacturing errors. In order to overcome these design constraints, compliant structures have been employed into parallel robots as investigated in Tanikawa, Arai, and Koyachi (1999) where notable precision is desirable at micron levels due to the fact that compliant structures possess simple configuration, do not suffer from backlash and have easy fabrication (Smith, 2000).

Typical development of motion platforms on macro scales comprise of kinematic configurations in which linear piston-like actuators are connected between the top (motion) plate and the bottom (fixed) plate through spherical or universal joints that are difficult to manufacture in MEMS fabrication technology. In literature there is a large body

of work (Tsai & Tahmasebi, 1993; Ben-Horin & Shoham, 1996; Ben-Horin, Shoham, & Djerassi, 1998; Honegger, 1998; Wenger & Chablat, 2000; Tsai & Tahmasebi, 1993; Ceccarelli, 1997) that investigates the design and development of such parallel structures which employ passive sub-mechanisms to connect the motion plate.

Through literature studies (Choi, Sreenivasan, & Choi, 2008; Kallio, Lind, Zhou, & Koivo, 1998; Tang, Pham, Li, & Chen, 2004; Wu, Chen, & Chang, 2008; Hudgens & Tesar, 1991) it has

Figure 1. Model of the proposed motion platform

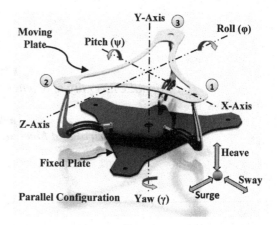

Figure 2. Kinematic configuration of the proposed architect

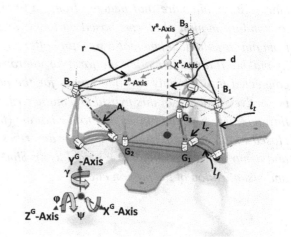

Figure 3. Views of the motion platform in different poses

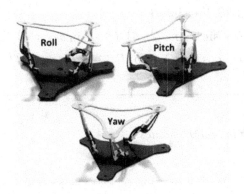

been realized that micromanipulators possess potential applications in fields such as optical fiber alignment, micro system assembly and micro-scale experiments in biological research. Therefore more and more micro-robotic systems are developed based upon parallel structures to realize systems with appreciable performance. However, conventional parallel mechanisms found in literature feature fewer degrees of freedom (Choi, Sreenivasan, & Choi, 2008; Kallio, Lind, Zhou, & Koivo, 1998; Tang, Pham, Li, & Chen, 2004; Wu, Chen, & Chang, 2008) that confines their motion envelope. Therefore, design and realization of micro-manipulators with 6-DOF motion capabilities with a reasonable workspace still require further research.

1.2 Literature Review and Related Work

Motion transmission is achieved in compliant mechanisms using flexure joints through elastic deformation of the material as observed in Ouyang, Tjiptoprodjo, Zhang, and Yang (2008). In a research study (Ryu, Kim, & Kim, 1997), it has been realized that, stress stiffening in compliant mechanisms corresponds to the stiffening of the structure itself due to its stress state in the presence of axial stresses. As investigated in Xu and

Li (2008), this phenomenon should be avoided to compensate the effect of amplified forces and reduced strokes to the mechanism which as a consequence causes nonlinearities to the actuation procedures.

Initial development of specialized joints based on Micro-fabricated hinges was proposed by Pister, Judy, Burgett, and Fearing (1992) to realize relative motion (Piyawattanametha, Fan, Lee, Su, & Wu, 1998) between the connecting links. Later, micro-fabricated hinges developed in surface micromachining to provide vertical motion in motion platforms was proposed by Fan, Wu, Choquette, and Crawford (1997). Another breakthrough in this context was the development of flexible joints (Piyawattanametha, Fan, Lee, Su, & Wu, 1998) that enabled the fabrication of spatial mechanisms with linear and revolute flexible joints as demonstrated in Liu & DeVoe (2001).

Wang, Hikita, Kubo, Zhao, Huang, and Ifukube (2003) investigated the motion analysis of a 6 degree-of-freedom parallel manipulator comprising elastic joints. Another research study (Liu, Xu, & Fei, 2003) proposed the design and development of

Figure 4. Representation of the platform orientation using the pitch-yaw-roll Euler angles

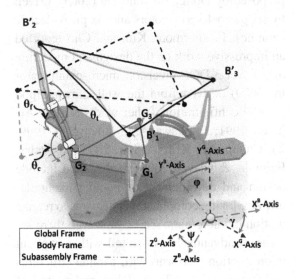

Box 1.

$$T_{xi}^{G} = \cos^3 \cos \ddot{o} B_{xi}^{B} - \cos^3 \sin \ddot{o} B_{yi}^{B} + \sin^3 B_{zi}^{B}$$

$$T_{yi}^{G} = \left(\cos \phi \sin \ddot{o} + \sin \phi \sin^3 \cos \ddot{o}\right) B_{xi}^{B} + \left(\cos \phi \cos \ddot{o} - \sin \phi \sin^3 \sin \ddot{o}\right) B_{yi}^{B} - \sin \phi \cos^3 B_{xi}^{B}$$

$$T_{zi}^{G} = \left(\sin \phi \sin \ddot{o} - \cos \phi \sin^3 \cos \ddot{o}\right) B_{xi}^{B} + \left(\sin \phi \cos \ddot{o} + \cos \phi \sin^3 \sin \ddot{o}\right) B_{yi}^{B} + \cos \phi \cos^3 B_{xi}^{B}$$

(1)

Figure 5. Physical model of the proposed motion platform in SimMechanics

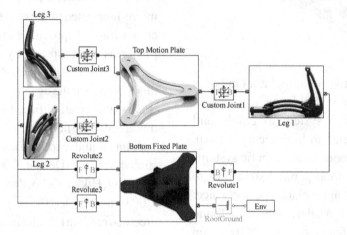

micro-manipulator with appreciable performance based upon 3-PRS (prismatic-revolute-spherical) in-parallel kinematic configuration. A compliant parallel-structure with high precision was further proposed by Dong, Sun, and Du (2007), driven by six piezoelectric motors and six piezoelectric ceramics. Furthermore, Kim and Cho reported an impressive work on the design and modeling of a novel 3-DOF precision micro-manipulator in (2009) derived from the well-known tripod parallel configuration. Other works (Lee & Arjunan, 1991; Tanikawa, Arai, & Koyachi, 1999; Yi, Chung, Na, Kim, & Suh, 2003; Niaritsiry, Fazenda, & Clavel, 2004) in this field include design and development of parallel manipulators employing compliant mechanisms to realize motion in micro/nano scales. However, there are some disadvantages associated with compliant micro-motion platforms with parallel structures and flexure joints. The modeling and control of

compliant micro-motion platforms with appreciable performance and precision demands more effort compared to conventional mechanisms due to the fact that the hinge compliance in the undesired directions cause unintended motions. Most of the 6-DOF parallel manipulators described in the literature constitute a kinematic configuration of six actuators as discussed in Wu, Chen, and Chang (2008), Hudgens and Tesar (1991), and Gao, Zhang, Chen, and Jin (2003).

However, there are certain disadvantages associated with compliant motion platforms with parallel structures and flexure joints to achieve micromanipulation. The modeling and control of a motion platform with significant precision at micron-scale is more difficult in using a compliant parallel structure than compared to the conventional mechanisms due to the reasons that include: the complexity of kinematic and dynamic modeling of parallel mechanism and

Figure 6. Visualization of the SimMechancis model in 3D animation viewer

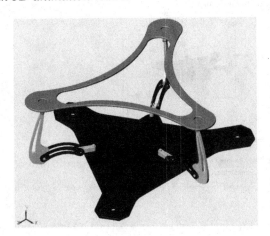

joint compliance in the undesired directions that cause unintended motions. Furthermore, most of the existing micromanipulators can provide only a planar 3-DOF motion, or spatial 3-DOF combined motions of translation and rotation.

An actuator design inspired from an anthropological kinematic structure may offer higher dexterity, large motion envelope, a possibility to realize any posture within the workspace and an ability to displace payload with large accelerations and velocities along complex coordinated trajectories, which serves a strong motivation of the work done presented in this paper. However, designing a motion platform based on actuator design inspired from anthropological kinematic structure involves several challenges of which

the greatest is to conceive a suitable kinematical model able to cope with the user inputs.

The objective here is to present a new architect of a micro-scaled 6-DOF motion platform which can achieve a reasonable workspace with minimum possible hinges and links. In order to realize this objective, the proposed architect is first modeled in Pro Engineer and then a kinematic model is determined through inverse and forward kinematic equations to define the pose of the motion plate in terms of its angular orientation and linear position with reference to global coordinate system. To validate the kinematic model, the second part of the paper presents a simulation model to conduct kinematic analysis of the proposed design through real-time simulations and visualize the motion cues during the simulation. The paper ends with a discussion of the significance of the proposed design and anticipated shortcomings when realizing such mechanisms in micro scale.

2. DESIGN AND MODELING

Parallel robots provide higher rigidity due to the fact that several actuators altogether provide support to the mechanism which also enables them to sustain high payloads. In serial manipulators, the geometrical and control errors accumulate however, in parallel configurations the errors are distributed between the actuators. Through several research studies found in literature, it has been

Figure 7. Physical model of an individual leg

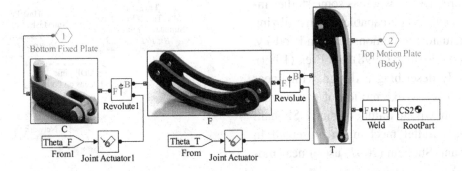

Figure 8. Structure of the SimMechancis simulation model

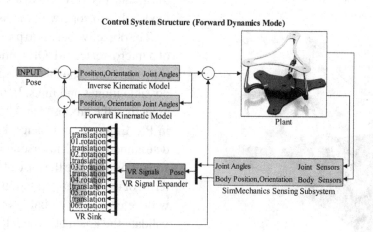

understood that, in serial configurations several active elements like motors/actuators move with the arm and a significant part of motor power is consumed in moving the motors themselves in contrast to the motion of the effective payload. On the other hand, parallel robots are designed with such a kinematic architecture that the motors, control computers and associated electrical equipment are located at the fixed base. This investigation concludes that parallel mechanisms are more suitable for MEMS fabrication in comparison to serial manipulators.

2.1 Architecture of Motion Platform

In literature there are research studies that deal with the design of motion platforms with different kinematic architectures, some of which have linear motors at the base moving along a triangle as presented in Daniel, Murky, and Angeles (1993) providing three DOFs, whereas some studies investigates 6-DOF motion capable designs utilizing rotational actuators as demonstrated in Brodsky, Shoham, and Glozman (1998) and Bieg (1999). Another study describing a design with linear motion and a screw-like motion of the moving platform is shown in Bamberger and Shoham (2004). The parallel mechanism presented in Bamberger and Shoham (2007) uses linear mo-

tors as actuators and 15 links connected through 21 revolute hinges that have their axes within the substrate plane concluding an architect suitable for MEMS fabrication. However, linear actuators with a large stroke should be developed in such parallel mechanisms to achieve a reasonable workspace. Such a demand of linear actuators with a large stroke may offer obstacles in MEMS fabrications. Moreover, the hinge reliability during fabrication and operation may cause low yield due to the presence of large number of hinges in the mechanism.

Figure 9. A block diagram of real-time simulation setup

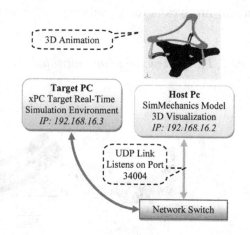

Figure 10. Visualization of a ±25 degrees pitch

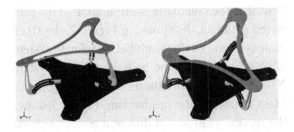

Figure 11. Visualization of a ±25 degrees roll

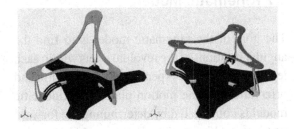

Figure 12. Visualization of a ± 50 degrees yaw

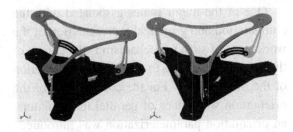

Figure 13. Visualization of a ±5micro-m vertical heave

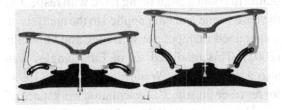

Based on these design considerations investigated here, this paper proposes an architect resembling a parallel robot to achieve a reasonable workspace with minimum possible hinges and links. Therefore, the architect is composed of three

sub-assemblies connected between the motion and fixed plates through passive and active joints that fits MEMS fabrication demands. Figure 1 shows the model presenting the proposed design.

To accomplish this feat, the consideration of appropriate joints in the development of micro-scaled mechanisms is of prime importance due to the fabrication constraints in MEMS fabrication technology. Joints are generally classified as flexible or rigid joints. In the field of MEMS fabrication technology, a study (Liu & DeVoe, 2001) investigates the modeling and fabrication of flexible joints to enable motion in multi channels. Another research work (Kota, 1999) studies the flexible joints in terms of their motion capabilities to enable linear or rotational motions while constraining the others. Through various research studies found in literature, it has been inferred that flexible joints provide smooth motion however their motion capabilities suffer from nonlinear restitution forces due to inherent stress in a limited motion envelope. Furthermore, the kinematic model of a flexible joint is not completely defined as observed in Bamberger and Shoham (2007). On the other hand, rigid joints enable motions in large ranges without suffering from inherent restitution forces. Though, the micro rigid joints are subjected to noise, impacts, friction and wear, still they are preferable for the development of micro scale parallel mechanisms with reasonable workspace. In the light of these investigations about the advantages and disadvantages of rigid and flexible joints, the kinematic architect proposed in this paper constitutes a hybrid structure containing both the rigid and flexible joints.

A motion plate (Body) represented by $B_1B_2B_3$ is connected to the bottom (fixed) plate $G_1G_2G_3$ through three subassemblies acting as legs to form a closed-loop parallel mechanism (Wang, Zhang, Chen, Liu, & Chen, 2007) as shown in Figure 2. Each leg consists of three links C, F & T represented by lengths l_c, l_f & l_t inter connected through two revolute joints θ_f & θ_t. Each

Figure 14. Visualization of a ±7micro-m horizontal sway

Figure 15. Visualization of a ±7mm horizontal surge

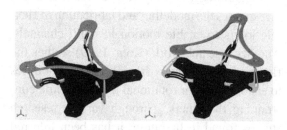

Figure 16. A comparative representation of the reference data and simulation results in terms of attitude angles in degrees

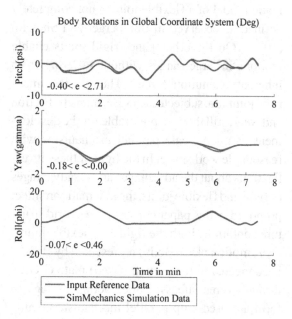

subassembly (leg) is connected between the fixed plate and the motion plate through a revolute joint represented by θ_c and three revolute hinges

resembling a 3-DOF joint respectively. Figure 2 describes the kinematic configuration of the proposed architect. As shown in Figure 2, the fixed plate is aligned with the global coordinate system X^G-Y^G-Z^G whereas, the body coordinate system X^B-Y^B-Z^B moves with the motion of the Centre of Mass (CoM) of the mechanism. Figure 3 shows the motion platform at different poses modeled in Pro Engineer WildFire 5.0.

2.2 Kinematic Model

The goal of the kinematic model is to find the angular rotations of the revolute joints within each leg for a given pose (specified by its position and orientation) of the motion plate. The kinematic model is constructed by determining the forward and inverse kinematic equations using Denavit-Hartenberg convention as investigated by Asif and Iqbal (2010).

One of the major issues associated with the computation and graphical representation of the workspace of such a mechanism is the choice of the set of Euler angles to describe the orientation of the motion plate. For the computation of the orientation workspace of parallel manipulators, an orientation parameterization was introduced in Bonev and Ryu (1999). Orientation workspace analysis of 6-DOF parallel manipulators) by defining a set of all attainable orientations of the platform about a point being fixed with respect to the base frame and later applied to the analysis of constrained manipulator in Bonev and Gosselin (2002). This modified set of Euler angles was also used in Huang, Wang, and Gosselin (1999) for representing the 2-D orientation workspace of Gough-Stewart parallel manipulators.

In this paper, we impose the first requirement that Euler angles φ and ψ must describe exactly the direction of the y-axis with respect to the fixed plate frame of reference that is aligned with the global coordinate system. This requirement is adequately met by choosing the Euler angles (ψ, γ, φ) shown in Figure 4. In this orientation

Figure 17. Tracking errors in terms of attitude angles in degrees

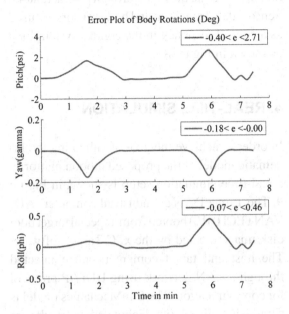

Figure 18. A comparative representation of the reference data and simulation results in terms of linear displacements in mm

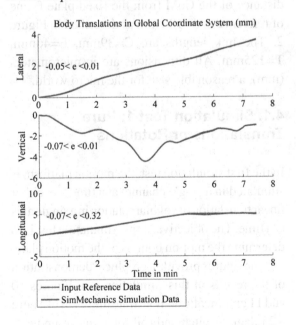

representation, the motion plate is first rotated about the base x-axis by an angle ψ, then about the new y-axis by an angle γ and finally about the z-axis by an angle φ.

Thus, the transformation from the body frame to the global frame is defined by the Equation 1 (see Box 1) using general homogeneous transformation matrices.

3. SIMULATION MODEL

The mathematical derivation for a simulation model is complex because of the complexity of parallel mechanisms (Huang, Chen, & Li, 2003; Wohlhart & Karl, 2003). A problem of visualization occurs when only numerical results are produced during the simulation. Furthermore, the behaviour of the mechanical joints and body of the designed motion platform as they operate in real is difficult to be visualised. Thus a solution using SimMechanics is suggested here.

SimMechanics software is a block diagram modelling environment for the engineering design and simulation of rigid body machines and their motions, using the standard Newtonian dynamics of forces and torques. It also presents the visual representation of the robot during the simulation (Ng, Ong, & Nee, 2006; Liu, Wang, Gao, & Wang, 2002). In this context, Ng, Ong, and Nee (2006) analysed 3-DOF micro Stewart platform using the SimMechanics. Wang, Deng, Zhang, and Meng (2006) analysed Assistant Robotic Leg using the SimMechanics. Furthermore, the work done in Qi, McInroy, and Jafari (2007), Huang, Hiller, and Fang (2007), and Dong, Zhang, and Lu (2005) demonstrates the modelling and simulation of manipulators in macro-world using SimMechanics. In the light of these investigations, a SimMechanics model of the proposed motion platform is created using a CAD translation utility provided by SimMechanics. The block diagram shown in Figure 5 is the Plant subsystem of the simulation model which describes the physical model of the proposed motion platform. The mechanical assembly is defined by rigid bodies with their mass properties, motions, kinematic constraints and coordinate systems. The actuat-

Figure 19. Tracking errors in terms of linear displacements in mm

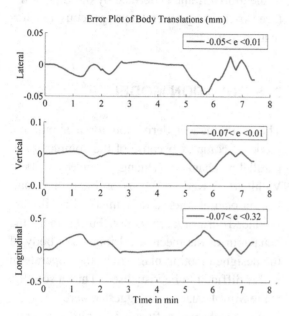

ing elements are the joint actuators which can be actuated by angular rotations determined by the kinematic model for a particular motion. The sensing elements constitute body and joint sensors to record the rotation angles of joints and pose of rigid bodies (positions and rotations).

Figure 6 shows the visualization of the SimMechanics model through the 3D Animation viewer.

Figure 7 represents the SimMechanics model of an individual leg in terms of its rigid bodies, joints and their actuators. The link "C" is connected to the fixed plate through a revolute joint, the link "F" is interconnected between link "C" & link "T" through a revolute joint and link "T" is finally connected to the moving plate through a custom joint shown in Figure 7 and Figure 5.

The overall simulation model is shown in Figure 8 which incorporates the forward and inverse kinematic equations described in the modeling section. The input data describes the pose of the motion plate in terms of its orientation and position. For each particular pose, respective joint

angles are computed by the inverse kinematic model to actuate the Plant through joint actuators. Sensors data from joint and body sensors are used as a control feedback in the control system loop as shown in Figure 8.

4. REAL-TIME SIMULATION

In order to achieve the goal of validating the kinematic model of the proposed motion platform, a real-time simulation setup is shown in Figure 9. The Target PC is an industrial computer (ADVANTECH 610) booted from a special target boot disk image created by the xPC Target software. The host and target computers are connected through a LAN network using UDP/IP protocol for communication. The SimMechanics model is simulated with a defined input and the results are validated by comparing the results of kinematic model and sensors data.

The platform dimension is r = 84 mm and the distance of the CoM from the fixed plate frame of reference is d=106 mm as described in Figure 2. The link lengths are C=39mm, F=46mm, T=125mm. All dimensions are in micrometers (mm), a reasonable size for the micro world.

4.1. Simulation Test 1: Pure Translations or Rotations

In the first simulation test, the motion platform is simulated in a single channel at a time, i.e. either linear translation or angular rotation in a single axis at a time. The objective of this simulation test is to determine the motion bounds of the motion plate.

The figures provide a graphical demonstration of the results of this simulation test. Figures 10 and 11 graphically demonstrate the result of a pure ±25 degrees pitch and roll rotations around z & x axes respectively. Figure 12 shows the result of a pure ±50 degrees yaw rotation around y axis. Similarly, Figures 13, 14, and 15 show the results of ±5mm vertical, and ±7mm horizontal linear translations of the motion platform along y, x & z axes respectively.

4.2. Simulation Test 2: Combined Motions

In the second test, the motion platform is simulated with an input data constituting of motions in more than one channel simultaneously, i.e. combined motions constituting of both linear translations and angular rotations. The main objective of this simulation test is to validate the execution of the workspace determination algorithm described in the earlier sections.

The reference data is represented by the solid red lines in Figures 16 and 18 that demonstrate a comparative representation of the simulation results and the reference data in terms of angular rotations and linear translations respectively. As apparent from the figures, the simulation data either matches or trails the input data that validates the ability of the kinematic model to keep the mechanism within its motion envelope. Figures 17 and 19 show the error plots to present the tracking errors during the simulation.

CONCLUSION AND FUTURE WORK

A new architecture of a 6-DOF motion platform is presented in this paper. The architecture consists of only revolute hinges as connecting elements, which are already well established in MEMS. The architect consists of only 9 links in form of three legs distributed around the body in a symmetric fashion. Each leg is comprised of three revolute joints of which two (θ_f & θ_t) have their axes within a common plane. Furthermore, each leg is actuated with three angular actuators of which two are located at the base. The inverse kinematic solution is given along with the direct kinematics to orient the motion plate at a pose specified by its position and orientation in global frame of reference with reasonable precision. Through real-time simulations it has been concluded that the proposed architect achieves a reasonable workspace both in single and combined motions. In single motions it achieves ±25 degrees of roll & pitch and ±50 degrees of yaw rotations. Similarly, ±5mm vertical and ±7mm horizontal motion bounds have been verified in linear translations in single channels. In combined motions, the kinematic model is capable of keeping the mechanism within its motion envelope to avoid the singular positions that are characteristic of parallel mechanisms.

The realization of this micro-parallel mechanism however still faces some major hindrances: the rotation axis of the revolute joint θ_c of each leg which connects the leg with the fixed plate is perpendicular to the axis of (θ_f & θ_t) that may cause difficulty in the fabrication at micron-level. Furthermore, the actuator which actuates θ_t is not located at the base and is carried by the moving leg resulting in a power compromise at (θ_c & θ_f) during the motion. Finally, the frictions and clearances at the revolute joints can reduce the positioning accuracy of the moving platform in real world. All these considerations are now the subject of future investigations.

REFERENCES

Asif, U., & Iqbal, J. (2010, November 24-26). Modeling and Simulation of Biologically Inspired Hexapod Robot using SimMechanics. In *Proceedings of the IASTED International Conference on Robotics (Robo 2010)*, Phuket, Thailand.

Bamberger, H., & Shoham, M. (2004). A new configuration of a six degrees-of-freedom parallel robot for MEMS fabrication. In *Proceedings of the IEEE International Conference on Robotics and Automation*, New Orleans, LA (pp. 4545-4550).

Bamberger, H., & Shoham, M. (2007). A novel six degrees-of-freedom parallel robot for MEMS fabrication. *IEEE Transactions on Robotics, 23*(2). doi:10.1109/TRO.2006.889493.

Ben-Horin, R., & Shoham, M. (1996). Construction of a six degrees-of-freedom parallel manipulator with three planarly actuated links. In *Proceedings of the ASME Design Conference.*

Ben-Horin, R., Shoham, M., & Djerassi, S. (1998). Kinematics, dynamics and construction of a planarly actuated parallel robot. *Robotics and Computer-integrated Manufacturing, 14*(2), 163–172. doi:10.1016/S0736-5845(97)00035-5.

Bieg, L. F. X. (1999). *U.S. Patent No. 5,901,936: Six degree-of-freedom multi-axes positioning apparatus.* Washington, DC: U. S. Patent and Trademark Office.

Bonev, I. A., & Gosselin, C. M. (2002). Advantages of the modified Euler angles in the design and control of PKMs. In *Proceedings of the International Conference on Kinematic Machines,* Chemnitz, Germany (pp. 171-188).

Bonev, I. A., & Ryu, J. (1999). Orientation workspace analysis of 6-DOF parallel manipulators. In *Proceedings of the ASME Design Engineering Technical Conference.*

Brodsky, V., Shoham, M., & Glozman, D. (1998). Double circular-triangular six degrees-of-freedom parallel robot. In *Proceedings of the 6th International Symposium on Advances in Robotic Kinematics,* Salzburg, Austria (pp. 155-164).

Ceccarelli, M. (1997). A new 3 DOF spatial parallel mechanism. *Mechanism and Machine Theory, 32*(8), 89–902. doi:10.1016/S0094-114X(97)00019-0.

Choi, Y., Sreenivasan, S. V., & Choi, B. J. (2008). Kinematic design of large displacement precision XY positioning stage by using cross strip flexure joints and over-constrained mechanism. *Mechanism and Machine Theory, 43*(6), 724–737. doi:10.1016/j.mechmachtheory.2007.05.009.

Daniel, H. R. M., Somber-Murky, P., & Angeles, J. (1993). The kinematics of 3-DOF planar and spherical double-triangular parallel manipulator. In Angeles, J., & Kovacs, P. (Eds.), *Computational kinematics* (pp. 153–164). Norwell, MA: Kluwer Academic.

Dong, W., Sun, L. N., & Du, Z. J. (2007). Design of a precision compliant parallel positioner driven by dual piezoelectric actuators. *Sensors and Actuators. A, Physical, 135,* 250–256. doi:10.1016/j.sna.2006.07.011.

Dong, Y., Zhang, L., & Lu, D. (2005). Mechanism Modelling and Simulation of Cobot Based on Simulink. *Machine Design and Research, 21*(4), 33–36.

Fan, L., Wu, M. C., Choquette, K. D., & Crawford, M. H. (1997). Self-assembled microactuated XYZ stages for optical scanning and alignment. In *Proceedings of the International Solid State Sensors Actuators Conference,* Chicago, IL (pp. 319-322).

Gao, F., Zhang, J., Chen, Y., & Jin, Z. (2003). Development of a new type of 6-DOF parallel micro-manipulator and its control system. In *Proceedings of the IEEE International Conference on Robotics, Intelligent Systems and Signal Processing* (Vol. 2, pp. 715-720).

Gosselin, C., & Angeles, J. (1991). A global performance index for the kinematic optimization of robotic manipulators. *ASME Journal of Mechanical Design, 113*(3), 220–226. doi:10.1115/1.2912772.

Hollar, S., Bergbreiter, S., & Pister, K. S. J. (2003). Bidirectional inchworm motors and two-DOF robot leg operation. In *Proceedings of the 12th International Solid State Sensors Actuators Conference*, Boston, MA (pp. 262-267).

Honegger, M. (1998). Nonlinear adaptive control of a 6 DOF parallel manipulator. In *Proceedings of the MOVIC Conference*, Zurich, Switzerland (Vol. 3, pp. 961-966).

Huang, J., Hiller, M., & Fang, S.-Q. (2007). Simulation Modeling of the Motion Control of a Two Degree of Freedom, Tendon Base, Parallel Manipulator in Operational Space Using MAT-LAB. *Journal of China University of Mining and Technology*, *17*(2), 179–183. doi:10.1016/S1006-1266(07)60067-4.

Huang, T., Wang, J., & Gosselin, C. (1999). Determination of closed form solution for the 2-D orientation workspace of Gough–Stewart parallel manipulators. *IEEE Transactions on Robotics and Automation*, *15*, 1121–1125. doi:10.1109/70.817675.

Huang, Z., Chen, L. H., & Li, Y. W. (2003). The singularity principle and property of Stewart parallel manipulator. *Journal of Robotic Systems*, 163–176. doi:10.1002/rob.10078.

Hudgens, J., & Tesar, D. (1991). Analysis of a fully-parallel six degree-of freedom micromanipulator. In *Proceedings of the 5th International Conference on Advanced Robotics: Robots in Unstructured Environments* (Vol. 1, pp. 814-820).

Jokiel, B., Benavides, B. L., Bieg, L. F., & Allen, J. J. (2001). Planar and spatial three-degree-of-freedom micro-stages in silicon MEMS. In *Proceedings of the Annual Meeting of the American Society of Precision Engineering*, Crystal City, VA (pp. 32-35).

Kallio, P., Lind, M., Zhou, Q., & Koivo, H. N. (1998). A 3-DOF piezohydraulic parallel micro-manipulator. In *Proceedings of the 1998 IEEE International Conference on Robotics and Automation* (Vol. 2, pp. 1823-1828).

Kim, H. S., & Cho, Y. M. (2009). Design and modeling of a novel 3-DOF precision micro-stage. *Mechatronics*, *19*, 598–608. doi:10.1016/j.mechatronics.2009.01.004.

Kota, S. (1999). Design of compliant mechanisms: Applications to MEMS. In *Proceedings of the SPIE Conference on Smart Electronics*, Newport Beach, CA (LNCS 3673, pp. 45-54).

Lee, K. M., & Arjunan, S. (1991). A three-degrees-of-freedom micromotion in-parallel actuated manipulator. *IEEE Transactions on Robotics and Automation*, *7*(5), 634–641. doi:10.1109/70.97875.

Lee, T. T., Liao, C. M., & Chen, T. K. (1988). On the stability Properties of Hexapod Tripod Gait. *IEEE Transactions on Robotics and Automation*, *4*(4).

Liu, D., Xu, Y., & Fei, R. (2003). Study of an intelligent micromanipulator. *Journal of Materials Processing Technology*, *139*, 77–80. doi:10.1016/S0924-0136(03)00185-7.

Liu, X.-J., Wang, J., Gao, F., & Wang, L.-P. (2002). Mechanism design of a simplified 6-DOF 6-RUS parallel manipulator. *Robotica*, 81–89.

Liu, Z., & DeVoe, D. L. (2001). Micromechanism fabrication using silicon fusion bonding. *Robotics and Computer-integrated Manufacturing*, *17*(1), 131–137. doi:10.1016/S0736-5845(00)00046-6.

Mehregany, M., Gabriel, K. J., & Trimmer, W. S. N. (1988). Integrated fabrication of polysilicon mechanisms. *IEEE Transactions on Electron Devices*, *35*(6), 719–723. doi:10.1109/16.2522.

Merlet, J. P. (2000). *Parallel Robots*. London, UK: Kluwer.

Ng, C. C., Ong, S. K., & Nee, A. Y. C. (2006). Design and development of 3-DOF modular micro parallel kinematic manipulator. *International Journal of Advanced Manufacturing Technology, 31*(1-2), 188–200. doi:10.1007/s00170-005-0166-y.

Niaritsiry, T. F., Fazenda, N., & Clavel, R. (2004). Study of the sources of inaccuracy of a 3DOF flexure hinge-based parallel manipulator. In *Proceedings of the IEEE International Conference on Robotics and Automation* (pp. 4091-4096).

Ouyang, P. R., Tjiptoprodjo, R. C., Zhang, W. J., & Yang, G. S. (2008). Micromotion devices technology: The state of arts review. *International Journal of Advanced Manufacturing Technology, 38*(5-6), 463–478. doi:10.1007/s00170-007-1109-6.

Pister, K. S. J., Judy, M. W., Burgett, S. R., & Fearing, R. S. (1992). Microfabricated hinges. *Sensors and Actuators. A, Physical, 33*(3), 249–256. doi:10.1016/0924-4247(92)80172-Y.

Piyawattanametha, W., Fan, L., Lee, S. S., Su, J. G. D., & Wu, M. C. (1998). MEMS technology for optical crosslink for micro/nano satellites. In *Proceedings of the International Conference on Integrated Nano/Microtechnology for Space and Biomedical Applications*, Houston, TX.

Qi, Z., McInroy, J. E., & Jafari, F. (2007). Trajectory tracking with parallel robots using low chattering, fuzzy sliding mode controller. *Journal of Intelligent & Robotic Systems, 48*(3), 333–356. doi:10.1007/s10846-006-9084-y.

Ryu, K. (1997). A criterion on inclusion of stress stiffening effects in flexible multibody dynamic system simulation. *Computers & Structures, 62*(6), 1035–1048. doi:10.1016/S0045-7949(96)00285-4.

Smith, S. T. (2000). *Flexures: Elements of Elastic Mechanisms*. Boca Raton, FL: CRC Press.

Tabib-Azar, M. (1998). *Microactuators: Electrical, magnetic, optical, mechanical, chemical & smart structures*. Norwell, MA: Kluwer Academic.

Tang, X., Pham, H. H., Li, Q., & Chen, I.-M. (2004). Dynamic analysis of a 3-DOF flexure parallel micromanipulator. In *Proceedings of the 2004 IEEE Conference on Robotics, Automation and Mechatronics* (Vol. 1, pp. 95-100).

Tanikawa, T., Arai, T., & Koyachi, N. (1999). Development of small-sized 3 DOF finger module in micro hand for micro manipulation. In *Proceedings of the IEEE/RSJ International Conference on Intelligent Robots and Systems* (pp. 876-881).

Tsai, L.-W., & Tahmasebi, F. (1993). Synthesis and analysis of a new class of six-degree-of-freedom parallel minimanipulator. *Journal of Robotic Systems, 10*(5), 561–580. doi:10.1002/rob.4620100503.

Wang, G., Zhang, L., Chen, D., Liu, D., & Chen, X. (2007). Modeling and Simulation of multi-legged walking machine prototype. In *Proceedings of the International Conference on Measuring Technology and Mechatronics Automation*.

Wang, L., Deng, Z., Zhang, L., & Meng, Q. (2006). Analysis of Assistant Robotic Leg on MATLAB. In *Proceedings of the 2006 IEEE International Conference on Mechatronics and Automation* (pp.1092-1096).

Wang, S. C., Hikita, H., Kubo, H., Zhao, Y.-S., Huang, Z., & Ifukube, T. (2003). Kinematics and dynamics of a 6 degree-of-freedom fully parallel manipulator with elastic joints. *Mechanism and Machine Theory, 38*, 439–461. doi:10.1016/S0094-114X(02)00132-5.

Wenger, P., & Chablat, D. (2000). Kinematic analysis of a new parallel machine tool: The orthoglide. In *Proceedings of the 7th International Symposium on Advances in Robot Kinematics*, Portoroz, Slovenia (pp. 305-314).

Wohlhart, K. (2003). Mobile 6-SPS parallel manipulators. *Journal of Robotic Systems*, *20*(8), 509–516. doi:10.1002/rob.10101.

Wu, T.-L., Chen, J.-H., & Chang, S.-H. (2008). A six-DOF prismatic-spherical-spherical parallel compliant nano-positioner. *IEEE Transactions on Ultrasonics, Ferroelectrics, and Frequency Control*, *55*(12), 2544–2551. doi:10.1109/ TUFFC.2008.970.

Xu, Q., & Li, Y. (2008). Structure improvement of an XY flexure micromanipulator for micro/ nano scale manipulation. In *Proceedings of the 17th IFAC World Conference* Seoul, Korea (pp. 12733-12738).

Yi, B. J., Chung, G. B., Na, H.-Y., Kim, W. K., & Suh, I. H. (2003). Design and experiment of a 3-DOF parallel micromechanism utilizing flexure hinges. *IEEE Transactions on Robotics and Automation*, *19*(4), 604–612. doi:10.1109/ TRA.2003.814511.

This work was previously published in the International Journal of Intelligent Mechatronics and Robotics, Volume 1, Issue 3, edited by Bijan Shirinzadeh, pp. 1-17, copyright 2011 by IGI Publishing (an imprint of IGI Global).

Chapter 11

Bond Graph Modeling and Computational Control Analysis of a Rigid–Flexible Space Robot in Work Space

Amit Kumar
Indian Institute of Technology Roorkee, India

Pushparaj Mani Pathak
Indian Institute of Technology Roorkee, India

N. Sukavanam
Indian Institute of Technology Roorkee, India

ABSTRACT

The combination of a rigid and a flexible link in a space robot is an interesting field of study from modeling and control point of view. This paper presents the bond graph modeling and overwhelming trajectory control of a rigid-flexible space robot in its work space using the Jacobian based controller. The flexible link is modeled as Euler Bernoulli beam. Bond graph modeling is used to model the dynamics of the system and to devise the control strategy, by representing the dynamics of both rigid and flexible links in a unified manner. The scheme has been verified using simulation for a rigid-flexible space manipulator with two links.

1. INTRODUCTION

The flexible manipulators will be useful for space application due to their light weight, less power requirement, ease of maneuverability and ease of transportability. Because of the light weight, they can be operated at high speed. For flexible manipulators flexibility of manipulator have considerable influence on its dynamic behaviors. The flexibility of the link as well as flexibility of joint affects the overall performance of the system. The control of such flexible manipulator is very much

DOI: 10.4018/978-1-4666-3634-7.ch011

influenced by the non-linear coupling of large rigid body motions and small elastic vibrations. In case of space robots the position and orientation of the satellite main body will change due to manipulator motion. The motion of the space robots also induces vibrating motions in structurally flexible manipulators. The combination of a rigid and a flexible link in a space robot is an interesting field of study from modeling and control point of view. A free-floating space robotic system is one in which the spacecraft's position and attitude are not actively controlled using external jets/thrusters. It does not interact dynamically with the environment during manipulator activity. For such systems, the linear and angular momentum is conserved. Thus, due to conserved linear and angular momenta, the spacecraft moves freely in response to the dynamical disturbances caused by the manipulator's motion. This disturbance of the base results in deviation of the end-effector from the desired trajectory. Moreover, the angular momentum conservation constraints are non-integrable rendering the system to be non-holonomic (Nakamura & Mukherjee, 1991).

Cartesian coordinates offer a better choice when it comes to specifying a task or defining an obstacle in workspace. The obstacles and the trajectory followed by the end-effector can be expressed in terms of simple homogeneous transformations in Cartesian space. In this method the user specifies the desired position and orientation of the end-effector with respect to the robot base frame, as a function of time t in terms of robot parameters. Taking time derivative of such equations gives us the velocity equations.

Kane (1961) presented a general method for obtaining the differential equations governing motion of both holonomic and nonholonomic systems. Yamada and Tsuchiya (1987) derived the equation of motion of multibody system such as space structure, whose base is free to move. They derived the equations of motion of a single rigid body by using position of the center of mass (*CM*) of the body as generalized coordinate. These equations were derived based on the Kane's method. Passarello and Huston (1973) improved upon the Kane's method. The advantage was automatic elimination of nonworking constraint forces and avoiding computation of the vector components of acceleration. The method also provides the arbitrary choice of dependent variables so that it may be applied to a variety of nonholonomic systems. Hemami and Weimer (1981) developed a feedback model of nonholonomically constrained dynamic system. The model was used for analysis, control and understanding of the nonholonomic constrained systems under impulsive and friction forces. In this model, the forces of constraint are explicit function of states and inputs. Fuyang, Hongtao, and Hongli (2009) used fast efficient integration method to solve complex differential equation of the dynamics of space flexible robots. Zhang and Yu (2004) developed the dynamic equations of planar cooperative manipulators with link flexibility in absolute coordinate with the help of Timoshenko beam theory and the finite element method.

Umetani and Yoshida (1989) developed a control method for space manipulators based on the concept of resolved motion rate control to continuously control the end-effector of a space manipulator mounted on a space vehicle. Yoshida (2003) demonstrated the concept of generalized Jacobian in ETS–VII. Yokokohji, Toyoshima, and Yoshikawa (1993) presented efficient computational algorithms for the trajectory control of multi arm free-flying space robots. The motion of the aircraft during manipulation is considered in order to obtain an accurate trajectory control using generalized Jacobian. Watanabe and Nakamura (1998) proposed a free-flying space robot having a special kinematic structure and mass distribution such that the *CM* of the total system remains fixed to the base body regardless of the manipulators configuration. They claimed that for a free-flying space robot with the *CM* invariance, an experimental system on the ground is simple to build. Further, the computational cost of Generalized Jacobian matrix is reduced, and the motion in a two dimensional plane is holonomic. Nenchev (1993) used resolved acceleration type controller for free-flying space robots with a kinematically redundant manipulator arm. The arm is controlled to track a desired end effector trajectory and at the same time to change the attitude of the system in a desired manner and without actuating jet thrusters or reaction wheels. The formulation is based on the fixed-attitude restricted (FAR) Jacobian matrix. He showed that when the FAR Jacobian is applied, the joint acceleration will be mainly derived from the null space of the manipulator link inertia matrix. Wang and Xie (2009) solved the problem of position/force tracking control of a free-flying space manipulator with uncertain kinematics and dynamics. They devised an adaptive Jacobian to cope with the uncertainties arising from free-flyer's kinematics, dynamics and surface stiffness and position. Stieber, Vukovich, and Petriu (1997) addressed the stability and control problems arising in vision based control of flexible space robot where the motion of a robot payload relative to the work space is measured at a considerable distance from the control actuator in the robot joints. Murotsu, Tsujio, Senda, and Hayashi (1992) proposed control schemes for flexible space manipulator using a virtual rigid manipulator concept i.e., "pseudo resolved acceleration control (RAC) for flexible manipulator", and "composite control of pseudo RAC and reduced order modal control". The validity of the proposed control schemes has been explained through the singular perturbation method. Licheng, Fuchun, Zengqi, and Wenjing (2002) worked on dynamic modeling, control and simulation of flexible dual arm space robot based on the Lagrange method and described the elastic

deflection by the assumed mode method. The inversion dynamic control method is performed to solve the tracking problem. Vaz and Samanta (1991) proposed a new approach called Travelling Independent Modal Space Control for the active control of flexibility in light weight manipulator systems using the concept of Intelligent Structures. This approach enhances actuator utilization and limits control spillover effects. Senda and Murotsu (2000) proposed a methodology for designing stable manipulation variable feedback control of a space robot with flexible links for positioning control to a static target and continuous path tracking control. Dwivedy and Eberhard (2006) carried out a survey of the literature related to dynamic analyses of flexible robotic manipulators. Both link and joint flexibility are considered. Jiang (1992) developed a concept of compensability for free floating flexible space robot arms. Using this concept end-effector behavior caused by the link flexural behavior and the satellite motion in response to the arm motion is considered as errors in the end-effector motion and this error is compensated by the joint behavior. Ueno, Xu, and Yoshida (1991) discussed modeling and control strategy of a 3-D flexible space robot, by considering mass of joint and neglecting mass of link. Lee (2004) developed a new trajectory control of a flexible link robot based on a distributed parameter dynamic model. Wang, Meng, and Liu (1999) studied influence of shear, rotary inertia on the dynamic characteristics of flexible manipulators. Luo and Guo (1997) have discussed the shear force feedback control of a single-link flexible robot with a revolute joint. Jiang (1992) presented kinematics and dynamics of flexible space robot arms. Samanta and Devasia (1988) have discussed modeling and control of flexible terrestrial manipulates using distributed actuator. The nonlinear coupling of large rigid body motion and small elastic vibration of the flexible arms has been taken into consideration in the model. The concept of using distributed piezoelectric

Figure 1. Schematic representation of two link rigid-flexible space robot

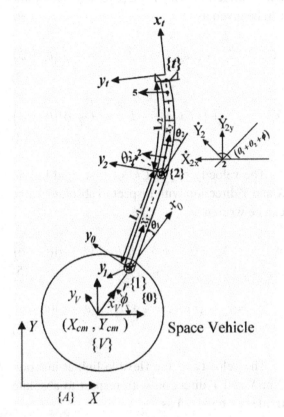

transducers for controlling elastic vibration of arms has been incorporated in the analysis. Sueur and Dauphin-Tanguy (1992) have elaborated bond-graph modeling of flexible robots. They proposed a solution for improvement of the slow subsystem. Majda, Borutzky, and Damic (2009) have provided good comparison of different formulations of 2D beam elements based on Bond Graph technique. Wu, Sun, Sun, and Wu (2004) proposed an optimal trajectory planning method with vibration reduction for a dual arm space robot with front flexible links using particle swarm optimization. Bond graphs are used to model the dynamics of the space robot as it offers flexibility in modeling and formulation of system equations. SYMBOLS Shakti (Mukherjee, 2006) software is used for bond graph (Mukherjee, Karmarkar, & Samantaray 2006) modeling and simulation.

In this paper we present the bond graph modeling and overwhelming trajectory control of a rigid-flexible space robot in its work space using the Jacobian based controller. Bond graphs are used to represent both rigid body and flexible dynamics of the link in a unified manner. To illustrate the methodology, an example of a two degree of freedom (DOF) rigid-flexible space robot is considered. The advantage of the controller is that it is provided with only tip velocity information of manipulator. To simplify the analysis we have considered Jacobian of rigid manipulator only.

2. MODELING OF TWO LINK RIGID-FLEXIBLE SPACE ROBOT

Modeling of the space robot dynamics involves the modeling of linear and rotational motion of the links and the base of the space robot. It is assumed that the space robot system has a single manipulator with two revolute joints and is in open kinematic chain configuration. Figure 1 shows the schematic sketch of a two link rigid-flexible space robot. The notations used in Figure 1 are described in Table 1. The flexible link is uniform with a known flexible rigidity EI_2, here E is the Modulus of Elasticity and I_2 is the inertia of the second link. The flexible link of the space robot has been modeled as Euler-Bernoulli beam.

3. BOND GRAPH MODELING

The kinematic analysis of the flexible link is performed as follows in order to draw the bond graph as shown in Figure 2. The flexible link is divided into four segments of equal length. The segment angles with respect to absolute frame $\{A\}$ are $\psi_1, \psi_2, \psi_3,$ and ψ_4 respectively. To draw the bond graph model two types of motion of the links are considered, namely the motion perpendicular to the link and the rotational motion of the link. The

Table 1. Notations and their description

Notations	Description
{A}	Absolute frame
{V}	Vehicle frame
{0}	Frame is located in space vehicle at the base of the robot
{1}	Frame located in first link at the base of first link
{2}	Frame located at second joint at the base of second link
{t}	Tip frame
L_1	Length of the rigid link
L_2	Length of flexible link
r	Distance between the robot base and center of mass (*CM*) of the space vehicle.
m_1 and m_2	Mass of first and second link
ϕ	Rotation of space vehicle
θ_1 and θ_2	Joint angles
τ_1 and τ_2	Torque on first and second link
A_2	Cross section area of second link
ρ_2	Density of second links
I_1 and I_2	Inertia of first and second link respectively
E	Modulus of Elasticity of link material
M_V	Mass of space vehicle
I_V	Inertia of space vehicle
X_{cm} and Y_{cm}	Location of *CM* of the space vehicle from absolute frame {A}

motion perpendicular to the link is resolved into *X* and *Y* components and link segment inertia are attached there. Pads are used for computational simplicity i.e., to avoid the differential causality.

The displacement relation for the tip of first rigid link in *X* and *Y* direction with respect to absolute frame can be expressed as

$$X_{t\_1} = X_{cm} + (r\cos\phi) + L_1\cos(\theta_1 + \phi) \quad (1)$$

$$Y_{t\_1} = Y_{cm} + (r\sin\phi) + L_1\sin(\theta_1 + \phi) \quad (2)$$

The velocity of the tip of first rigid link in *X* and *Y* direction with respect to absolute frame can be given as

$$\dot{X}_{t\_1} = \dot{X}_{cm} - (r\sin\phi)\dot\phi - L_1\sin(\theta_1 + \phi)(\dot\theta_1 + \dot\phi) \quad (3)$$

$$\dot{Y}_{t\_1} = \dot{Y}_{cm} + (r\cos\phi)\dot\phi + L_1\cos(\theta_1 + \phi)(\dot\theta_1 + \dot\phi) \quad (4)$$

The velocity of the *CM* of first rigid link in *X* and *Y* direction with respect to absolute frame can be written as

$$\dot{X}_{t\_1\_cm} = \dot{X}_{cm} - (r\sin\phi)\dot\phi - 0.5L_1\sin(\theta_1 + \phi)(\dot\theta_1 + \dot\phi) \quad (5)$$

$$\dot{Y}_{t\_1\_cm} = \dot{Y}_{cm} + (r\cos\phi)\dot\phi + 0.5L_1\cos(\theta_1 + \phi)(\dot\theta_1 + \dot\phi) \quad (6)$$

The velocity of the flexible link at junction 2 in *X* and *Y* direction with respect to absolute frame can be given as

$$\dot{X}_{2x} = \dot{X}_{t\_1} + \dot{Y}_2\cos\left(\frac{\pi}{2} + \psi_1\right) \quad (7)$$

$$\dot{Y}_{2y} = \dot{Y}_{t\_1} + \dot{Y}_2\sin\left(\frac{\pi}{2} + \psi_1\right) \quad (8)$$

where $\psi_1 = \phi + \theta_1 + \theta_2$

Similarly the velocities of the other segment junction can be evaluated. The tip (end-effector) velocities in *X* and *Y* direction can be written as

$$\dot{X}_{t\_2} = \dot{X}_{5x} - \left(\frac{L_2}{8}\sin\psi_4\right)\dot\psi_4 \quad (9)$$

Figure 2. Bond graph model of a two link rigid-flexible space robot with Jacobian based controller

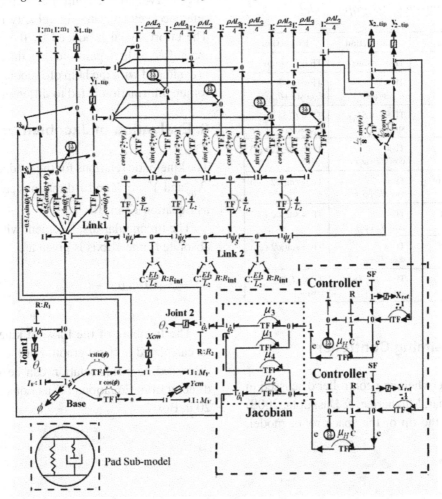

$$\dot{Y}_{t\_2} = \dot{Y}_{5y} + \frac{L_2}{8}\cos\left(\psi_4\right)\dot{\psi}_4 \qquad (10)$$

where $\dot{X}_{5x}$ and $\dot{Y}_{5y}$ are velocities at junction 5 of flexible link in X and Y direction. The complete bond graph is shown in Figure 2. From the kinematic relations the different transformer moduli used in the bond graph are derived and are shown in Table 2 and it is with the assumption that at the tip there is no force or moment acting. This bond graph is appended with the evaluation of the Ja-

cobian and overwhelming controller and is shown in Figure 2. Next section presents evaluation of the Jacobian and overwhelming controller concept.

4. CONTROLLER DESIGN

Controller consists of two parts

1. Overwhelming controller,
2. Evaluation of Jacobian assuming that the link is rigid.

Table 2. Transformer moduli for finding velocities at various points in bond graph model

Description	X direction	Y direction
Robot base	TF = -r sin(ϕ)	TF = r cos(ϕ)
First link tip	TF = -L_1sin($\phi+\theta_1$)	TF = L_1cos($\phi+\theta_1$)
First link CM	TF = -(0.5L_1) sin($\phi+\theta_1$)	TF = (0.5) L_1cos($\phi+\theta_1$)
First segment of second link	TF = cos($\pi/2+\psi_1$)	TF = sin($\pi/2+\psi_1$)
Second segment of second link	TF = cos($\pi/2+\psi_2$)	TF = sin($\pi/2+\psi_2$)
Third segment of second link	TF = cos($\pi/2+\psi_3$)	TF = sin($\pi/2+\psi_3$)
fourth segment of second link	TF = cos($\pi/2+\psi_4$)	TF = sin($\pi/2+\psi_4$)
Second link tip	TF = -(L_2/8) sin(ψ_4)	TF = (L_2/8) cos(ψ_4)

1. Overwhelming Controller

The linear overwhelming controller discussed in (Pathak, Kumar, Mukherjee, & Dasgupta, 2008) is applied to the tip of the space robot model.

The overwhelming controller is provided with reference velocity and tip velocity information. The effort signal produced by the controller is magnified by high gain and then the joint torques is evaluated with the help of Jacobian. This joint torque information is fed to different joints.

2. Evaluation of Jacobian

The kinematic relations for the tip displacements X_{tip} and Y_{tip} in X and Y directions can be written as Equation 17 in Box 1.

The tip angular displacement with respect to absolute frame X axis is given as,

$$\theta_{tip} = \varphi + \theta_1 + \theta_2 \tag{18}$$

The Jacobian of the forward kinematics can be calculated in bond graph. For the planar case discussed here this relation can be worked out directly from Equation (17) as Equations 19 and 20 in Box 2.

Box 1.

$$\begin{bmatrix} X_{tip} \\ Y_{tip} \end{bmatrix} = \begin{bmatrix} X_{CM} \\ Y_{CM} \end{bmatrix} + \begin{bmatrix} r\cos\phi + L_1\cos(\phi+\theta_1) + L_2\cos(\phi+\theta_1+\theta_2) \\ r\sin\phi + L_1\sin(\phi+\theta_1) + L_2\sin(\phi+\theta_1+\theta_2) \end{bmatrix} \tag{17}$$

Box 2.

$$\begin{bmatrix} \dot{X}_{tip} \\ \dot{Y}_{tip} \end{bmatrix} = \begin{bmatrix} \dot{X}_{CM} \\ \dot{Y}_{CM} \end{bmatrix} + \begin{bmatrix} -r\dot{\phi}\sin\phi - L_1(\dot{\phi}+\dot{\theta}_1)\sin(\phi+\theta_1) - L_2(\dot{\phi}+\dot{\theta}_1+\dot{\theta}_2)\sin(\phi+\theta_1+\theta_2) \\ r\dot{\phi}\cos\phi + L_1(\dot{\phi}+\dot{\theta}_1)\cos(\phi+\theta_1) + L_2(\dot{\phi}+\dot{\theta}_1+\dot{\theta}_2)\cos(\phi+\theta_1+\theta_2) \end{bmatrix} \tag{19}$$

$$\begin{bmatrix} \dot{X}_{tip} \\ \dot{Y}_{tip} \end{bmatrix} = \begin{bmatrix} \dot{X}_{CM} - r\dot{\phi}\sin\phi \\ \dot{Y}_{CM} + r\dot{\phi}\cos\phi \end{bmatrix} + \begin{bmatrix} -L_1\sin(\phi+\theta_1) - L_2\sin(\phi+\theta_1+\theta_2) & -L_2\sin(\phi+\theta_1+\theta_2) \\ L_1\cos(\phi+\theta_1) + L_2\cos(\phi+\theta_1+\theta_2) & L_2\cos(\phi+\theta_1+\theta_2) \end{bmatrix} \begin{bmatrix} \dot{\theta}_1 \\ \dot{\theta}_2 \end{bmatrix} +$$

$$\begin{bmatrix} -L_1\sin(\phi+\theta_1) - L_2\sin(\phi+\theta_1+\theta_2) \\ L_1\cos(\phi+\theta_1) + L_2\cos(\phi+\theta_1+\theta_2) \end{bmatrix} [\dot{\phi}] \tag{20}$$

Table 3. Parameters and values used for simulation

Parameter	Value
Modulus of Elasticity	$E = 70 \times 10^9$ N/m^2
Link length	$L_1 = 0.4$ *m*; $L_2 = 0.5$ *m*
Link Inertia	$I_1 = 0.2153$ Kg.m^2; $I_2 = 6.609 \times 10^{-13}$ m^4
Location of robot base from vehicle *CM*	$r = 0.1$ *m*
Mass of the first link	$M_1 = 1.0$ kg
Cross section area of second link	$A_2 = 6.288 \times 10^{-5}$ m^2
Joint resistances	$R_1 = 0.001$ Nm/(rad/s), $R_2 = 0.001$ Nm/(rad/s)
Mass of base	$M = 5.0$ kg
Moment of Inertia of space vehicle	$I_V = 1.0$ kg-m^2
Gain value	$\mu_H = 2.0$
Density of Aluminum	$\rho = 2700$ kg/m^3
Controller parameters	$M_C = 0.005$, $K_C = 0.01$, $R_C = 0.005$
Internal damping coefficient for material	$R_{int} = 0.0001$ N/(m/s)

For the evaluation of joint torques if we assume that space vehicle rotational velocity $\dot{\phi}$, and *CM* velocity $\dot{X}_{CM}, \dot{Y}_{CM}$ are small, then neglecting the first and last term of Equation (20) we get Equations 21 and 22 in Box 3, where [*J*] is the Jacobian. The Jacobian transpose maps Cartesian forces acting at the robot tip into equivalent joint torques. So the controller can be placed in the work space and Jacobian can be used to get the joint torques. However, this controller will be suitable for slow speed actuation only as it does not involves any inertia term in it. With the help of Equation (22) various transformer modulli shown in bond graph can be evaluated as

$$\mu_1 = -L_1 \sin(\phi + \theta_1) - L_2 \sin(\phi + \theta_1 + \theta_2),$$

$$\mu_2 = L_1 \cos(\phi + \theta_1) + L_2 \cos(\phi + \theta_1 + \theta_2),$$

$$\mu_3 = -L_2 \sin(\phi + \theta_1 + \theta_2),$$

$$\mu_4 = L_2 \cos(\phi + \theta_1 + \theta_2)$$

5. SIMULATION AND RESULTS

The initial configuration of the space robot is shown in Figure 3. It is assumed that initially base rotation; first and second joint angles are 0 radian, 0.1 radian and 0 radian respectively.

Let the reference trajectory in *X* and *Y* direction be assumed to be a polynomial of third degree (i.e., a cubic polynomial) say,

$$x(t) = b_{0x} + b_{1x} t + b_{2x} t^2 + b_{3x} t^3, \tag{23}$$

$$y(t) = b_{0y} + b_{1y} t + b_{2y} t^2 + b_{3y} t^3, \tag{24}$$

where b_{0x}, b_{1x}, b_{2x}, b_{3x}, b_{0y}, b_{1y}, b_{2y} and b_{3y} are constants.

Box 3.

$$
\begin{bmatrix} \dot{X}_{tip} \\ \dot{Y}_{tip} \end{bmatrix} = \begin{bmatrix} -L_1 \sin(\phi + \theta_1) - L_2 \sin(\phi + \theta_1 + \theta_2) & -L_2 \sin(\phi + \theta_1 + \theta_2) \\ L_1 \cos(\phi + \theta_1) + L_2 \cos(\phi + \theta_1 + \theta_2) & L_2 \cos(\phi + \theta_1 + \theta_2) \end{bmatrix} \begin{bmatrix} \dot{\theta}_1 \\ \dot{\theta}_2 \end{bmatrix} \tag{21}
$$

$$
\begin{bmatrix} \dot{X}_{tip} \\ \dot{Y}_{tip} \end{bmatrix} = [J] \begin{bmatrix} \dot{\theta}_1 \\ \dot{\theta}_2 \end{bmatrix} \tag{22}
$$

Figure 3. Initial configuration of rigid–flexible space robot

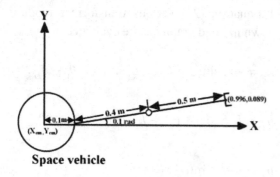

Space vehicle

Then reference velocities in *X* and *Y* direction will be given as

$$\dot{x}(t) = b_{1x} + 2b_{2x}t + 3b_{3x}t^2 \tag{25}$$

$$\dot{y}(t) = b_{1y} + 2b_{2y}t + 3b_{3y}t^2 \tag{26}$$

Let us assume that,

- The initial and final tip position be given as, $x(0) = x_0$, $x(t_f) = x_f$, $y(0) = y_0$, and $y(t_f) = y_f$.

- The initial and final velocities are zero i.e., $\dot{x}(0) = 0$, $\dot{x}(t_f) = 0$, $\dot{y}(0) = 0$, and $\dot{y}(t_f) = 0$.

Figure 4. (a) Reference trajectory in x direction, (b) reference trajectory in y direction, (c) tip trajectory in x direction, (d) tip trajectory in y direction

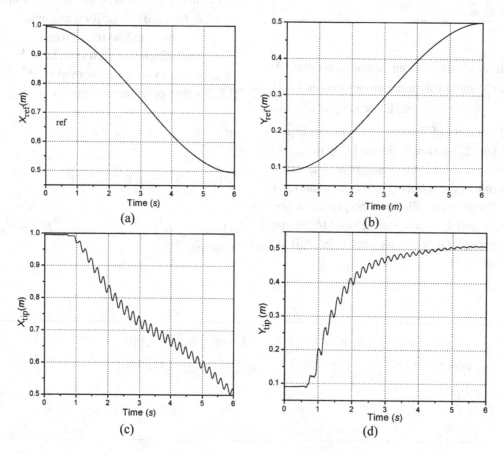

Figure 5. (a) Plot of base rotation versus time, (b) plot of first joint angle versus time, (c) plot of second joint angle versus time, (d) centre of mass trajectory in x direction, (e) centre of mass trajectory in y direction, (f) centre of mass trajectory of space vehicle

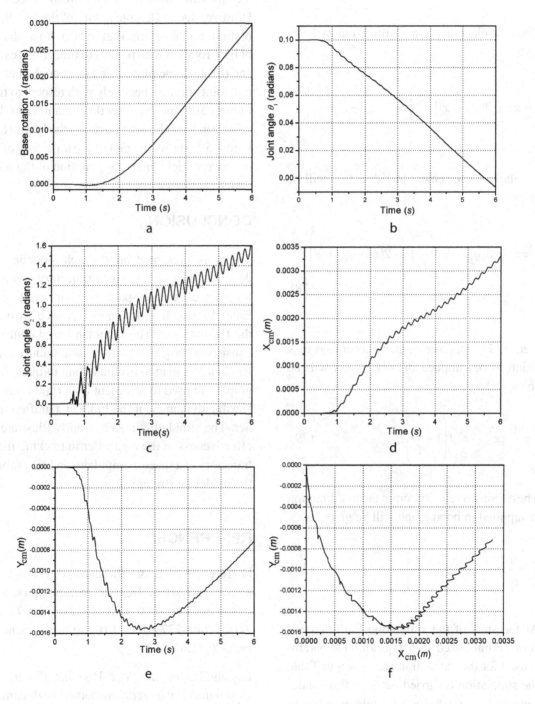

Applying these four constraints to Equations (23), (24), (25) and (26) and solving for the constants $b_{0x}, b_{1x}, b_{2x}, b_{3x}, b_{0y}, b_{1y}, b_{2y}$ and b_{3y} one gets,

$$b_{0x} = x_0; \; b_{1x} = 0; \; b_{2x} = 3(x_f - x_0) \, / \, t_f^2;$$

$$b_{3x} = -2(x_f - x_0) \, / \, t_f^3$$

$$b_{0y} = y_0;\ b_{1y} = 0;\ b_{2y} = 3(y_f - y_0)\,/\,t_f^2;$$
$$b_{3y} = -2(y_f - y_0)\,/\,t_f^3$$

Thus displacement in X direction will be given by,

$$x(t) = x_0 + 3(x_f - x_0)\left(\frac{t}{t_f}\right)^2 - 2(x_f - x_0)\left(\frac{t}{t_f}\right)^3 \tag{27}$$

Similarly displacement in Y direction will be given by,

$$y(t) = y_0 + 3(y_f - y_0)\left(\frac{t}{t_f}\right)^2 - 2(y_f - y_0)\left(\frac{t}{t_f}\right)^3 \tag{28}$$

Hence, the reference velocity command in X direction to be supplied in bond graph will be given as,

$$\dot{x}(t) = \frac{6}{t_f^2}(x_f - x_0)t\left(1 - \frac{t}{t_f}\right). \tag{29}$$

The reference velocity command in Y direction to be supplied in bond graph will be given as,

$$\dot{y}(t) = \frac{6}{t_f^2}(y_f - y_0)t\left(1 - \frac{t}{t_f}\right) \tag{30}$$

At the start of the simulation the reference trajectory is initialized to tip trajectory. The parameters used for the simulation are shown in Table 3. The simulation is carried out for 6.0 seconds.

Figures 4(a) and (b) show the reference trajectory in X and Y directions respectively. Figures 4(c) and (d) show the variation of tip trajectory in X and Y directions respectively. From Figures 4(a), and (c) and Figures (b) and (d) it is seen that the tip follows the reference command effectively. However due to flexible nature of the link tip vibrations are observed. Figure 5(a) shows the plot of base rotation with respect to time. Figures 5(b) and (c) show the plot of first joint angle and second joint angle respectively with respect to time. Figures 5(d) and (e) show the variation of CM of vehicle in X and Y direction respectively with time. Figure 5(f) shows the plot of the trajectory of CM of space vehicle, indicating motion of the base.

CONCLUSION

The work presented bond graph modeling and design of the overwhelming trajectory controller for a rigid-flexible space robot in work space. Euler Bernoulli beam model is used to model the flexible link. The Jacobian based controller consists of two parts (1) an overwhelming controller (2) a Jacobian to evaluate joint torques. Bond graphs are used to represent both rigid body and flexible dynamics of the links in a unified manner. The simulation results clearly illustrate the effectiveness of the controller in tracking the tip trajectory of a space manipulator and also validate the modeling strategy.

REFERENCES

Dwivedy, S. K., & Eberhard, P. (2006). Dynamic analysis of flexible manipulators, a literature review. *Mechanism and Machine Theory*, *41*(7), 749–777. doi:10.1016/j.mechmachtheory.2006.01.014.

Fuyang, T., Hongtao, W., & Hongli, S. (2009). Efficient numerical integration method for dynamic of flexible space robots system. In *Proceedings of the International Asia Conference on Informatics in Control, Automation and Robotics* (pp. 102-106).

Hemami, H., & Weimer, F. C. (1981). Modeling of nonholonomic dynamic systems with applications. *ASME Journal of Applied Mechanics, 48*, 177–182. doi:10.1115/1.3157563.

Jiang, Z. H. (1992). Compensability of end-effector motion errors and evaluation of manipulability for flexible space robot arms. In *Proceedings of the IEEE International Conference on System, Man, and Cybernetics* (pp. 486-491).

Jiang, Z. H. (1992). Kinematics and dynamics of flexible space robot arms. In *Proceedings of the IEEE/RSJ International Conference on Intelligent Robots and Systems Raleigh* (pp. 1681-1688).

Kane, T. R. (1961). Dynamics of nonholonomic systems. *ASME Journal of Applied Mechanics*, 574-578.

Lee, H. H. (2004). A new trajectory control of a flexible-link robot based on a distributed-parameter dynamic model. *International Journal of Control*, 546–553. doi:10.1080/00207170410 001695656.

Licheng, W., Fuchun, S., Zengqi, S., & Wenjing, S. (2002). Dynamic modeling control and simulation of flexible dual arm space robot. In *Proceedings of the IEEE TENCON* (pp. 1282-1285).

Luo, Z. H., & Guo, B. Z. (1997). Shear force feedback control of a single-link flexible robot with a revolute joint. *IEEE Transactions on Automatic Control, 42*(1).

Majda, C., Borutzky, W., & Damic, V. (2009). Comparison of different formulations of 2D beam elements based on bond graph technique. *Simulation Modelling Practice and Theory, 17*(1), 107–124. doi:10.1016/j.simpat.2008.02.014.

Mukherjee, A. (2006). *Users manual of SYMBOLS Shakti*. Retrieved from http://www.htcinfo.com/

Mukherjee, A., Karmarkar, R., & Samantaray, A. K. (2006). *Bondgraph in modeling simulation and fault identification*. New Delhi, India: I. K. International Publishing House Pvt.

Murotsu, Y., Tsujio, S., Senda K., & Hayashi, M. (1992). Trajectory control of flexible manipulators on a free flying space robot. *IEEE Control Systems*, 51-57.

Nakamura, Y., & Mukherjee, R. (1991). Nonholonomic path planning of space robots via a bidirectional approach. *IEEE Transactions on Robotics and Automation, 7*(4), 500–514. doi:10.1109/70.86080.

Nenchev, D. N. (1993). A controller for a redundant fee-flying space robot with spacecraft attitude/manipulator motion coordination. In *Proceedings of the IEEE/RSJ International Conference on Intelligent Robots and Systems*, Yokohama, Japan (pp. 2108-2114).

Passarello, C. E., & Huston, R. L. (1973). Another look at nonholonomic systems. *ASME Journal of Applied Mechanics*, 101-104.

Pathak, P. M., Kumar, P., Mukherjee, A., & Dasgupta, A. (2008). A scheme for robust trajectory control of space robots. *Simulation Modelling Practice and Theory, 16*, 1337–1349. doi:10.1016/j.simpat.2008.06.011.

Samanta, B., & Devasia, S. (1988). Modeling and control of flexible manipulators using distributed actuators: A bond graph approach. In *Proceeding of the IEEE International Workshop on Intelligent Robots Systems* (pp. 99-104).

Senda, K., & Murotsu, Y. (2000). Methodology for control of a space robot with flexible links. *IEEE Proceedings on Control Theory Applications, 147*(6), 562–568. doi:10.1049/ip-cta:20000870.

Stieber, M. E., Vukovich, G., & Petriu, E. (1997). Stability aspects of vision based control for space robots. In *Proceedings of the IEEE International Conference on Robotics and Automation* (pp. 2771-2776).

Sueur, C., & Dauphin-Tanguy, G. (1992). Bond-graph modeling of flexible robots: The residual flexibility. *Journal of the Franklin Institute, 329*(6), 1109–1128. doi:10.1016/0016-0032(92)90006-3.

Ueno, H., Xu, Y., & Yoshida, T. (1991). Modeling and control strategy of a 3-D flexible space robot. In *Proceedings of the IEEE/RSJ International Workshop on Intelligent Robots and Systems* (pp. 978-983).

Umetani, Y., & Yoshida, K. (1989). Resolved motion rate control of space manipulators with generalized jacobian matrix. *IEEE Transactions on Robotics and Automation, 5*(3), 303–314. doi:10.1109/70.34766.

Vaz, A., & Samanta, B. (1991). Digital controller implementation for flexible manipulators using the concept of intelligent structures. In *Proceedings of the IEEE/RSJ International Workshop on Intelligent Robots and Systems*, Osaka, Japan (pp. 952-958).

Wang, D., Meng, M., & Liu, Y. (1999). Influence of shear, rotary inertia on the dynamic characteristics of flexible manipulators. In *Proceedings of the IEEE Pacific Rim Conference on Communications, Computers and Signal Processing* (pp. 615-618).

Wang, H., & Xie, Y. (2009). Adaptive jacobian position/force tracking control of free-flying manipulators. *Robotics and Autonomous Systems, 57*, 173–181. doi:10.1016/j.robot.2008.05.003.

Watanabe, Y., & Nakamura, Y. (1998). A space robot of the center-of-mass invariant structure. In *Proceedings of the IEEE/RSJ International Conference on Intelligent Robots and Systems*, Victoria, BC, Canada (pp. 1370-1375).

Wu, H., Sun, F., Sun, Z., & Wu, L. (2004). Optimal trajectory planning of a flexible dual-arm space robot with vibration reduction. *Journal of Intelligent & Robotic Systems*, 147–163. doi:10.1023/B:JINT.0000038946.21921.c7.

Yamada, K., & Tsuchiya, K. (1987). Formulation of rigid multibody system in space. *JSME International Journal, 30*(268), 1667–1674.

Yokokohji, Y., Toyoshima, T., & Yoshikawa, T. (1993). Efficient computational algorithms for trajectory control of free-flying space robots with multiple arms. *IEEE Transactions on Robotics and Automation, 9*(5), 571–579. doi:10.1109/70.258050.

Yoshida, K. (2003). Engineering test satellite VII flight experiments for space robot dynamics and control: Theories on laboratory test beds ten years ago, now in orbit. *The International Journal of Robotics Research, 22*(5), 321–335. doi:10.1177/0278364903022005003.

Zhang, C., & Yu, Y. Q. (2004). Dynamic analysis of planar cooperative manipulators with link flexibility. *Transactions of ASME, 126*, 442–448. doi:10.1115/1.1701875.

This work was previously published in the International Journal of Intelligent Mechatronics and Robotics, Volume 1, Issue 3, edited by Bijan Shirinzadeh, pp. 18-30, copyright 2011 by IGI Publishing (an imprint of IGI Global).

Chapter 12
Experimental Study of Laser Interferometry Based Motion Tracking of a Flexure–Based Mechanism

Umesh Bhagat
Monash University, Australia

Bijan Shirinzadeh
Monash University, Australia

Yanling Tian
Tianjian University, China

ABSTRACT

This paper presents an experimental study of laser interferometry-based closed-loop motion tracking for flexure-based four-bar micro/nano manipulator. To enhance the accuracy of micro/nano manipulation, laser interferometry-based motion tracking control is established with experimental facility. The authors present and discuss open-loop control, model-based closed-loop control, and robust motion tracking closed-loop control for flexure-based mechanism. A comparative error analysis for closed-loop control with capacitive position sensor and laser interferometry feedback is discussed and presented. Model-based closed-loop control shows improvement in position and motion tracking over open-loop control. Robust control demonstrates high precise and accurate motion tracking of flexure-based mechanism compared to the model-based control. With this experimental study, this paper offers evidence that the laser interferometry-based closed-loop control can minimize positioning and tracking errors during dynamic motion, hence realizing high precision motion tracking and accurate position control.

DOI: 10.4018/978-1-4666-3634-7.ch012

INTRODUCTION

Ultra-precision manipulation is one of the very important techniques for the micro/nano engineering and for many applications in manufacturing and medical sciences. Precise control and accurate motion tracking at micro/nano level is a requirement for present and future automation systems. Such precise actuation and control can be achieved with piezoelectric actuator driven flexure-based mechanisms. These mechanisms are highly appropriate platforms for micro/nano manipulation (Chung, Choi, & Kyung, 2006; Li & Xu, 2010; Spanner & Vorndran, 2003; Speich & Goldfarb, 1998; Tian, Shirinzadeh, & Zhang, 2009). Flexure-based mechanisms have no backlash, zero friction and negligible hysteresis, and can offer unlimited motion resolution (Paros & Weisbord, 1965; Mohd Zubir & Shirinzadeh, 2009; Smith, 1997; Tian, Shirinzadeh, & Zhang, 2010). However, piezoelectric actuators possess non-linearities including the hysteresis and creep/drift effects. The presence of such non linearities cannot guarantee positioning accuracy and precise motion tracking of the flexure-based mechanism. However, in the past decade, increased demand for high-precision micro/nano motion tasks has generated high level of research interests in precise positioning and accurate motion tracking of micro/nano manipulators. Many appropriate closed-loop control strategies have been proposed to achieve the desired motion tracking of piezoelectric actuator-driven flexure-based mechanisms (Bashash & Jalili, 2008; Liaw & Shirinzadeh, 2008, 2009; Motamedi et al., 2010; Rakotondrabe, Haddab, & Lutz, 2009; Saeidpourazar & Jalili, 2006; Shiou et al., 2010; Xu & Li, 2009, 2010). An important component associated with closed-loop control of micro/nano multi-axis manipulator is the sensing and measurement of motion characteristics. Use of interferometry-based sensing to measure changes in position, length, distance and optical length is well demonstrated in recent past (Som-

margren, 1987; Speckle 2010; Optical Metrology, 2010). Laser interferometry-based sensing and measurement system is capable of delivering sub nanometer accuracy when used for displacement measurement (Lee, Yoon, & Yoon, 2011; Minoshima, 2010; Schott, 2010; Schuldt et al., 2010; Shelley, 2007; Zeng & He, 2009; Zhou, Zhang, & Cheng, 2009). There has also been a number of research studies carried out on laser interferometry-based motion control for flexure-based mechanisms (Qi, Zhao, & Lin, 2007; Yeh, Ni, & Pan, 2005; Zhang & Menq, 2007) and it shows immense potential for further research.

Research proposed in this work is motivated by our previous efforts in the control of the flexure-based mechanism driven by piezoelectric actuators (Liaw, Shirinzadeh, & Smith, 2007, 2008b), as well as model-based control, and a robust motion tracking control (Liaw, Shirinzadeh, & Smith, 2008a). In model-based closed-loop control, we use supposed knowledge of system parameters to design the motion controller. The robust motion tracking control is employed in such a way that, it adapts the unknown system parameters, piezoelectric actuators nonlinearities, and external disturbances in the micro/nano manipulation system. In this study, the primary research objective is to track a specified motion trajectory using laser interferometry-based motion control. The secondary research objective is to validate experimental results for positioning accuracy and tracking performance with error analysis for capacitive position sensor and laser interferometry-based sensing and measurement technique.

The model of a piezo driven flexure-based micro/nano manipulator is described before presenting model-based and the robust motion tracking controller designs. Further, measurement error analysis and experimental study is detailed and results are presented and discussed ahead of conclusion.

MODEL OF FLEXURE-BASED MICRO/NANO MANIPULATOR

In our study the flexure-based manipulator mechanism is constructed by using flexure hinges. Therefore, the manipulator is monolithic structure, and hinges are assumed to be compliant in bending about one axis but rigid about the cross axes. The flexure hinge used is a notch-type hinge and the schematic of such hinge is shown in Figure 1.

The flexure hinge is simple in shape and operation and the design of four-bar flexure-based mechanism is detailed in our previous study (Liaw et al., 2008a). A lumped parameter dynamic model which combines the flexure-based mechanism and the piezoelectric actuator is formulated for the purpose of high-precision motion control. This is achieved by extending the model of a piezoelectric actuator, which is shown in Figure 2, and is described by Equation (1),

$$m_z \ddot{x}_z + b_z \dot{x}_z + k_z x_Z + f_m = T_{em}(v_{in} - v_h) \quad (1)$$

In Equation (1), x_z is the actuator displacement; m_z, b_z, and k_z are the mass, damping, and stiffness respectively; T_{em} is electromechanical transformer ratio; v_{in} is the applied (input) voltage; and v_h is the hysteresis voltage of the piezoelectric actuator model. The lumped parameter

Figure 1. Notch type flexure based hinge

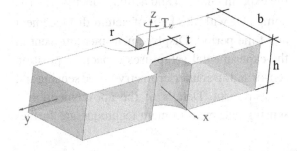

dynamic model of the piezoelectric actuator driven mechanism is given by Equation (2),

$$m_{lp} \ddot{x}_m + b_{lp} \dot{x}_m + k_{lp} x_m + v_h = v_{in} \quad (2)$$

where,

$$m_{lp} = \frac{1}{T_{em}} (m_z l_{zm} + m_a + \frac{I_b + I_c}{l_3^2}) \quad (3a)$$

$$b_{lp} = \frac{1}{T_{em}} b_z l_{zm} \quad (3b)$$

$$k_{lp} = \frac{1}{T_{em}} (k_z l_{zm} + \frac{4k_{\tau_z \alpha_z}}{l_3^2}) \quad (3c)$$

I_b, and I_c are the moment of inertia of link B and C of the flexure-based mechanism, respectively (Liaw et al., 2008a). By using the lumped parameter dynamic model given in Equation (3), the motion equation of the micro/nano manipulator results in Equation (4).

$$m_{lp} \ddot{x}_m + b_{lp} \dot{x}_m + k_{lp} x_m + v_n + v_d = v_{in} \quad (4)$$

where v_n and v_d represent all the non-linear effects and external disturbances, respectively.

MODEL-BASED CONTROL, ROBUST MOTION TRACKING CONTROL METHODOLOGY

In this section, we present brief information about closed-loop controller: model-based controller and for robust motion controller.

Figure 2. Model of piezoelectric actuator

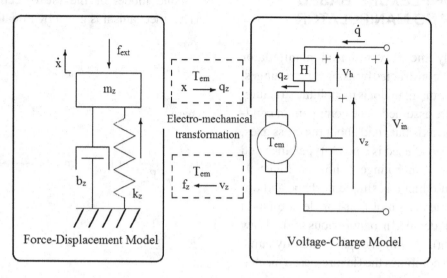

Model-Based Control

In model-based control, we will make use of the knowledge of the system parameters (i.e., m_d, b_d, and k_d) in controller design. A model-based closed-loop controller is established with the known system parameters and is shown in Equation (5),

$$m_{lp}\ddot{x}_s + b_{lp}\dot{x}_{act} + k_{lp}x_{act} = v_{in} \qquad (5)$$

In model-based control methodology, new input acceleration $\ddot{x}_s$ is expressed as,

$$\ddot{x}_s = \ddot{x}_{des} - \frac{1}{m_d}[(b_d\dot{x}_{err}) + (k_dx_{err})]. \qquad (6)$$

To realize, model-based controller, the values for m_d, b_d and k_d are calculated with considering lowest structural resonance of the mechanism $(\omega_{rm}, \omega_{rm})$, undamped natural frequency (ω_n), and damping ration (ξ), and is shown in the Equation (7),

$$m_d = 1,\ b_d = (2\omega_n) \cong (4\pi f_n)\ \text{and}$$
$$k_d = (\omega_n^2) \cong (2\pi f_n)^2 \qquad (7)$$

In Equation (5), x_{act} is actual position of mechanism, $\dot{x}_{act}$ is actual velocity, and $\ddot{x}_{des}$ is desired acceleration. Velocity $(\dot{x}_{err})$ and Position errors (x_{err}) are computed as,

$$\dot{x}_{err} = \dot{x}_{des} - \dot{x}_{act} \qquad (8)$$

$$x_{err} = x_{des} - x_{act} \qquad (9)$$

Actual position (x_{act}) of the mechanism is derived from the position sensor system used in the experimentation and actual velocity $(\dot{x}_{act})$ is computed with the help of actual displacement and time period. Displacement measurement in the experimental setup uses capacitive position sensor and laser interferometry-based sensing and measurement. Details of the above mentioned sensing and measurement technique are further explained in experimental setup. Position error and velocity errors are computed by using actual position measured by position sensing system and desired position (x_{des}) and velocity $(\dot{x}_{des})$ given by desired motion trajectory.

Robust Motion Control

An advanced control methodology is established for the purpose of high-precision tracking of the specified motion trajectory $x_{md}(t)$ for flexure-based micro/nano manipulator. With the tested and validated robust motion tracking control approach (Liaw et al., 2008a), the physical parameters of the system are assumed to be unknown or uncertain. There exist bounded non-linear effects and external disturbances within such a closed-loop system and this uncertainty of the motion system is expressed as Equation (10).

$$\left| \nabla m_{lp} \right| = \left| m_{lp} - \hat{m}_{lp} \right| \leq \delta m_{lp} \tag{10a}$$

$$\left| \nabla b_{lp} \right| = \left| b_{lp} - \hat{b}_{lp} \right| \leq \delta b_{lp} \tag{10b}$$

$$\left| \nabla k_{lp} \right| = \left| k_{lp} - \hat{k}_{lp} \right| \leq \delta k_{lp} \tag{10c}$$

where, ∇m_{lp}, ∇b_{lp} and ∇k_{lp} represent the parametric errors, $\hat{m}_{lp}$, $\hat{b}_{ln}$ and $\hat{k}_{lp}$ represent the estimated parameters, and δm_{lp}, δb_{lp} and δk_{lp} denote the bounds of the system parameters. It is assumed that the exact values of the parameters in Equation (4) are unknown, but the estimated values and their corresponding bounds of the system parameters are known and available. With this assumption gain values are chosen as,

$$k_p = m_d^{-1} k_d \tag{11a}$$

$$k_v = m_d^{-1} b_d - \alpha. \tag{11b}$$

With the introduction of the saturation function in the control law, the accuracy of the switching function is guaranteed to stay within the boundary layer. From the closed-loop dynamics of the control law, the steady-state switching function σ_{ss} within the boundary layer is obtained as

$$\sigma_{ss} = \frac{k_d e_{pss}}{m_d \alpha} \tag{12}$$

where, e_{pss} is the steady-state position error. With the knowledge of damping ration (ξ) and undamped natural frequency (ω_n), the desired parameters m_d, b_d and k_d are obtained with the help of Equation (7).

MEASUREMENT ERROR ANALYSIS

All mechanism and systems designed for micro/nano manipulation involve some level of uncertainties. To accomplish highest possible precision in motion of micro/nano manipulation mechanism, an investigation to identify the sources of errors and their contribution to the overall measurement uncertainty is essential. With the analysis of measurement subsystem, general expression for uncertainty estimation and the total error in measurement can be derived. Let us assume $'\nabla m'$ is a total measurement error in a system and it is a function of all independent error sources, resulting subsystem errors as

$$(\pm \nabla x_1, \pm \nabla x_2, \pm \nabla x_3, \ldots \ldots \pm \nabla x_n),$$

and $(x_1, x_2, x_3, \ldots \ldots x_n)$ are the system factors which causes measurement errors. With this assumption, general expression for uncertainty is derived as shown in Equation (13).

$$m \pm \nabla m = f(x_1 \pm \nabla x_1, x_2 \pm \nabla x_2, \ldots \ldots x_n \pm \nabla x_n)$$
$$(13)$$

In order to model, evaluate and analyze the general expression for uncertainty in our experimental system setup, we need to consider associated subsystems and the errors coupled with it.

The real time dynamic positioning measurement with the use of laser interferometry-based sensing is demonstrated in earlier research(Shirinzadeh et al., 2010). However, there exists some degree of uncertainty in the measurements obtained with the help of laser interferometry. The overall accuracy of any laser interferometry system is affected by a number of factors. A few of them are discussed below.

1. **Environmental errors** (E_{err}): An error of approximately 1ppm occurs in the index of the laser used due to each of the following environmental changes;
   a. 2.8 mmHg change in air pressure
   b. 10% change in the relative humidity
   c. 1° change in the air temperature

Controlling environmental conditions can help in minimizing the environmental errors. Environmental errors can be compensated with the Edlen's equation (Zygo, 2010) in Box 1, where, P, T and RH are pressure, temperature and relative humidity respectively.

2. **Geometric errors** (G_{err}): Geometric errors in measurement are caused mainly due to misalignment of the optics, which includes errors such as Cosine error, Abbe error and Polarization error. Cosine error is caused by an angular misalignment between the laser and the axis of motion. An offset between the measurement laser beam and the axis of motion causes Abbe error and polarization mixing error is due to the misalignment of the laser head relative to the interferometer and the imperfections in the polarization beam splitter.

3. **Instrumentation errors** (I_{err}): Instrumentation errors are based on the assumed and/or determined system parameters, such as system parameters supplied by manufacturer of the equipment. Zygo provided instrumentation errors are specified in Zygo error manual (Zygo, 2010). With the additional knowledge of laser interferometry subsystem and its error sources, the general expression for measurement uncertainty can be written as,

Table 1. Mechanism physical parameters

Flexure hinge width (b)	16 mm	Flexure hinge radius (r)	5 mm
Flexure hinge thickness (t)	0.5 mm	Flexure hinge link length (l_s)	50 mm
Angular Stiffness $\left(4k_{\tau_z \alpha_z}\right)$	6.44 Nm/rad	Lowest structural resonance $\left(\omega_{rm}\right)$	$326.64 \dfrac{rad}{s} \approx 54Hz$
Static linear Stiffness $\left(k_{f_m x_m}\right)$	0.01N/m^2	Undamped Natural Frequency $\left(\omega_n\right)$	$163.32 \dfrac{rad}{s} \approx 25Hz$

Box 1.

$$n = 3.836391.P \left[\frac{1 + P\left(0.817 - 0.0133.T\right).10^{-6}}{1 + 0.0036617} \right] - 3.033.10^{-3} RH.e^{0.057627T} \tag{14}$$

$$m \pm \nabla m = f(E_{err}, G_{err}, I_{err}) \tag{15}$$

In our experimental study, capacitive position sensors are also used as position feedback device. These capacitive sensors are analog non-contact devices, which consist of two RF-driven plates. The capacitance is directly related to the distance between the plates and it can be expressed as,

$$C = \frac{\varepsilon_0 \times \varepsilon_r \times A}{d} \; and \; \frac{\Delta C}{C} = \frac{\Delta d}{d} \tag{16}$$

where, ΔC and Δd, are changes in capacitance and position, ε_0 is permittivity of air or free space (in vacuum), given as $\varepsilon_0 = 8.85 \times 10^{-12}$ F/m or $\frac{1}{(36.10^9)\pi}$ F/m, ε_r is dielectric constant of the insulating medium, for air $\varepsilon_r \approx 1$, A is the overlapping area of plates (m²), and d is the distance between the plates (m).

Electronics which convert the capacitance information into a signal proportional to the distance may introduce some errors into measurement. Signal amplifier unit, electrical noise and misalignment are some of the factors which contribute towards the errors induced in capacitive position sensor system. Theoretically, in the capacitive position sensor based system, the measurement resolution is limited only by quantum noise. But in practical applications there are a few factors which affects the overall accuracy of capacitive position sensor system, such as (PHYSIK, 2010),

- Stray radiation (S_{err}),

- Electronics-induced noise (N_{err}),
- Geometric effects (G_{err1}).

The general expression for measurement uncertainty for capacitive position sensor system can be derived as function of error sources shown in Equation (17),

$$m \pm \nabla m = f(S_{err}, N_{err}, G_{err1}) \tag{17}$$

EXPERIMENTAL SETUP AND STUDY

An experimental research facility is established to investigate and characterize the performance in the tracking control with laser interferometry-based sensing. A VMEbus computer system is used to route laser interferometry-based signals into the feedback control – i.e., the closed-loop control. A detailed block diagram of the experimental set-up is shown in Figure 3 (Schematic block diagram of experimental setup).

Experimental set-up consists of a flexure-based four-bar mechanism based on a monolithic flexure-hinged structure. This mechanism is driven by piezoelectric actuators with an amplifier module, from Physik Instrumente (PI) capable of 45 µm displacement corresponding to a range of operating voltage from 0 to 100 V. In the capacitive position sensor system, a signal processing unit and a capacitive position sensor with measurement range of up to 50 µm from PI is used.

In laser interferometry-based sensing and measurement system, PCI to VME interface kit from the National Instruments is utilized in the

Figure 3. Schematic block diagram of experimental setup

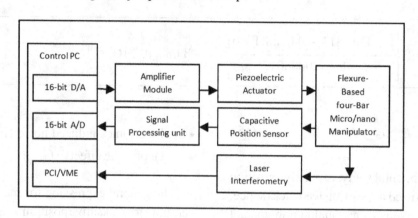

control PC. PCI/VME interface kit consists of two boards; a PCI board (PCI-MXI-2 board) and a 6U VME bus board (VME-MXI-2 module). We utilize a Zygo laser system in interferometry setup, which is composed of High Stability Plain Mirror interferometer (HSPMI) and the schematic of HSPMI system, is presented in Figure 4 (Laser interferometer schematic of HSPMI).

The Zygo Laser head uses a Helium-Neon frequency stabilized laser to generate a laser beam with two frequency components (F1 and F2) that are orthogonally polarized, and offset by 20 MHz. Control PC is running on an operating system capable of hard real-time control. Digital-to-analogue (D/A) board and an analogue-to-digital

(A/D) board in the control PC are of 16-bit resolution, and are used to generate the control signal and to acquire the position data of the micro/nano manipulator, respectively. In the experiments, the sampling frequency of the control loop is set at 1.25 kHz. For the micro/nano manipulator under study, 7000 series Aluminum alloy with $E = 72 \times 10^9 \ N \ / \ m^2$ is used as the material for the flexure-based manipulator. With the designed parameters presented in Table 1, angular stiffness $(4k_{\tau_z \alpha_z})$, static linear stiffness $(k_{f_m x_m})$, structural resonance (ω_{rm}) and undamped natural frequency (ω_n) were calculated.

Figure 4. Laser Interferometer schematic

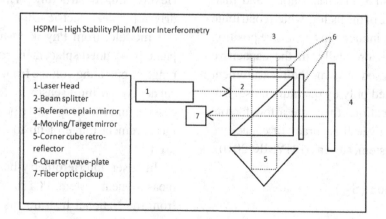

The closed-loop system is required to follow a jerk-free desired motion trajectory, which is shown in Figure 5.

This desired motion trajectory is formed by segments of higher-order 4-5-6-7 polynomials with zero acceleration at the beginning and the end. A low-pass second-order Butterworth filter (Ludeman, 1986) is employed to smooth the control signal prior to output as a commanding signal and the cutoff frequency is set at the upper limit of the undamped natural frequency, i.e., 25 Hz.

In order to study the effect of laser interferometry-based control methodology, an open-loop controller is established by eliminating the undesirable terms from Equation (4).

$$\hat{m}_{lp}\, \ddot{x}_m + \hat{b}_{lp}\, \dot{x}_m + \hat{k}_{lp}\, x_m = v_{in} \qquad (18)$$

In the Equation (18), $\hat{m}_{lp}$, $\hat{b}_{lp}$ and $\hat{k}_{lp}$ are estimated parameters, and this open loop controller is realized only for the comparison purpose. Lumped parameters of flexure-based four-bar mechanism are presented in Table 2.

The experiments are performed in a stable environment where the variations of room temperature and humidity are roughly controlled within 2° C and 10%, respectively. The experimentation is carried out for tracking 3 micrometer displacement with capacitive position sensor feedback and laser interferometry-based feedback. The tracking performances and experimental results of capacitive position sensor and Laser interferometry-based feedback systems are discussed further in result and discussion section.

RESULT AND DISCUSSION

The piezo-driven flexure-based four-bar manipulator is commanded to travel in a range of 3 micrometers to track the desired motion trajectory as shown in Figure 5. The resulting position of the flexure-based micro/nano manipulator with the capacitive position sensor-based control is shown in Figure 6. Laser interferometry-based feedback control of the flexure-based micro/nano manipulator and its resulting motion tracking with position error is presented in Figure 7. In model-based closed-loop control, motion tracking errors

Figure 5. Desired motion trajectory

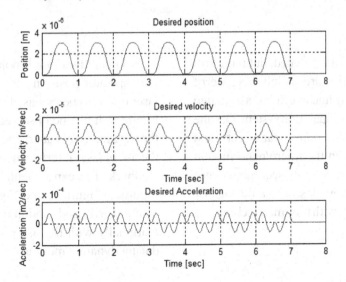

Table 2. Lumped parameters of piezo-driven flexure-based four-bar micro/nano mechanism

Lumped parameter	Desired values for Model-based control	Estimated values for Robust control	Bound for Robust control
Mass (V s2/m)	$m_d = 1$	$\hat{m}_{lp} = 1$	$\delta m_{lp} = 1$
Damping (V s/m)	$b_d = 314.1593$	$\hat{b}_{lp} = 1 \times 10^3$	$\delta b_{lp} = 1 \times 10^3$
Stiffness (V/m)	$k_d = 24674.0110$	$\hat{k}_{lp} = 1 \times 10^6$	$\delta k_{lp} = 1 \times 10^6$
Non-linearities and external disturbances (V)			$\delta v_{nd} = 30$

Figure 6. Capacitive position sensor based actual position tracking in open and closed-loop

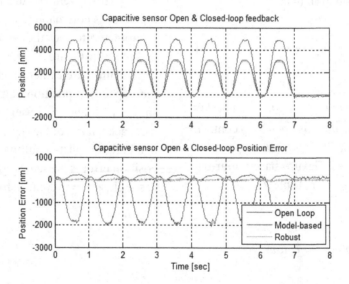

and positioning errors are reduced, but there still exist some errors which are mainly because of uncertainties and non-linear effects associated with the micro/nano manipulation system. Despite parametric uncertainties, non-linear effects, and external disturbances in the micro/nano motion system, laser interferometry-based robust motion tracking control demonstrates a reasonably precise tracking ability with minimal tracking and position errors.

Comparative closed-loop tracking results with the capacitor position sensor based system and laser interferometry-based system is presented in Figure 8. It can be observed from Figure 8, that the tracking errors are reduced considerably with the use of laser interferometry-based sensing and feedback. The experimental results presented in Figure 6, Figure 7, and Figure 8 show that the position error and motion tracking error reduced marginally with the use of laser interferometry during dynamic motion.

Figure 7. Laser interferometry based actual position tracking in open and closed-loop control

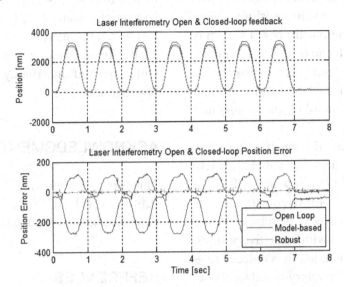

Figure 8. Tracking error: capacitive sensor based vs laser interferometry based control

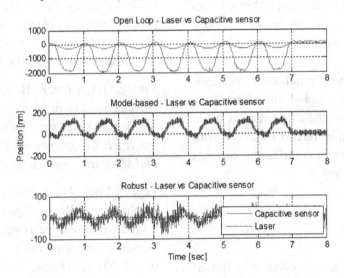

Table 3. Position error summary

Position Error in nanometer (nm)			
System – control strategy		**Laser Interferometry system**	**Capacitive position Sensor**
Steady state Error	Model-based control	< ± 20	< ± 100
	Robust motion control	< ± 20	< ± 80
Open Loop tracking		< – 300	< – 2000
Model-based closed-loop tracking		-20 < error < +100	-80 < error < +175
Robust Motion closed-loop tracking		< ± 20	< ± 80

Comparative error analysis for open and closed-loop control with the capacitive position sensor system and laser interferometry-based system is presented in Table 3. It can be inferred from the experimental data that, laser interferometry based robust control motion tracking results in more accurate and precise control of micro/nano manipulator.

In model-based closed-loop, position error with capacitive position sensor is approximately in between -80nm to +175 nm, which is dropped down between -20nm to +100 nm with laser interferometry-based feedback. In robust motion control the position error with capacitive position sensor is within $\pm$ 80 nm, which is reduced to $\pm$ 20 nm with laser interferometry-based feedback. The position tracking errors were less than $\pm$ 20 nm at the steady-state for laser interferometry-based system. Robust motion control methodology with laser interferometry-based feedback delivers high tracking performance with enhanced position accuracy.

Variations in ambient conditions, mechanical vibrations, acoustic, and electrical noise are some of the sources of the measurement error. The measurement can be further improved with knowledge of the factors contributing towards measurement uncertainty.

CONCLUSION

Laser interferometry based system is proposed and investigated for the tracking of desired motion trajectory for flexure-based micro/nano manipulator. The proposed laser interferometry based system shows improvement in motion tracking, and reduces tracking and positioning errors. An improvement in the position tracking with the help of laser interferometry is demonstrated with the help of experimentation. Stability and error analysis of laser interferometry based control system can be utilized to improve tracking performance. The experimental study offers evidence of high-precision

motion tracking of micro/nano manipulator with laser interferometry-based control. Future research based on laser interferometry-based closed-loop control of micro/nano multi-axis manipulators with improved uncertainty expression is under investigation.

ACKNOWLEDGMENT

This work is supported by ARC LIEF grant (LE0347024, LE0775692) and ARC Discovery grant (DP0450944, DP0666366, and DP0986814).

REFERENCES

Bashash, S., & Jalili, N. (2008). Adaptive robust control strategy for coupled parallel-kinematics piezo-flexural micro and nano-positioning stages. *IEEE/ASME Transactions on Mechatronics, 14*(1), 11–20. doi:10.1109/TMECH.2008.2006501.

Chung, G. J., Choi, K. B., & Kyung, J. H. (2006). Development of precision robot manipulator using flexure hinge mechanism. In *Proceedings of the IEEE Conference on Robotics, Automation and Mechatronics*, Bangkok, Thailand (pp. 1-6).

Lee, J., Yoon, H., & Yoon, T. H. (2011). High-resolution parallel multipass laser interferometer with an interference fringe spacing of 15 nm. *Optics Communications, 284*(5), 1118–1122. doi:10.1016/j.optcom.2010.10.073.

Li, Y., & Xu, Q. (2010). A totally decoupled piezo-driven XYZ flexure parallel micropositioning stage for micro/nanomanipulation. *IEEE Transactions on Automation Science and Engineering, 8*(2), 265–279. doi:10.1109/TASE.2010.2077675.

Liaw, H. C., & Shirinzadeh, B. (2008). Enhanced adaptive motion tracking control of piezo-actuated flexure-based four-bar mechanisms for micro/nano manipulation. *Sensors and Actuators. A, Physical, 147*(1), 254–262. doi:10.1016/j.sna.2008.03.020.

Liaw, H. C., & Shirinzadeh, B. (2009). Neural network motion tracking control of piezo-actuated flexure-based mechanisms for micro-/nanomanipulation. *IEEE/ASME Transactions on Mechatronics, 14*(5), 517–527. doi:10.1109/TMECH.2009.2005491.

Liaw, H. C., Shirinzadeh, B., & Smith, J. (2007). Enhanced sliding mode motion tracking control of piezoelectric actuators. *Sensors and Actuators. A, Physical, 138*(1), 194–202. doi:10.1016/j.sna.2007.04.062.

Liaw, H. C., Shirinzadeh, B., & Smith, J. (2008a). Robust motion tracking control of piezo-driven flexure-based four-bar mechanism for micro/nano manipulation. *Mechatronics, 18*(2), 111–120. doi:10.1016/j.mechatronics.2007.09.002.

Liaw, H. C., Shirinzadeh, B., & Smith, J. (2008b). Sliding-mode enhanced adaptive motion tracking control of piezoelectric actuation systems for micro/nano manipulation. *IEEE Transactions on Control Systems Technology, 16*(4), 826–833. doi:10.1109/TCST.2007.916301.

Ludeman, L. (1986). *Fundamentals of digital signal processing*. New York, NY: John Wiley & Sons.

Minoshima, K. (2010). High-precision absolute length metrology using fiber-based optical frequency combs. In *Proceedings of the International Conference on Electromagnetics in Advanced Applications*, Sydney, NSW, Australia (pp. 800-802).

Mohd Zubir, M. N., & Shirinzadeh, B. (2009). Development of a high precision flexure-based microgripper. *Precision Engineering, 33*(4), 362–370. doi:10.1016/j.precisioneng.2008.10.003.

Motamedi, M., Vossoughi, G., Ahmadian, M. T., Rezaei, S. M., Zareinejad, M., & Saadat, M. (2010). Robust adaptive control of a micro tele-manipulation system using sliding mode-based force estimation. In *Proceedings of the American Control Conference* (pp. 2811-2816).

Paros, J. M., & Weisbord, L. (1965). How to design flexure hinges. *Machine Design, 37*, 151–156.

PHYSIK. (2010). *INSTRUMENTE*. Retrieved from http://www.physikinstrumente.com/

Qi, Y. Y., Zhao, M. R., & Lin, Y. C. (2007). Study on a positioning and measuring system with nanometer accuracy. In *Proceedings of the First International Conference on Integration and Communication of Micro and Nanosystems* (pp. 1573-1577).

Rakotondrabe, M., Haddab, Y., & Lutz, P. (2009). Development, modeling, and control of a micro-/nanopositioning 2-DOF stickSlip device. *IEEE/ASME Transactions on Mechatronics, 14*(6), 733–745. doi:10.1109/TMECH.2009.2011134.

Saeidpourazar, R., & Jalili, N. (2006). Modeling and observer-based robust tracking control of a nano/micro-manipulator for nanofiber grasping applications. In *Proceedings of the ASME International Mechanical Engineering Congress and Exposition* (pp. 969-977).

Schott, W. (2010). Developments in homodyne interferometry. *Key Engineering Materials, 437*, 84–88. doi:10.4028/www.scientific.net/KEM.437.84.

Schuldt, T., Gohlke, M., Weise, D., Peters, A., Johann, U., & Braxmaier, C. (2010). High-resolution dimensional metrology for industrial applications. In *Proceedings of the 9th International Symposium on Measurement Technology and Intelligent Instruments*, Saint Petersburg, Russia (Vol. 437, pp. 113-117).

Shelley, T. (2007). Extreme measurements become routine. *Eureka, 27*(2), 29–30.

Shiou, F. J., Chen, C. J., Chiang, C. J., Liou, K. J., Liao, S. C., & Liou, H. C. (2010). Development of a real-time closed-loop micro-/nano-positioning system embedded with a capacitive sensor. *Measurement Science & Technology, 21*(5). doi:10.1088/0957-0233/21/5/054007.

Shirinzadeh, B., Teoh, P. L., Tian, Y., Dalvand, M. M., Zhong, Y., & Liaw, H. C. (2010). Laser interferometry-based guidance methodology for high precision positioning of mechanisms and robots. *Robotics and Computer-integrated Manufacturing*, *26*(1), 74–82. doi:10.1016/j.rcim.2009.04.002.

Smith, S. T., V. G. B., Dale, J. S., & Xu, Y. (1997). Elliptical flexure hinges. *The Review of Scientific Instruments*, *68*(3), 1474–1483. doi:10.1063/1.1147635.

Sommargren, G. E. (1987). A new laser measurement system for precision metrology. *Precision Engineering*, *9*(4), 179–184. doi:10.1016/0141-6359(87)90075-4.

Spanner, K., & Vorndran, S. (2003, July 20-24). Advances in piezo-nanopositioning technology. In *Proceedings of the IEEE/ASME International Conference on Advanced Intelligent Mechatronics* (pp. 1338-1343).

Speich, J. E., & Goldfarb, M. (1998). A three degree-of-freedom flexure-based manipulator for high resolution spatial micromanipulation. In *Proceedings of the SPIE Conference on Microbiotics and Micromanipulation*.

Tian, Y., Shirinzadeh, B., & Zhang, D. (2009). A flexure-based five-bar mechanism for micro/nano manipulation. *Sensors and Actuators. A, Physical*, *153*(1), 96–104. doi:10.1016/j.sna.2009.04.022.

Tian, Y., Shirinzadeh, B., & Zhang, D. (2010). Design and dynamics of a 3-DOF flexure-based parallel mechanism for micro/nano manipulation. *Microelectronic Engineering*, *87*(2), 230–241. doi:10.1016/j.mee.2009.08.001.

Xu, Q., & Li, Y. (2009). Global sliding mode-based tracking control of a piezo-driven XY micropositioning stage with unmodeled hysteresis. In *Proceedings of the IEEE International Conference on Intelligent Robots and Systems* (pp. 755-760).

Xu, Q., & Li, Y. (2010). Precise tracking control of a piezoactuated micropositioning stage based on modified Prandtl-Ishlinskii hysteresis model. In *Proceedings of the IEEE International Conference on Automation Science and Engineering* (pp. 692-697).

Yeh, H. C., Ni, W. T., & Pan, S. S. (2005). Digital closed-loop nanopositioning using rectilinear flexure stage and laser interferometry. *Control Engineering Practice*, *13*(5), 559–566. doi:10.1016/j.conengprac.2004.04.019.

Zeng, Y., & He, G. (2009). Study on interferometric measurements of a high accuracy laser-diode interferometer with real-time displacement measurement and it's calibration. In *Proceedings of the 4th International Symposium on Advanced Optical Manufacturing and Testing Technologies: Optical Test and Measurement Technology and Equipment*.

Zhang, Z., & Menq, C. H. (2007). Laser interferometric system for six-axis motion measurement. *The Review of Scientific Instruments*, *78*(8). doi:10.1063/1.2776011.

Zhou, X. Y., Zhang, T., & Cheng, Y. H. (2009). Study of high precision and high measurement speed dual-longitudinal thermal frequency stabilization laser interferometer. *Dianzi Keji Daxue Xuebao/Journal of the University of Electronic Science and Technology of China*, *38*(3), 451-454.

Zygo. (2010). *Corporation - Zygo error sources.* Retrieved from http://www.zygo.com/library/papers/ZMI_Error_Sources.pdf

This work was previously published in the International Journal of Intelligent Mechatronics and Robotics, Volume 1, Issue 3, edited by Bijan Shirinzadeh, pp. 31-45, copyright 2011 by IGI Publishing (an imprint of IGI Global).

Chapter 13
Analysis and Implementation for a Walking Support System for Visually Impaired People

Eklas Hossain
University of Wisconsin-Milwaukee, USA

Riza Muhida
International Islamic University of Malaysia, Malaysia

Md Raisuddin Khan
International Islamic University of Malaysia, Malaysia

Ahad Ali
Lawrence Technological University, USA

ABSTRACT

Visually impaired people are faced with challenges in detecting information about terrain. This paper presents a new walking support system for the blind to navigate without any assistance from others or using a guide cane. In this research, a belt, wearable around the waist, is equipped with four ultrasonic sensors and one sharp infrared sensor. Based on mathematical models, the specifications of the ultrasonic sensors are selected to identify optimum orientation of the sensors for detecting stairs and holes. These sensors are connected to a microcontroller and laptop for analyzing terrain. An algorithm capable of classifying various types of obstacles is developed. After successful tests using laptop, the microcontroller is used for the walking system, named 'Belt for Blind', to navigate their environment. The unit is also equipped with a servo motor and a buzzer to generate outputs that inform the user about the type of obstacle ahead. The device is light, cheap, and consumes less energy. However, this device is limited to standard pace of mobility and cannot differentiate between animate and inanimate obstacles. Further research is recommended to overcome these deficiencies to improve mobility of blind people.

INTRODUCTION

Orientation, navigation and mobility are perhaps three of the most important aspects of human life. Significant features of information to aid navigation for active mobility are passed to human through the most complex sensory system, the vision system. This visual information forms the basis for navigational tasks; as such an individual with impaired vision is at a disadvantage because appropriate information about the environment is not available. The term blindness refers to people

DOI: 10.4018/978-1-4666-3634-7.ch013

who have no sight at all as well as to those considered as blind have limited vision, which cannot be said to be severely visually impaired, (WHO, 1998). The major causes of blindness are age-related macular degeneration, cataracts, glaucoma, diabetic retinopathy, trachoma, onchocerciasis, by birth, lack of eye care and accident (Times of India, 2000). The development of assisting devices to aid visually impaired people in their everyday life has been increasing. In some cases, solutions to providing sensory supplementation such as Braille through electronic reading machines have been very effective. However, truly adequate solutions for navigation assistance for visually impaired have not yet been achieved. A number of devices have already been developed to address some of the difficulties faced by visually impaired people with regard to travel (Baldwin, 1998). This study therefore aims at examining the viability of different types of devices for mobility aid of blind; either using sensors or cameras, but some of them used both.

Visual impairment is one of the most common disabilities worldwide. WHO reported that due to the lack of epidemiological data, especially form the developing and under developed countries, the exact number of blind persons in the world is not known. In 1994, WHO estimated that it was around 38 million with a further 110 million cases of low vision, which are at risk of becoming blind? In 1998, the total population of visual impairment was more than 150 million people (WHO, 1998). Currently, there is a total of about 45 million blind people in the world and a further 135 million have low vision and this number is expected to double by 2020 (Times of India, 2000). The number of people who become blind each year is estimated to be 7 million. Over 70% of the people with vision problem receive treatment and their vision is restored. Thus the number of blind persons worldwide is estimated to increase by up to 2 million per year (WHO, 1997). Eight percent of these cases are ageing-related.

There are a quarter of a million people in the UK who are registered as visually impaired. However, in UK, actually there have nearly one million people entitled to register as a visually impaired person, and 1.7 million with the vision difficult. This represents over three percent of the UK population (NFB, 2002). In Britain, more than twenty thousand children grow up with visual impairment, and there are two hundred vision-related accidents per day in the UK alone (Leonard & Gordon, 1999; Viisola, 1995). There are approximately 10 million visually impaired people in the United States (AFB, 2001). In addition, statistics state that for every seven minutes, someone in America is becoming visually impaired (Blasch, 1999). In Malaysia, alarming increment in blind population is noted with about 46.9% from 1990 to 1999. By September 2000, there were about 13,835 registered in Blind Associations and it is predicted that, it might be less than 50% of the total blind population in the country (JKM, 2000).

Usually, to work outdoor as well as indoor, blind people face difficulties. This is why many of them use a guide cane that is considered as cheap and helpful to them. That purely mechanical device is used to detect the surface of the ground and any obstacles around him. A guide cane is so economical and light in weight that it can be folded and can be brought to any places without or with any difficulty. However, a guide cane also has its drawback. It must be used many times in order for the user to detect any change to the ground or to avoid obstacle. Therefore, only trained users will be able to use the guide cane defiantly. Moreover, it cannot differential among various obstacles on the way of the user's life. Besides that, blind person needs to scan the walking area continuously while walking. Another drawback is that the guide cane cannot detect any obstacle within the range of 2 to 3 meters. A guide cane can only detect an object when it has a contact with it. If there is no contact, the user will eventually bump to it. Another example is that the guide cane is not a device that can detect objects from above the waist such as obstacles close to the head. And lastly, it cannot detect any moving object and therefore are exposed to dangers of hitting vehicles or even moving animals. Another option for a blind person is to use a guide dog or horse. However, a both dog and horse will cost a lot as it needs an extensive training to help the blind person. In Islam, all Muslims are prohibited to have any kind of contact with dogs. Therefore, it is not applicable for a blind Muslim. Usually, blind people are not self dependent and this will be difficult for them to take care of the dog besides themselves. In an unknown environment, either a mobile robot or a blind person has a common point. That is they cannot detect an object without any help from a device. Both of them have the ability to walk or move to a certain point but a mobile robot uses sensors in order to avoid any obstacle. If this technology is sued on the blind person, it can be a big help for them to walk safely and reduce the danger when walking without a guide cane.

When multiple sensors are installed on the blind person, then they do not need to scan their area to walk in front. The transfer of mobile robot technology is actually a new development in order to help this type of community. In the past, robots have been used to aid the blind person to walk. But this new technology assists the user to walk without having any difficulty. Besides that, it is more economical to apply the technology directly to the person rather than buying a complicated robot that will cost a lot. In this case, it becomes difficult to mimic nature in its entirety of human vision system. A blind user will utilize a guide cane to scan obstacles in front of him/her, waving the guide cane from left from right and vice versa continuously. This scanning is not effective because there is a possibility for the guide cane to miss a spot as the user is waving the guide cane, it may lead to a collision. Besides, the blind user may reach his destine by taking help from healthy vision person. However, the blind needs

to adjust his time according to the availability of his companion who will guide him. So this will restrict his freedom to go other places according to his will. To overcome this problem, an array of sensors with different orientations is used so that mapping of the environment could be done easily. Having with modern technology, the walking support systems for blind are still not sophisticated in terms of mobility, safety and cost; this problems lead to motivation of designing a prototype of smart walking support system for visually impaired people. The walking support system will help the blind user to avoid obstacles in the way of his destination.

Electric Assistive Technologies (EATs) provide the blind people spatial information about the environment in assisting them for navigation. ETAs must satisfy three requirements: (1) accuracy of detecting the presence of a stationary object; (2) accuracy of detecting the inter-space between stationary objects; and (3) accuracy of tracking a moving object The progress of the ETA development can be divided into two stages (Wong, 2000; Sainarayanan, 2002) for Blind Navigation Aids: Early ETAs and Modern ETAs. Russsel Path sounder (Duen, 1998) was invented in 1965 and it is one of the most early ETAs. Chest mounted board with two ultrasonic sensors produce three levels of click sound to indicate the distance. The Sonic guide has wide beam ultrasonic transmitter mounted in the centre of the spectacle and a receiver on each side of the transmitter. Depending on the distance of the obstacle the sound tones are fed back to the right and left ear of the blind. Nottingham obstacle detector works in an analog technique as that of MOWAT sensor, but produces an audible note when the obstacle is within the device range, the maximum range being around two meters (Dodds, 1981). C-5 Laser Cane was introduced in 1973. It has three transmitters and three photodiodes as receives. It produces two different tones and vibrations upon detection of obstruction within 4 meter range (Lofving, 1998;

Benjamin, 1973). Some researchers attempted to solve this problem, in the late 1960s and the early 1970s, by building an array of tactile feedback devices and attaching them through standard camera (Snaith, 1998).

By the early 1990s the focus has switched from mobility and obstacle detection to orientation and location. With the advanced development of the high sensitive sensors and computing devices, the research had been focused to new directions. MIMS infrared mobility aid uses emitters and receives built into spectacle frames to provide obstacle detection. NavChair is a navigation system that is developed to provide mobility to individuals that find difficult to use a powered wheelchair due to cognitive, perceptual or motor impairments. The user will basically control the path of navigation and the wheelchair's motion along that path, and the NavChair restricts itself to ensure collision free travel. The Guide cane does not block the users hearing with audio feedback and since the computer automatically analyzes the situation and guides the user without requiring user to manually scan the area, there is no need for extensive training. Cardin (2007) proposed and developed a wearable system that can detect obstacle that surrounds the user by using multi-sonar system and sending appropriate vibro-tactile feedback so that the mobility of visually impaired can be increased. Sonic Eye works with the concept of mapping of image to sound (Reid, 1998). It works by scanning the device's thin window over the object of interest from left to right. Some of the students from University of Guelph, in Canada, developed an inexpensive, built with off-the-shelf components, wearable and low power device. It will transform depth information into tactile or sound information for use by blind people during navigation. The electron-neural vision system (ENVS) aims to achieve obstacle avoidance and navigation in outdoor environments with the aid of visual sensors, GPS, and electro-tactile simulation (Meers, 2005). Johnson and Higgins (2006)

described a compact, wearable device that converts visual information into a tactile signal and to create a wearable device that converts visual information into tactile signal to help visually impaired people self-navigate through obstacle avoidance with commercially available parts. The preservation and analysis of important image information needed for blind people for safe navigation in an indoor environment are presented in Tyflos navigation system (Dakopoulos & Bourbakis, 2008). A computerized travel aid was developed for blind based on mobile robotics technology (Shoval et al., 1993, 1994, 1998).

The paper is organized as follows: the next section presents a detail overview on experiments for terrain detection and data analyses. Algorithm and hardware development are then discussed. Finally, conclusions are drawn.

EXPERIMENTS FOR TERRAIN DETECTION AND DATA ANALYSES

This section presents experimental data on different terrain and their analyses. Experiments conducted for terrain detection mainly uses the sensory system designed and developed in our project, and concentrates on critical obstacles like stairs up, stairs down, drop off, overhang etc. These obstacles are considered as critical for the blind people following an interview with MAB personnel.

Over Head Obstacle

A blind man generally carries a white cane to scan obstacles at the ground level. Thus, overhead obstacle is only detected when his head hits against such obstacle. In this research, to address of this type of obstacles, a sharp IR sensor is attached to the belt facing upward direction making an angle with a line perpendicular to the ground. This arrangement can detect overhead obstacles before hitting the obstacle. Two different types of such obstacles are shown in Figure 1(a) and (b), and corresponding sensor data is shown in Figure 1(c). Assuming a clearance of 12 inches above the head the distance of an over head obstacle from the IR sensor on the belt at waist level can be about 50 inches. Therefore, less than 50 inches reading of the IR sensor is considered as overhead obstacle.

Wall/Front Obstacle

Walls and pillars are the common obstacles for both indoor and outdoor. Figure 2(a) and (b) illustrates two such cases. It is obvious readings of the sensors S3 and S4 will be decreasing once such obstacles are detected. Once readings of both the sensors show less than 25 inches, it will be considered obstacle in front of the person. In such a situation a blind person will have the op-

tion to move left or right to confirm nature of the obstacle i.e., wall or pillar.

Hole in Front

Holes are common obstacles on walk ways. Figure 3(a) and (b) depict a small hole and corresponding data read by the sensors S1 and S2. Unlike drop off in front, in the case of a hole high value of data appears for a very short duration. In the current state of the hardware developed for terrain detection, this scenario was put in the category of drop off in front. However, it is expected that proper training of a blind person will make him feel the difference between drop off in front and small hole in front. Figure 3(c) shows a double hole in front of a person and Figure 3(d) shows sensor data by sensors S1 and S2.

Figure 1. (a) Photograph of a slanted over head obstruction, (b) photograph of an over head obstruction coming out of a wall, (c) sensor data showing detection of over head obstruction

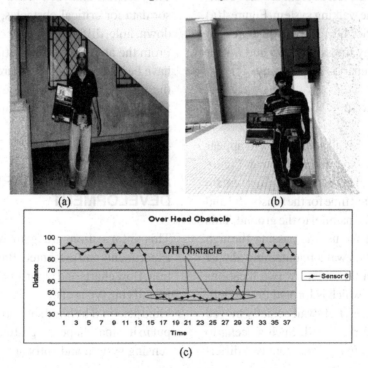

(a)   (b)

(c)

Figure 2. (a) Wall in front of a person, (b) column in front of a person, (c) sensor data shown by sensors S3 and S4

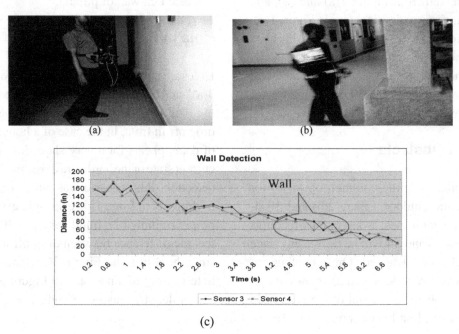

(a) (b)

(c)

Obstacle on Left or Right

Obstacles either on the left or right are also common scenarios for the walking system. Figure 4(a) shows such a scenario. 4 (b) and (c) illustrates the trend of data read by the sensors S3 and S4 for the left and right scenarios respectively.

Complex Scenario

In real-life it is rear that all scenarios will appear as discrete maps as experimented above. Figure 5(a) shows readings of lower distances that remain constant for significant time for the sensors S3 and S4, which are directed parallel to the ground. This is a scenario where the user is passing through a passage between two walls maintaining almost equal distance from the walls. At two locations suddenly sensor S3, which is located toward left, shows higher distances. Left wall actually moved a bit away from the right wall. Such a scenario needs actuation data to the user from two differ-

ent actuators as well as training of the user. The sketch of the scenario is shown in Figure 5(b). This section analyzed trend of ultrasonic sensor data for critical obstacles, like stair up, stair down, hole, different types of drop offs and so on. From the above analyses distinguishing features have been identified which are later compiled in the form of flow chart as well algorithm in the following section for developing blind support system hardware.

ALGORITHM AND HARDWARE DEVELOPMENT

This information are gathered from different experiments and presented. Based on the information, flow charts and algorithms are developed for identifying types of obstacles on the walkway of a blind person. The blind support system hardware is put in its final shape integrating actuating system, sensing system and software.

Figure 3. (a) Photograph of a hole in front of a person, (b) sensor data shown by sensors S1 and S2, (c) photograph of a double hole in front of a person, (d) sensor data shown by sensors S1 and S2

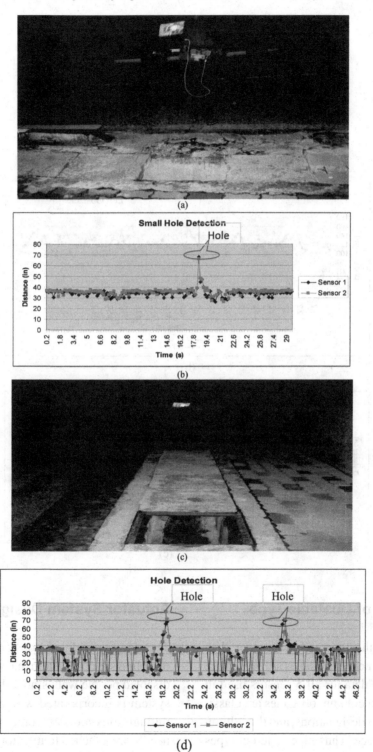

Figure 4. (a) Photograph of obstacle on the left side of a person, (b) sensor data detecting obstacle on left side, (c) sensor data detecting obstacle on right side

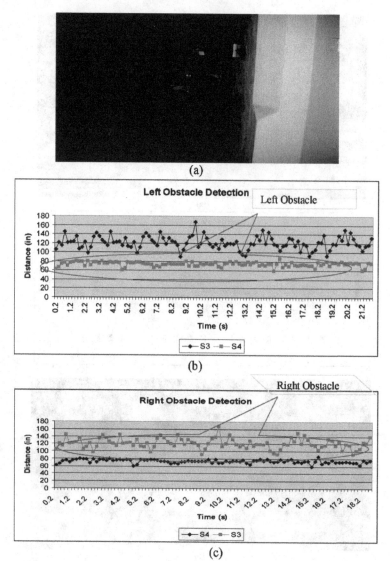

Classification of Obstacle Type

Experimental results on trends of sensor data against different types of obstacles are presented. Table 1 shows the data for all 5 sensors used in the experiments. In column, obstacles are classified with numerical designations, and the column is shown in bold face. Thus nine different types of obstacles are being represented through their respective sets of sensor data.

Actuator System Design

The study provides the design and development of a blind support system that would help visually disabled people walk smoothly. Proper actuating system is incorporated with the sensing system through software. A sevomotor with an indicator needle are attached to its rotor which is shown in Figure 6. It is selected as the actuating mechanism for conveying the information to its user. Based on

Figure 5. (a) Sensor data showing complex scenario, (b) sketch showing complex scenario

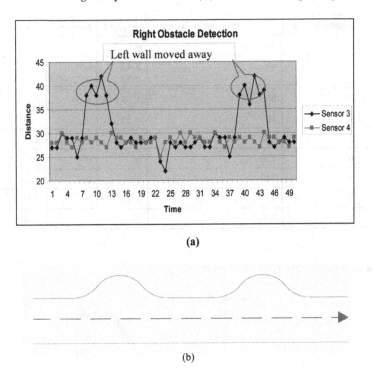

(a)

(b)

the scenario mapped in Table 1, the servomotor shows 9 positions. The servomotor assembly is attached to the pouch on the *"Belt for Blind"*. On top of that a buzzer is also attached to this system. The moment an obstacle is detected, the buzzer will beep and the needle will move to a position corresponding to the type of the obstacle detected. On touching the dial after hearing the beep the user will be able to identify type of the obstacle and accordingly plan his move.

Strategy for Control of the Smart Walking Support System for the Blind

This device is only used when the user intends to walk or move from one place to another. Therefore, a switch either to turn ON or OFF for the device will be available and located at the side of the user where it is easy to reach. Once the user turns the switch ON, the smart walking support system will be active and the ultrasonic sensor will start to scan the environment. When the ultrasonic sensor is active, it will emit a short burst of ultrasonic wave when it is "fired". If an object is located in front of the sensor, some of the ultrasonic waves will be reflected back to the sensor, which switches into a microphone mode immediately after firing. Once the echo from the object is received by the sensor, it is be converted to electrical signal and then sends the electrical signal to the controller that controls the output of the system (Figure 7). The controller measures the time that has elapsed between firing the ultrasound and receiving the echo. Because the velocity of ultrasound traveling through air is almost constant, the controller can easily compute the distance between the object and the sensor from the measured time-of-flight. The ultrasonic sensors used here have a maximum range of 3.3 meters.

Ultrasound waves propagate from the sensor in a cone-shaped propagation profile, in which

Table 1. Obstacle classification based on sensor data

Obstacle Name	Obstacle Type	Servo Position in degree	Sensor 1	Sensor 2	Sensor 3	Sensor 4	Sensor 5
Over Head	1	20	0	0	0	0	< 50
Stair Down	2	40	> 45 < 52	> 45 < 52	0	0	0
Hole/drop off in front	3	60	> 55	> 55	0	0	0
Stair Up	4	80	< 35	< 35	< 60	< 60	0
Wall	5	100	0	0	< 25	< 25	0
Front Left	6	120	0	0	< 15	< 35	0
Front Right	7	140	0	0	< 35	< 15	0
Drop off_ Left	8	160	0	0	> 55	< 40	0
Drop off_Right	9	180	0	0	< 40	> 55	0

Figure 6. Servo positions showing types of obstacles

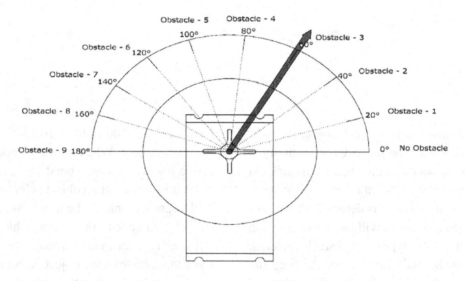

the opening angle of the cone is about 40°. This arrangement for each of the ultrasonic sensors assures coverage of the area in front of the user. One major difficulty in the use of multiple ultrasonic sensors is the fact that these sensors can cause mutual interference, called crosstalk. Crosstalk is a phenomenon in which the wave-front emitted by one of the ultrasonic sensor secularly reflects off smooth surfaces and is subsequently detected by another ultrasonic sensor. In the past, researchers had to employ slow firing schemes to allow each sensor's signal to dissipate before the next sensor was fired. Therefore, this problem is avoided by making sure that the ultrasound fired by each sensor does not interfere with each other. This problem is solved by taking into consideration the maximum distance for the neighboring sensor. The case for ultrasound wave overlapping with the neighboring sensor will not occur for this setup.

For the walking support system for the blind named as belt for blind, it will start off by simply

Figure 7. Block diagram representing the basic control system

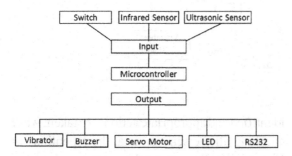

scanning the area of interest that is about 120° in front of the user and only 40 degree at the back of the user as shown in the flowchart (Figure 9). Once it detects an object, it will determine whether the object is constantly exists or not and are shown in the flowchart as in Figure 8. For example, the hand movement will cause a disturbance to the sensor because it will detect the hand as an object once the hand swings in front of the sensor. This problem is resolved by making a counter in order to determine the object really exist in front of the sensor. The counter will count for 3 milliseconds and after that, it will verify that the object exists in front of the sensor. After the verification of an obstacle whether it is on the front, right, or left it will classify the obstacle as stair, hole, drop off, and so on. The problem will arise when the user is walking fast. This will make the vibrator for the respective direction to vibrate and thus making the user to slow down the movement in order to make sure that he can follow the guidance from the vibrator to a safe path. Hence, a user will be able to follow the direction safely only if he walks slowly. This occurs because of the step distance when the user walks. Normally, for our case, the user has a step distance of 0.4 meter. Therefore, from the analysis, the sensor will suddenly move 0.4 meter in front. But for a mobile robot, the robot moves forward step by step and therefore it can move fast because the step distance can be less than 5cm due to the step distance, the sensor

needs to have a fast response in order to send the signal to the microcontroller to the output system.

Once the obstacle has been overcome, the entire servo position and buzzer will be turned off and the controller will loop back to scanning the area of interest. The block diagram in Figure 7 shows the basic control system. In brief, the belt

Figure 8. Flowcharts for walking support system and determining which sensor detected the obstacle

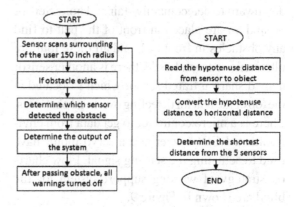

Figure 9. Flowchart for detecting various obstacles (over head, stair down, hole, stair up, wall, front left, front right, drop off left, and drop off right) in front of the user

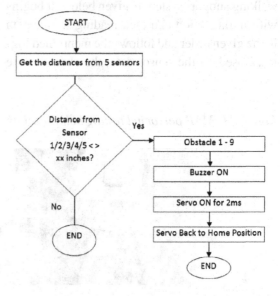

Figure 10. Setup of obstacle avoidance experiment

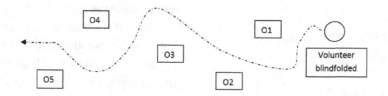

is consisting of five sensors placed in different position and orientation so that a satisfactory safety distance can be covered for the user to navigate his way. Sensor S1 and S2 are placed downward to detect mainly stair and hole. Sensor S3 and S4 are placed in front of the belt to find any obstacles in front of it and all these four are ultrasonic sensors. Lastly, there is another sensor named sharp infrared sensor which is located at the top of the belt for detecting overhead obstacles. There is a micro controller to get data from environment. LCD, buzzer and a servo motor have used to actuating the sensing signal. Flow charts on strategy of walking support system for the blind are shown in Figure 9.

Algorithm for the Walking Support System

The step by step procedures of the algorithm for walking support system is given below. It begins with initialization and then reading sensor data in the given order and follows the mentioned logics. Based on the sensor data, the obstacles are identified and appropriate actions are taken. After checking all the sensors data, the repetitions of logic are used for any walking path.

Obstacle Avoidance and Detection

In this experiment, the efficiency of the system was tested using obstacle placed randomly in front of the user. A volunteer is chosen randomly, a brief explanation was given on how to follow instructions from the smart walking support system. When the volunteer understands on how the system works, he is taken blindfolded. Obstacle was placed randomly in front of him and he is instructed to move forward with the guidance of the walking support system. The setup is shown in Figure 10.

The experiment is done with 3 volunteers and the results are satisfactory. A blindfolded user walks slowly due to the fear of collision with the obstacle, thus the volunteer has sufficient time to respond to the instructions given. The setup of the obstacle was changed to check for efficiency of the system. The user will walk forward, detect-

Figure 11. MAB personnel testing belt for blind

Table 2.

```
"Switch ON"
Begin
  Initialized: S1, S2, S3, S4, S5 ∈ {0}
  S ∈ {0° position with no obstacle}
  Repeat
  Begin
    If S5 < 50 then Obs1
    Begin
      B and S ∈ {20° position for 02 sec}
      S ∈ {0° position}
    End;
    If 45 < S1 < 52 and 45 < S2 < 52 then Obs2
    Begin
      B and S ∈ {40° position for 02 sec}
      S ∈ {0° position}
    End;
    If S1 > 55 and S2 > 55 then Obs3
    Begin
      B and S ∈ {60° position for 02 sec}
      S ∈ {0° position}
    End;

    If S1 < 35 and S2 < 35 and S3 < 60 and S4 < 60 then Obs4
    Begin
      B and S ∈ {80° position for 02 sec}
      S ∈ {0° position}
    End;
    If S3 < 25 and S4 < 25 then Obs5
    Begin
      B and S ∈ {100° position for 02 sec}
      S ∈ {0° position}
    End;
    If S3 < 15 and S4 < 35 then Obs6
    Begin
      B and S ∈ {120° position for 02 sec}
      S ∈ {0° position}
    End;
    If S3 < 35 and S4 < 15 then Obs7
    Begin
      B and S ∈ {140° position for 02 sec}
      S ∈ {0° position}
    End;
    If S3 > 55 and S4 < 40 then Obs8
    Begin
      B and S ∈ {160° position for 02 sec}
      S ∈ {0° position}
    End;
    If S3 < 40 and S4 > 55 then Obs9
    Begin
      B and S ∈ {180° position for 02 sec}
      S ∈ {0° position}
    End;
  End;
  Until "Switch OFF"
End.
```

ing O1, buzzer and servo will switch ON and the user moves to his left. He then walks forward and detects another obstacle O2 and obstacle O3, then again buzzer will turn ON and also servo will move some angle based on obstacle type and the user walks avoiding going to his left according to the servo instruction. Similarly, the last object can be avoided.

Various obstacles are being detected via actuation systems. The prototype of *"Belt for Blind"* was tested at the Malaysian Association for Blind (MAB) and they have recommended our device as a great work for blind people's navigation. Figure 11 shows MAB personnel using the system in an unknown environment. The validation of the algorithm for detecting various obstacles was verified and tested. Sensors, actuators and software are integrated to convey mapping of the surrounding for smooth movement of visually impaired person. Experiments conducted both with blind folded person and by birth blind person are conducted successfully to validate the functionality of the prototype of the walking support system (*Belt for Blind*) designed and developed in this research.

CONCLUSION

A new Walking Support System for the visually impaired people named as *"Belt for Blind"* is developed for detecting information about terrain where the environment consists of various obstacles such as stair, hole and so on. Algorithms are developed through extensive experimentations that are able to differentiate different obstacles around the walkway of a blind person. The walking support system is incorporated with actuation system that is able to convey interpretation of obstacles around the user successfully. The prototype of the system is successfully tested on a blind user with minimum training. The system passed the ultimate test of real-time implementation on a prototype. In addition to fulfilling the objectives, certain characteristics about the structure were also ascertained as expected; the device is equipped with a servo motor and a buzzer to generate outputs that inform the user about the type of the obstacle ahead. The configurations calculated at the Modeling and experimental stages were validated through the performance of the prototype. The successful execution of the sensing signal also affirmed the suitability of the control algorithm. This device is limited to standard pace of mobility of the user. It requires making more sophisticated system so that user can walk at a pace of a normal human. The used sensors sometimes are affected with temperature and humidity and also its accuracy and precision is not enough to get proper data. It is my recommendation that use more filter to eliminate all possible noise from the system. For complete walking support system, one should plan for more sensors to further enhance mapping capability of the system. In such case it is expected number obstacle scenario will increase significantly, which would need alpha numeric brail representations.

REFERENCES

Aguerrevere, D., Choudhury, M., & Barreto, A. (2004). *Portable 3D sound/sonar navigation system for blind individuals.* Paper presented at the 2nd LACCEI International Latin American Caribbean Conference on Engineering Technology, Miami, FL.

Akopoulos, D., Boddhu, S. K., & Bourbakis, N. (2007). A 2D vibration array as an assistive device for visually impaired. In *Proceedings of the 7th IEEE International Conference on Bioinformatics and Bioengineering,* Boston, MA (Vol. 1, pp. 930-937).

Audette, R., Balthazaar, J., Dunk, C., & Zelek, J. (2000). *A stereo-vision system for the visually impaired.* Guelph, ON, Canada: University of Guelph.

Belote, L. (2006). *Low vision education and training: Defining the boundaries of low vision patients. A personal guide to the VA Visual Impairment Services Program.* San Francisco, CA: Visual Impairment Service Team.

Benham, T., & Benjamin, J. (1963). Active energy radiating systems: An electronic travel aid. In *Proceedings of the International Congress on Technology and Blindness* (pp. 167-176).

Benjamen, J. M., Ali, N. A., & Schepis, A. F. (1973). A laser cane for the blind. *Proceedings of the San Diego Biomedical Symposium, 12,* 53–57.

Beurle, R. L. (1951). Electronic guiding aids for blind people. *The British Journal of Psychology, 42*(1-2), 164–171.

Blasch, B. B., Long, R. G., & Griffin-Shirely, N. (1999). National evaluation of electronic travel aids for blind and virtually impaired individuals: Implications for design. In *Proceedings of the 12th Annual Conference Rehabilitation Engineering Society of North America,* New Orleans, LA (pp. 133-134).

Blasch, B. B., Wiener, W. R., & Welsh, W. R. (1997). *Foundations of orientation and mobility* (2nd ed.). New York, NY: AFB Press.

Borenstein, J., & Koren, Y. (1985). Error eliminating rapid ultrasonic firing for mobile robot obstacle avoidance. *IEEE Transactions on Robotics and Automation, 11*(1), 132–138. doi:10.1109/70.345945.

Borenstein, J., & Koren, Y. (1985). Obstacle avoidance with ultrasonic sensors. *IEEE Journal on Robotics and Automation, 4*(2).

Borenstein, J., & Koren, Y. (1991). The vector field histogram – Fast obstacle avoidance for mobile robots. *IEEE Journal on Robotics and Automation, 7*(3), 278–288. doi:10.1109/70.88137.

Borenstein, J., & Ulrich, I. (1997). The GuideCane – A computerized travel aid for the active guidance of blind pedestrians. In *Proceedings of the IEEE International Conference on Robotics and Automation*, Albuquerque, NM (pp. 1283-1288).

Bourbakis, N. G., & Kavraki, D. (1996). Intelligent assistants for handicapped people's independence: Case study. In *Proceedings of the IEEE International Joint Symposium on Intelligent Systems* (pp. 337-344).

Bouzit, M., Chaibi, A., De Laurentis, K. J., & Mavroidis, C. (2004). Tactile feedback navigation handle for the visually impaired. In *Proceedings of the International Mechanical Engineering Congress and Exposition*, Anaheim, CA (pp. 13-19).

Brabin, J. A. (1982). New developments in mobility and orientation aids for the blind. *IEEE Transactions on Bio-Medical Engineering, 29*, 285–290. doi:10.1109/TBME.1982.324945.

Cardin, S., Thalmann, D., & Vexo, F. (2007). A wearable system for mobility improvement of visually impaired people. *The Visual Computer, 23*(2), 109–118. doi:10.1007/s00371-006-0032-4.

Dakopoulos, D., & Bourbakis, N. (2008). Preserving visual information in low resolution images during navigation for visually impaired. In *Proceedings of the 1st International Conference on Pervasive Technologies related to Assistive Environments*, Athens, Greece (pp. 5-19).

Davies, T. C., Burns, C. M., & Pinder, S. D. (2007). Mobility interfaces for the visually impaired: What's missing? In *Proceedings of the 7th ACM SIGCHI New Zealand Chapter's International Conference on Computer-Human Interaction: Design Centered HCI*, Hamilton, New Zealand (Vol. 254, pp. 41-47).

Dodds, A. G., Armstrong, J. D., & Shingledecker, C. A. (1981). The Nottingham obstacle detector: development and evaluation. In *Proceedings of the 7th ACM SIGCHI New Zealand Chapter's International Conference on Computer-Human Interaction: Design Centered HCI*, Hamilton, New Zealand (Vol. 75, pp. 203-209).

Easton, R. D. (1992). Inherent problems of attempts to apply sonar and vibrotactile sensory aid technology to the perceptual needs of the blind. *Optometry and Vision Science, 69*(1), 3–14. doi:10.1097/00006324-199201000-00002.

González-Mora, J. L., Rodríguez-Hernández, A., Rodríguez-Ramos, L. F., Díaz-Saco, L., & Sosa, N. (2009). Development of a new space perception system for blind people, based on the creation of a virtual acoustic space. In J. Mira & J. V. Sánchez-Andrés (Eds.), *Proceedings of the International Work-Conference on Engineering Applications of Bio-Inspired Artificial Neural Networks* (LNCS 1607, pp. 321-330).

Hub, A., Diepstraten, J., & Ertl, T. (2004). Design and development of an indoor navigation and object identification system for the blind. In *Proceedings of the 6th International ACM SIGACCESS Conference on Computers and Accessibility* (pp. 147-152).

Ifukube, T., Sasaki, T., & Peng, C. (1991). A blind mobility aid modeled after echolocation of bats. *IEEE Transactions on Bio-Medical Engineering, 38*(5), 461–465. doi:10.1109/10.81565.

Ito, K., Okamoto, M., Akita, J., Ono, T., Gyobu, I., Tagaki, T., et al. (2005). CyARM: An alternative aid device for blind persons. In Proceedings of Extended Abstracts on Human Factors in Computing Systems, Portland, OR (pp. 1483-1488).

Jabatan Kebajikan Masyarakat (JKM). (2000). *Buletin Perangkaan Kebajikan (1999)*. Kuala Lumpur, Malaysia: Percetakan Nasional Malaysia Berhad.

Johnson, L. A., & Higgins, C. M. (2006). A navigation aid for the blind using tactile-visual sensory substitution. In *Proceedings of the IEEE Annual International Conference of the Engineering in Medicine and Biology Society* (pp. 6289-6292).

Kulyukin, V. (2005). *Robotic guide for the blind*. Logan Hill, UT: Utah State University.

Levine, S., Koren, Y., & Borenstein, J. (1990). NavChair control system for automatic assistive wheelchair navigation. In *Proceedings of the 4th International Conference of the Computer for Handicapped Persons*, Vienna, Austria.

Lofving, S. (1998). Extending the cane range using laser technique. In *Proceedings of the 9th International Orientation and Mobility Conference*.

Malaysian Association for Blind (MAB). (n. d.). *Tun Sambanthan*. Kuala Lumpur, Malaysia. *MAB*.

Malvern, B. J., Jr. (1973). The new C-5 laser cane for the blind. In *Proceedings of the Carnahan Conference on Electronic Prosthetics*, Lexington, KY (pp. 77-82).

Meijer, P. B. L. (1992). An experimental system for auditory image representations. *IEEE Transactions on Bio-Medical Engineering*, *39*(2), 112–121. doi:10.1109/10.121642.

Panda and Edie. (1999). *The guide horse program*. Retrieved from http://guidehorse.org/

Sainarayanan, G., Nagarajan, R., & Yaacob, S. (2001). Interfacing vision sensor for real-time application in MATLAB. In Proceedings of Scored, Malaysia.

Shoval, S., Borenstein, J., & Koren, Y. (1993). The NavBelt – A computerized travel aid for the blind. In *Proceedings of the Rehabilitation Engineering and Assistive Technology Society of North America Conference*, Las Vegas, NV (pp. 240-242).

Shoval, S., Borenstein, J., & Koren, Y. (1994). Mobile robot obstacle avoidance in a computerized travel aid for the blind. In *Proceedings of the IEEE Conference on Robotics and Automation*, San Diego, CA (pp. 2023-2029).

Shoval, S., Borenstrin, J., & Koren, Y. (1998). The NavBelt – A computerized travel aid for the blind based on mobile robotics technology. *IEEE Transactions on Bio-Medical Engineering*, *45*(11), 1376–1386. doi:10.1109/10.725334.

Ulrich, I., & Borenstein, J. (2001). The guidecane – applying mobile robot technologies to assist the visually impaired people. *IEEE Transactions on Systems, Man, and Cybernetics. Part A, Systems and Humans*, *31*(2), 131–136. doi:10.1109/3468.911370.

Van den Doel, K., Smilek, D., Bodnar, A., Chita, C., Corbett, R., Nekrasovski, D., & McGrenere, J. (2004). Geometric shape detection with sound view. In *Proceedings of the Tenth Meeting of the International Conference on Auditory Display*, Sydney, Australia (pp. 1-8).

Wong, F., Nagarajan, R., Yaacob, S., Chekima, A., & Belkhamza, N.-E. (2001). An image segmentation method using fuzzy – based threshold. In *Proceedings of the 6th International Symposium on Signal Processing and its Applications*, Kuala Lumpur, Malaysia (pp. 144-147).

World Health Organization (WHO). (1997). *Blindness and visual disability, part i of vii: General information*. Geneva, Switzerland: WHO.

Yuan, D., & Manduchi, R. (2004). A tool for range sensing and environment discovery for the blind. In. *Proceedings of the Conference on Computer Vision and Pattern Recognition, 3*, 39.

This work was previously published in the International Journal of Intelligent Mechatronics and Robotics, Volume 1, Issue 3, edited by Bijan Shirinzadeh, pp. 46-62, copyright 2011 by IGI Publishing (an imprint of IGI Global).

Chapter 14
Which is Better?
A Natural or an Artificial Surefooted Gait for Hexapods

Kazi Mostafa
National Sun Yat-sen University, Taiwan

Innchyn Her
National Sun Yat-sen University, Taiwan

Jonathan M. Her
National Taiwan University, Taiwan

ABSTRACT

Natural multiped gaits are believed to evolve from countless generations of natural selection. However, do they also prove to be better choices for walking machines? This paper compares two surefooted gaits, one natural and the other artificial, for six-legged animals or robots. In these gaits four legs are used to support the body, enabling greater stability and tolerance for faults. A standardized hexapod model was carefully examined as it moved in arbitrary directions. The study also introduced a new factor in addition to the traditional stability margin criterion to evaluate the equilibrium of such gaits. Contrary to the common belief that natural gaits would always provide better stability during locomotion, these results show that the artificial gait is superior to the natural gait when moving transversely in precarious conditions.

INTRODUCTION

The topics of bio-mimetic robots have attracted much attention lately. There have been a variety of man-made bio-mimetic walking machines so far (Wang, 2010). Among them the six-legged types are popular designs, each often emulating some kind of crawling (McGhee & Iswandhi, 1979; Brooks, 1989), or running (Clark & Cutkosky, 2006), or even climbing insect (Palmer, Diller, & Quinn, 2009; Spenko, Haynes, Saunders, Cutkosky, Rizzi, Full, & Koditschek, 2008). Ad-

DOI: 10.4018/978-1-4666-3634-7.ch014

vantages for a robot to use the hexapod configuration (Figure 1A and 1B) are satisfactory walking efficiency and at the same time static stability. Hexapod gaits are much diversified, i.e., having a larger number of patterns than biped, quadruped or even myriapod gaits. A robot designer has to know which gait is the best for a specific use of the hexapod (Chu & Pang, 2002; Srinivasan & Ruina, 2006; Erden & Leblebicioglu, 2007; Starke, Robilliard, Weller, Wilson, & Pfau, 2009).

Natural Gaits

There are apparently two broad categories of multiped gaits: natural and artificial gaits. Natural gaits (Figure 1C) are those used by living creatures. For instance, Wilson (1966) has reported in his classic paper that the most common gaits for hexapods are the slow wave gait, the ripple gait and the tripod gait. Note that Wilson's ripple gait was also referred to as the metachronal gait (Ferrell, 1993; Schreiner, 2004). Song and Waldron (1987) have developed a mathematical formula to theorize the natural gaits. This is possible since in a natural gait the legs on either side always move successively from the rear end to the front end of the animal (Wilson, 1966). Recently, some researchers have studied the metachronal gait of a stinkbug *Mesocerus marginatus* (Frantsevicha & Cruse, 2006) and that of a stick insect *Aretaon asperrimus* (Jeck & Cruse, 2007). As to transversely moving animals, the walking, trotting, and galloping gaits of the ghost crab have been investigated thoroughly (Full & Weinstein, 1992). However, the crab used an alternating tetrapod gait when it walks and trots, and a dynamically stable gait when it gallops with aerial phases.

Figure 1. (a) A terrestrial hexapod model (P, Q, U are dimension parameters, L1, L2, L3, R1, R2, R3 are the legs, and the star denote CG.), (b) an inverted, hanging posture of the hexapod, (c) a natural surefooted gait: the type-1 paired metachronal gait, showing pairing sequence (steps M_I, M_{II}, and M_{III}) of the legs, (d) the artificial diametric gait and its pairing sequence (steps D_I, D_{II}, D_{III})

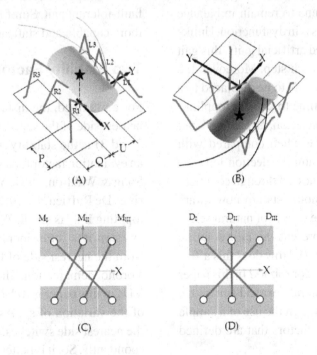

Artificial Gaits

On the contrary, artificial gaits (Figure 1D) are created in the laboratory, only to be used by walking machines. The leg sequences of some artificial gaits, being too different from the natural gaits, cannot be expressed by Song and Waldron's gait formula (Wang, 2010). The first artificial gait was a simulated wave gait developed by McGhee and Frank (1968) on a quadruped robot that crawled along a straight line. Shortly after that, a similar artificial gait for the hexapod machine was also developed (Bessonov & Umnov, 1973). As for the need for robots to conquer uneven terrain and obstacles, there have been the adaptive gait (McGhee & Iswandhi, 1979) and the FTL (follow-the-leader) gait (Ozguner, Tsai, & McGhee, 1984), which was pioneering obstacle-avoiding gaits. Recently, some non-periodic open-loop gaits were also proposed for climbing hexapods (Spenko, Haynes, Saunders, Cutkosky, Rizzi, Full, & Koditschek, 2008). These gaits, which require real-time feedbacks and much computing power, are obviously for high-end robotic products. Yet in a less sophisticated form but still interestingly, Yang and Kim (2000) have designed a periodic fault-tolerant gait, which allows a hexapod machine to remain in balance even when one of the legs is in dysfunction. Unlike all previously-mentioned artificial gaits, this gait moves a pair of legs in each step. Moreover, the center of gravity (CG) needs to be displaced to a specific location every time to prevent toppling.

By traditional interpretation, an artificial gait does not stand a chance when compared with a natural gait, for the natural selection tends to favor what has been winnowed through countless generation. However, biologists are now aware of the possibility that the results of natural selection not necessarily represent the *best* solution (Campbell & Reece, 2007). How does this affect our gait selection process for robots? In this paper we compare a typical natural hexapod gait with a typical artificial gait, through the use of a simple model and two stability factors that are defined in the following section.

MODELING AND STABILITY FACTORS

The Hexapod Computer Model

The standardized hexapod model we used is shown in Figure 1A. This symmetrical, rectangular model was actually suggested by Yang (Yang & Kim, 2000; Yang, 2009). Three legs are featured at each side, labeled as Li and Ri (i=1 to 3). L_1 and R_1 are the leading legs as the robot moves forward. P and Q define the size of the allocated reachable region of a leg, and U is the width of the body. In this paper, we simply let P=Q=U=1. The mass of the legs is small, and the CG (denoted as a star) is at the centroid of the body. In Figure 1C and 1D, we show the steady-state leg sequences of the two gaits to be discussed. Each gait cycle contains three steps. Steps M_I, M_{II}, and M_{III} constitute the type-1 paired metachronal gait, and steps D_I, D_{II}, D_{III}, the diametric gait. A step begins when one pair of legs leave the ground, and ends when they once again touch the ground. Throughout this paper, a factor C_{gc} is used to represent the percentage of completion of a full surefooted gait cycle. The concept of surefooted gait is not similar as fault-tolerant gait. Surefooted gait designed so as more capable and statically stable gait.

Two Stability Factors

For a multiped, the distance from the CG to the nearest side of the support polygon (SP) (Figures 2 and 3) is the stability margin s_m, and has been an evaluation factor for a gait (Chu & Pang, 2002; Song & Waldron, 1987; Yang & Kim, 2000; Harris & De Ruffieu, 1993). The hexapod is prone to toppling if s_m is small. When a hexapod starts in any step, its s_m often increases, as CG moves away from the nearest side of the SP until it reaches a Voronoi demarcation. The main purpose of using VDs in this paper is to facilitate the visualization of the variation of s_m. As the CG crosses a VD, the nearest side switches, and s_m decreases correspondingly. So, it is better to have a VD crossed at the middle of a step, since the variation of within the step will be smaller. In this way VD can be a useful reference for a robot designer. Note that for s_m, we watch particularly the minimum value within the whole cycle. This is when the hexapod becomes most susceptible to imbalance. Thus, the gait with a larger minimum s_m is better.

Besides the general-purpose stability margin s_m, in this paper we propose another stability criterion that is especially significant for precarious walking conditions. It is a measurement of the fault-tolerant property of a surefooted gait. We considered the distance from the CG to the intersection of the diagonals of the SP in every

step. That intersection is shown as a small box in Figures 2 and 3 and is called the critical point. As Yang and Kim designed their anti-toppling gait, they required that the CG of the robot always be moved to that point (Yang & Kim, 2000). If the CG is close to the critical point, the hexapod may less likely encounter a great tipping moment when it loses a foothold. We refer to this distance as the error margin e_m and used it to signify the unbalanced torque should the hexapod lose a foothold. However, it is the maximum values of e_m that are used in the evaluations to determine which gait is more stable.

Figure 2. Steps of the steady-state type-1 paired metachronal gait walking on a 45° line, where (a) corresponds to step M_I in Figure 1C, (b) step M_{II}, and (c) step M_{III}

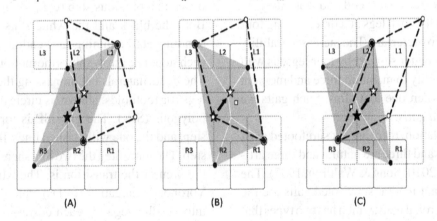

Figure 3. Steps of the steady-state diametric gait walking on a 45° line, where (a) corresponds to step D_I in Figure 1D, (b) step D_{II}, and (c) step D_{III}

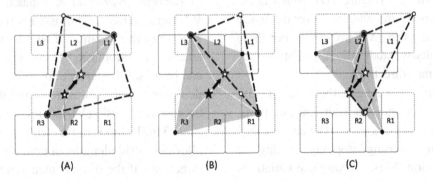

RESULTS

The Surefooted Gaits

This paper studies surefooted gaits. Proposed surefooted gaits statically stable because when it walks the center of gravity lies within its support polygon. Thus, they are statically stable, periodic (less computer-intensive than the open-loop gaits), hexapod (not like the gait used by ghost crabs) gaits. They are ideal for robots under difficult conditions. Consider Figure 1B, a hexapod crawling on the ceiling in an upside down posture. Most hexapod animals can move easily over non-level and even inverted surfaces. For hexapod robots, however, in a precarious situation like this, the fast, commonly-used tripod gait is not suitable. The failure of only one leg is doomed to disintegrate the equilibrium of the entire machine. Now a compromise between speed and stability is to use a gait that lifts two legs at a time, leaving four legs clinging to the terrain. The increased stability results from a better support by four legs, as well as from the ability to maintain force and moment equilibrium when one foot slips. Such gaits we call *surefooted gaits*.

By permutation, there are six surefooted gaits for hexapods, and three are natural and three artificial (Wang, 2010; Song & Waldron, 1987). The most significant natural surefooted gaits are the paired metachronal gaits, which has two types that are mirror images with respect to the Y-axis. Only one is discussed in detail here (Figure 1C). As for the artificial gait, we were particularly interested in the diametric gait (Figure 1D), which is the artificial hexapod surefooted gait ever discussed in the literature (Yang & Kim, 2000). For these two chosen gaits, we studied their stability as they move in arbitrary directions.

Even within a specific gait, there are a steady-state gait sequence and some transient variations. Transient gait sequences are used when an animal or robot is in a starting, stopping, turning, or curving condition. These fleeting gait variations are not discussed in this paper. On the other hand, the steady-state sequence represents the lasting, unique and recurring steps of the gait. In our analysis, we also let the hexapod model walk along a straight line with a constant speed and the contact between a leg and the ground is a point.

Planning of the Gaits

We detail these gaits moving in an oblique (45°) direction, which is also known as crab angle in Figures 2 and 3. For example, Figure 2A is step M_I, corresponding to the gait completion factor C_{gc} varying from 0% to 33% in the type-1 paired metachronal gait. The three boxes in solid gray lines on each side are leg workspaces when a step starts, and dotted gray boxes signify workspaces when a step ends. The solid boxes in a step (e.g., step M_{II} in Figure 2B) correspond to the dashed boxes in a previous step (e.g., M_I in Figure 2A). Both the black and the white stars are the CG, denoting its beginning and ending positions in each step. The arrow shows the motion of the body. The quadrilaterals encompassing the supporting legs (or footholds, shown as circles) are support polygons (SPs). The gray SP is for the current step, and the one with dashed lines, the following step. The more area the two SPs share in common, the securer the transition is. The white lines are Voronoi demarcations (VDs) which divide an SP into smaller regions, each corresponding to the area nearest to a side. Such VDs are constructed that any point on them is equidistant to the two nearest sides, and they are used in many branches of sciences (Kawasaki & Tanaka, 2010). So, it is better to have a VD crossed at the middle of a step, since the variation of within the step will be smaller. We introduced VD in this paper and which can be a useful reference for a robot designer. A designer can able to estimate the distance from nearest side by using VD as a reference. In Figures 2 and 3 the hexapod moves with the same greatest allowable stride lengths in both the X and the Y directions. If the displacements in the two axial directions are the same, the hexapod then walks along a t=+45° line, where (t) is the direction angle, measuring counterclockwise from the +X axis. Therefore, by adjusting the stride lengths in the two axial directions, we can direct the animal or robot to walk in any specific angle.

Let's compare Figure 1C with Figure 2. The M_I pair (legs R1-L3) is presently the lifted legs in Figure 2A; no footprints are shown in the solid gray boxes for these two legs. However, at the end of this M_I step, they will touch the ground at the farthest ahead positions, depicted as white circles in the corresponding workspace boxes (Figure 2A). As to the M_{II} pair (L2, R3), the legs will be at the farthest behind positions of their boxes at the end of this M_I step (dotted boxes in Figure 2A). Such legs are shown as black dots. Lastly, the M_{III} pair (L1, R2) has just moved to the farthest ahead positions of their boxes at the beginning of the M_I step. They are shown as double circles. Similar representation is used for M_{II} and M_{III} steps and the diametric steps D_I, D_{II}, D_{III} in Figure 3.

Walking in Normal and Inverted Surfaces

Consider again the two factors used to evaluate the gaits: the stability margin s_m and the error margin e_m. The stability margin s_m is a traditional measurement of the equilibrium of a legged robot under normal conditions. On the other hand, the new error margin e_m is for slippery, precarious walking conditions. In normal upright walking conditions (Figure 1A), each foot of a hexapod machine experiences a vertical compressional reaction force from the ground, so there is no need to use claws or suction cups (Palmer, Diller, & Quinn, 2009; Ferrell, 1993). However, when a foot slips or buckles, CG may fall outside the support polygon. In order to make the torques remain in equilibrium, the reaction force of a foot needs to become tensional. In that case a claw or suction cup has to be used to provide the tension, which is proportional to the error margin e_m. On

the other hand, when a hexapod is in a hanging position (Figure 1B), all reaction forces of the feet are at first tensional. If a foot loses the grip of its foothold and imbalance occurs, one of the reaction forces needs to become compressional. This is equally, if not more, dangerous since this compression force must be counterbalanced by increasing the tension of other feet. If the error margin e_m is sufficiently large, all feet can no longer hold and the entire hexapod creature or machine may fall.

Evaluating the Gaits

Figure 4 depicts the stability margin s_m and the error margin e_m for a hexapod that moves in an arbitrary direction (t) using the type-1 paired metachronal gait and the diametric gait. As shown in Figure 4, comprehensive results are presented in the inserted 3D graphs (marked as M for metachronal and D for diametric gait). In 3D graph, the t axis signifies the direction angle, ranging from –90° to +90°. Stability factors s_m and e_m are represented as the vertical axes. And the C_{gc} axis is the completion of the gait cycle, denoted in percentages. The three C_{gc} ranges correspond to the three steps of the gaits.

The larger 2D plots demonstrate the curves of s_m and e_m for axial directions only. Subscript M means metachronal and D means diametric. The minimum position of each s_m and the maximum of each e_m curve are labeled on the right hand side. Using s_m as a ranking criterion (where higher minimum value is favored), we find from Figure 4A that the type-1 paired metachronal gait prefers most to walk in –90° (–Y), following by 0° (X), and lastly +90° (+Y direction). As for the diametric gait, the ±Y (±90°) directions are the same, with the +X (0°) direction being somewhat inferior.

Our second evaluating criterion, e_m, is designed for hexapods when losing a foothold seems unavoidable. From Figure 4B we see that two thirds of the e_m value of the $0°_D$ curve is greater than

Figure 4. Comparison of the two surefooted gaits (M for metachronal and D for diametric) in the axial directions: (a) showing the s_m, and (b) showing the e_m (In this figure, t, sm, em, and C_{gc} in the legend are direction angle, stability margin, error margin, and normalized completion factor, respectively.)

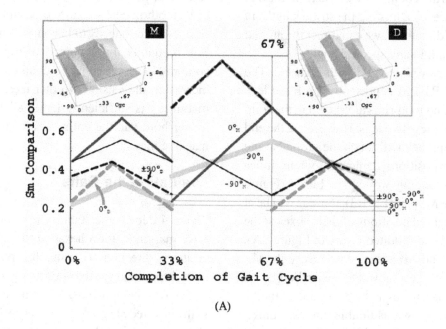

(A)

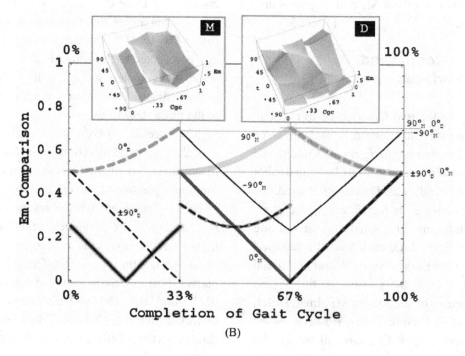

(B)

even the maximum value of the $0°_M$ curve. This reveals that when the hexapod moves in the X direction, using the diametric gait is more perilous if the hexapod loses a foothold. However, when it comes to moving in the Y direction, we observed that the greatest e_m value of the $\pm90°_D$ curve lies well below that of the $90°_M$ and the $-90°_M$ curves, meaning in this rare case, the artificial gait wins.

Veering Preference of a Gait

From Figure 4, we see that type-1 paired meta-chronal gait has a slight veering preference; turning left and turning right does not have the same stability. This is because its leg sequences are not symmetrical with respect to the X-axis. This peculiarity is like that of the canter gait of the horse. Horses using a canter gait do have a preference regarding turning left or right (Harris & De Ruffieu, 1993). Figure 4A and 4B confirm that the type-1 metachronal gait prefers the $-Y$ direction to the $+Y$ direction. As for type-2 metachronal gait, which is a perfect reflection of type-1 metachronal gait with respect to the Y-axis, we know by analogy that it would love the $+Y$ and loathe the $-Y$ direction. However, for the artificial diametric gait there is no such turning preference concerning the left or the right, since its pairing sequences are symmetrical with respect to the X-axis.

DISCUSSION

In normal conditions, when a hexapod moves primarily in the $+X$ direction, either the diametric gait or the metachronal gait is acceptable. The minimal s_m values of the $0°_M$ curve and the $0°_D$ curve are about the same (Figure 4A). Although roughly two thirds of the $0°_M$ curve is within a higher (0.45, 0.7) interval, and two thirds of the $0°_D$ curve is in a lower (0.2, 0.45) interval, the CGs of both gaits stay safely in the SPs. However, considering the possibility of losing a foothold, i.e., when e_m needs to be taken into account, things are not the same. The $0°_D$ artificial diametric gait is prone to instability. Therefore, it is recommended for a hexapod to always use the natural metachronal gait when it moves in the X direction.

When an animal or a robot moves sideways in the Y direction, the smallest s_m values of the $\pm 90°_D$ and the $-90°_M$ curves are equal, and that of the $+90°_M$ curve is the lowest, not by a wide margin, though. This suggests that in safe and secure conditions, using either the diametric gait or the metachronal gaits seems to be okay. However, if the hexapod is to walk in a difficult situation, when losing a foothold is predictable, we have to consider the trends of the error margin e_m. Figure 4B shows that the greatest e_m value of the $\pm 90°_D$ curve is well below that of the $90°_M$ and the $-90°_M$ curves. Hence the artificial diametric gait produces least imbalanced torque should the hexapod lose a foothold on a slippery terrain. As a result, the $\pm 90°$ artificial diametric gait is the better choice in this condition.

SUMMARY

This paper showed the comparison between two surefooted gaits for six-legged walking machines. A standardized hexapod model was evaluated by using computer simulation. In our study, we focused mainly on selection of gait between artificial and natural. Our simulated result and stability factors can be used easily to robots of different geometry. Moreover, we proposed two factors such as stability margin and error margin to evaluate static and dynamic capabilities of hexapod robots. In precarious walking conditions, static stability criteria are more significant then dynamic capabilities. For this kind of working environment, walking speed is less important than static stability. Thus, we emphasized more on static stability criteria to evaluate our standardized hexapod model.

By analyzing the stability in terms of two factors s_m and e_m, this paper studied whether an artificial gait could prevail under certain circumstances when compared with a natural counterpart. For most biological locomotion, our results seem only theoretical. Animal even insects rarely move along an oblique path without first turning their head and aligning their body with the direction of travel. Not to mention how difficult it is to teach them a new gait. However, for man-made walking machines, since the location of the sensing devices are not necessarily constrained in a forward-looking head (Roennau, Kerscher, Ziegenmeyer, Zollner, & Dillmann, 2009) and the functions of a robot are usually very programmable, our results provide a practical insight and quick guidance as to which gait to use particularly when sideway walking is needed in a mission.

ACKNOWLEDGMENT

We thank Messrs L. W. Liu and D. W. Wu for creating the figures and Mr. D. Li for miscellaneous assistance. Financial support by the National Science Council of Taiwan under grant NSC 98-2221-E110-023 is also acknowledged.

REFERENCES

Bessonov, A. P., & Umnov, N. V. (1973). The analysis of gaits in six-legged vehicle according to their static stability. In *Proceedings of the Symposium on Theory and Practice of Robots and Manipulators*, Udine, Italy (pp. 1-9).

Brooks, R. A. (1989). A robot that walks: Emergent behavior from a carefully evolved network. *Neural Computation, 1,* 253–262. doi:10.1162/neco.1989.1.2.253.

Campbell, N. A., & Reece, J. B. (2007). *Biology.* Upper Saddle River, NJ: Pearson.

Chu, S. K., & Pang, G. K. (2002). Comparison between different models of hexapod robot gait. *IEEE Transactions on Systems, Man, and Cybernetics. Part A, Systems and Humans, 32*(6), 752–756. doi:10.1109/TSMCA.2002.807066.

Clark, J. E., & Cutkosky, M. R. (2006). The effect of leg specialization in a biomimetic hexapedal running robot. *ASME. Journal of Dynamic Systems, Measurement, and Control, 128,* 26–35. doi:10.1115/1.2168477.

Erden, M. S., & Leblebicioglu, K. (2007). Analysis of wave gaits for energy efficiency. *Autonomous Robots, 23*(3), 213–230. doi:10.1007/s10514-007-9041-z.

Ferrell, C. (1993). *Robust agent control of an autonomous robot with many sensors and actuators.* Unpublished master's thesis, MIT, Cambridge, MA.

Frantsevicha, L. I., & Cruse, H. (2005). Leg coordination during turning on an extremely narrow substrate in a bug *Mesoceruse marginatus* (Heteroptera, Coreidae). *Journal of Insect Physiology, 51*(10), 1092–1104. doi:10.1016/j.jinsphys.2005.05.008.

Full, R. J., & Weinstein, R. B. (1992). Integrating the physiology, mechanics and behavior of rapid running ghost crabs: Slow and steady doesn't always win the race. *Integrative and Comparative Biology, 32*(3), 382–395. doi:10.1093/icb/32.3.382.

Harris, S. E., & De Ruffieu, F. L. (1993). *Horse gaits, balance and movement.* New York, NY: Howell Book House.

Jeck, T., & Cruse, H. (2007). Walking in Aretaon asperrimus. *Journal of Insect Physiology, 53*(7), 724–733. doi:10.1016/j.jinsphys.2007.03.010.

Kawasaki, T., & Tanaka, H. (2010). Formation of a crystal nucleus from liquid. *Proceedings of the National Academy of Sciences of the United States of America, 107*(32), 14036–14041. doi:10.1073/pnas.1001040107.

McGhee, R. B., & Frank, A. A. (1968). On the stability properties of quadruped creeping gaits. *Mathematical Biosciences, 3,* 331–351. doi:10.1016/0025-5564(68)90090-4.

McGhee, R. B., & Iswandhi, G. I. (1979). Adaptive locomotion of a multi-legged robot over rough terrain. *IEEE Transactions on Systems, Man, and Cybernetics*, *9*(4), 176–182. doi:10.1109/TSMC.1979.4310180.

Ozguner, F., Tsai, S. J., & McGhee, R. B. (1984). An approach to the use of terrain-preview information in rough terrain locomotion by a hexapod walking machine. *The International Journal of Robotics Research*, *3*(2), 134–146. doi:10.1177/027836498400300211.

Palmer, L. R., Diller, E. D., & Quinn, R. D. (2009). Design of a wall-climbing hexapod for advanced maneuvers. In *Proceedings of the International Conference on Intelligent Robots and Systems*, St. Louis, MO (pp. 625-630).

Roennau, A., Kerscher, T., Ziegenmeyer, M., Zöllner, J. M., & Dillmann, R. (2009). Adaptation of a six-legged walking robot to its local environment. In Kozlowski, K. R. (Ed.), *Robot motion and control* (pp. 155–164). Berlin, Germany: Springer-Verlag. doi:10.1007/978-1-84882-985-5_15.

Schreiner, J. N. (2004). *Adaptations by the locomotor systems of terrestrial and amphibious crabs walking freely on land and underwater*. Unpublished master's thesis, Louisiana State University, Eunice, LA.

Song, S. M., & Waldron, K. J. (1987). An analytical approach for gait study and its applications on wave gaits. *The International Journal of Robotics Research*, *6*(2), 60–71. doi:10.1177/027836498700600205.

Spenko, M. J., Haynes, G. C., Saunders, J. A., Cutkosky, M. R., Rizzi, A. A., Full, R. J., & Koditschek, D. E. (2008). Biologically inspired climbing with a hexapedal robot. *Journal of Field Robotics*, *25*(4-5), 223–242. doi:10.1002/rob.20238.

Srinivasan, M., & Ruina, A. (2006). Computer optimization of a minimal biped model discovers walking and running. *Nature*, *439*, 72–75. doi:10.1038/nature04113.

Starke, S. D., Robilliard, J. J., Weller, R., Wilson, A. M., & Pfau, T. (2009). Walk-run classification of symmetrical gaits in the horse: a multidimensional approach. *Journal of the Royal Society of London*, *6*(33), 335–342.

Wang, H. P. (2010). *Mathematic model and trajectory of multiped gaits*. Unpublished master's thesis, National Sun Yat-sen University, Kaohsiung, Taiwan.

Wilson, D. M. (1966). Insect walking. *Annual Review of Entomology*, *11*, 103–122. doi:10.1146/annurev.en.11.010166.000535.

Yang, J. M. (2009). Fault-tolerant gait planning for a hexapod robot walking over rough terrain. *International Journal of Intelligent & Robotic System*, *54*(4), 613–627. doi:10.1007/s10846-008-9282-x.

Yang, J. M., & Kim, J. H. (2000). A fault tolerant gait for a hexapod robot over uneven terrain. *IEEE Transactions on Systems, Man, and Cybernetics. Part B, Cybernetics*, *30*(1), 172–180. doi:10.1109/3477.826957.

This work was previously published in the International Journal of Intelligent Mechatronics and Robotics, Volume 1, Issue 3, edited by Bijan Shirinzadeh, pp. 63-72, copyright 2011 by IGI Publishing (an imprint of IGI Global).

Chapter 15
Design and Validation of Force Control Loops for a Parallel Manipulator

Giuseppe Carbone
University of Cassino, Italy

Enrique Villegas
University of Cassino, Italy

Marco Ceccarelli
University of Cassino, Italy

ABSTRACT

This paper addresses problems for design and validation of force control loops for a 3-DOF parallel manipulator in drilling applications. In particular, the control design has been investigated for a built prototype of CaPaMan2bis at LARM (Laboratory of Robotics and Mechatronics of Cassino). Two control loops have been developed, each one with two types of controllers. The first one is a Constrained Control Loop, which limits the force that is applied to an object to stay below a given value. The second one is a Standard Control Loop with external force feedback, which keeps the force at a given value. The control loops have been implemented on CaPaMan2bis by a Virtual Instrument in LABVIEW Software. CaPaMan2bis has been attached to a serial robot to make dynamic tests. The results of the experimental tests show the effectiveness and quick response of both algorithms after a careful calibration process.

1. INTRODUCTION

Parallel manipulators have been used for many applications because of their better characteristics with respect to those of serial manipulators, such as high stiffness, high payload, high precision and high velocity and acceleration (Merlet, 2006). For these reasons parallel manipulators are suitable for applications such as manipulation, packing, assembly and disassembly processes, motion simulation and milling machines. Some examples are Delta Robot, Orthoglide, Stewart Platform, Tricept and

DOI: 10.4018/978-1-4666-3634-7.ch015

3-UPU (Merlet, 2006, 2009; Clavel, 1988; Bonev, 2009 Wenger & Chablat, 2002; Stewart, 1965; Gosselin, 1988; Zabalsa & Ros, 2007).

At LARM: Laboratory of Robotics and Mechatronics, a three D.O.F (degrees of freedom) spatial parallel manipulator named as CaPaMan has been built and studied (Ceccarelli, 1997). For a first version several analysis and tests have been studied and reported (Ceccarelli, 1997, 1998; Ceccarelli & Decio, 1999; Mendes & Ceccarelli, 2001; Ottaviano & Ceccarelli, 2002; Ceccarelli, Decio, & Jimenez, 2002; Wolf, Ottaviano, Shoham, M., & Ceccarelli, 2004; Ottaviano & Ceccarelli, 2006). Later two prototypes have been built named CaPaMan2 and CaPaMan2bis which also have been studied and reported (Aguirre, Acevedo, Carbone, & Ottaviano, 2003; Carbone, Ceccarelli, Ottaviano, Checcacci, Frisoli, Avizzano, & Bergamasco, 2003; Carbone & Ceccarelli, 2005a, 2005b; Hernández-Martínez, Ceccarelli, Carbone, & López-Cajún, 2008). In addition CaPaMan2bis has been implemented as a part of a hybrid robotic architecture for surgical tasks (Carbone & Ceccarelli, 2005b) as a trunk module in a humanoid robot design that has been named as CALUMA (CAssino Low-cost hUMAnoid robot) (Nava, Carbone, & Ceccarelli, 2006) and as an intelligent wrist for milling applications (Briones-Leon, Carbone, & Ceccarelli, 2009).

This paper proposes a further development for a force control for the manipulator robot CaPaMan2bis as a terminal tool for a serial parallel robotic architecture for drilling applications. A force-feedback control for parallel manipulators is discussed in Merlet (1988), two other reference methods of force control are presented in Erlbacher (2000), another one has been applied for the LARM Hand III in Iannone, Carbone, and Ceccarelli (2008) and several methods are explained in Yiu (2002).

Two force control loops has been developed specifically for CaPaMan2bis and its application in drilling operations. The first one is a constrained control loop that actuates when the force is higher than the desired and the second is a Standard control loop that actuates at all times taking the force to the desired level. For each control loop PD and PID controllers have been implemented and experimentally tested.

2. A SERIAL-PARALLEL MANIPULATOR AT LARM

The experimental setup at LARM is composed by the serial robot SCARA Adept Cobra and the parallel robot CaPaMan2bis as shown in Figure 1. The serial robot is a SCARA Adept Cobra

Figure 1. a) SCARA adept cobra adapted

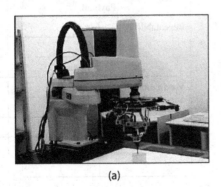
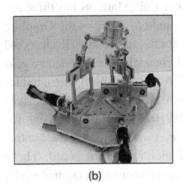

(a)   (b)

Adapted (Figure 2) (Adept Inc., 2009). We can see the characteristic of this robot in Table 1. For the application presented in this article it was used just the first two joints that allowed a movements in the horizontal plane. The prototype of CaPaMan2bis showed in Figure 2 is composed of a movable platform MP, which is connected to a fixed base FP by means of three leg mechanisms. Each leg mechanism is composed of an articulated parallelogram AP whose coupler carries a revolute joint RJ, a connecting bar CB that transmits the motion from AP to MP through RJ, and a spherical joint BJ, which is installed on MP at point J. The revolute joint RJ, which is installed on the coupler of AP has the rotation axis coinciding with the parallelogram plane.

Each leg mechanism is rotated of $\pi/3$ with respect to the neighboring one so that the leg planes lie along two vertices of an equilateral triangle, giving symmetry properties to the mechanism. Design parameters of the mechanism are depicted in Figure 3.a, and they can be identified for the k-th leg mechanism (k=1, 2, 3) as: a_k is the length of the frame link; b_k is the length of input crank; c_k is the length of the coupler link; d_k is the length of follower crank and h_k is the length of connecting bar CB. The size of the movable platform MP and fixed base FP is given by the distance r_p and r_f as depicted in Figure 3.a. Points H and O are the center points of MP and FP, respectively. Points O_k is the middle point of the frame link a_k, J_k is the connecting points between the k-th leg mechanism and platform MP.

The manipulator CaPaMan2bis has three actuators MAXON motors RE25 with Graphite Brushes. Each motor has an encoder HEDL-5540 of 500 ppr with 3 channels and a planetary gear head MAXON 32 A (0.75-4.5 Nm) with a reduction ratio of 1:66. It also has three servo-drives MAXON ADS 50/10, Model 4-Q-DC, one for each motor, that are used to amplify the control signal of the NI motion control board. These amplifiers work in current or torque control mode.

The UMI accessory is use to make the connections of the amplifiers and encoders to the motion control board. The control signal for the motors is perform by a PCI motion control board that sends a signal from -10V to 10 V depending on the spin direction and can control 4 axes at the same time with a PID control loop. It also acquires the currents from each motor to determine the position of each leg. A Measurement & Automation Explorer (MAX) software is use to tune up the motors with an auto-tuning tool that also estimated the PID parameters to the close loop position control. The PID parameters are: Kp=76, Kd = 224 and Ki = 35.

Figure 2. CaPaMan2bis prototype

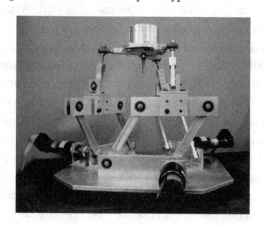

Table 1. General characteristics of SCARA adept cobra adapted

Payload		5.5kg
Repeatability	(x,y)	±0.020 mm
	(z)	±0.01 mm
	Theta	±0.03°
Maximum Joint Speed	Joint 1	360°/sec.
	Joint 2	672°/sec.
	Joint 3	1,100 mm/sec.
Weight	Joint 4	1200°/sec.
		34 kg

Figure 3. Force control loops: a) constrained force control loop; b) standard force control loop

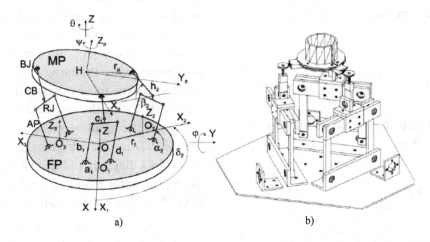

a)   b)

The estimation of these parameters is obtained by inducing little impulse signals in the system, and then using the controlled system's frequency response to design the PID parameters values.

As regards of the sensor equipment a load cell was used to measure the force applied to the object (Figure 4). This type of sensor was used because of its small size and low cost, as reported in LAUMAS (2009).

3. PROPOSED CONTROL LOOPS

The way to control the force applied to the end effector is adding a external force feedback to the position control loop through a measurement of the force, to accomplish this, two control loops were develop. Namely the two control loops were made for force control; one is a constrained control loop and the other a Standard control loop as shown in Figure 5(Villegas, 2009).

The first one has the objective of limit the force so this one doesn´t surpass a given desired force, so if the actual force is lower than the desired, the force control doesn´t react. The force control loop with a PD controller is shown in Figure 1.a. $F_d(t)$ stands for the desired force and $F_{out}(t)$ stands for the current force measure for the pressure sensor. The force error $\varepsilon_F(t)$ is obtained as:

$$E_f(t)=F_{out}(t)-F_d(t) \tag{1}$$

Figure 4. The used load cell: a) picture; b) size

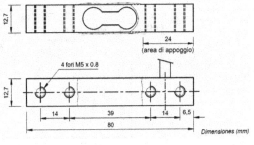

1   A)   B)

As we want the control to react just when $F_{out}(t) \geq F_d(t)$ the true force error will be:

$$e_F(t) = \begin{cases} \varepsilon_F(t) & if\ F_{out}(t) \geq F_d(t) \\ 0 & if\ F_{out}(t) < F_d(t) \end{cases} \quad (2)$$

which is the input to the PD force controller. The force control signal:

$$u_F(t) = Kpe_F(t) + Kd\frac{\Delta e_F(t)}{\Delta t} \quad (3)$$

will be added with the desired position $P_d(t)$ and subtract with the current position, obtaining the position error $e_p(t)$. This will be the input to the PID position controller to originate the position control signal $u_p(t)$ that will be the input to the motors that will move the CaPaMan2bis.

The second control loop is shown in Figure 5.b. The objective of this control is to reach the desired force limiting the position to not surpass the desired one, we can noticed that now the true force error $e_F(t)$ is the same as the previous force error $\varepsilon_F(t)$ so the control will be present in the whole process and when the current position is equal to the desired, the process will end. The notations for Figure 5.a are the same as for Figure 5.b.

Also both control loop have been implemented with a PID force controller. This one uses the same block diagrams of Figure 5.a and Figure 5.b but replacing the PD force controller with a PID controller. This way the control signal will be:

$$U_F(t) = Kpe_F(t) + Kd\frac{\Delta e_F(t)}{\Delta t} + Ki\int_0^t e_F(\tau)d\tau \quad (4)$$

The mathematical and identify models of the robot CaPaMan2bis have been difficult to find because of the complexity of the revolute joins and the dynamic variables of the system. For this matter using mathematical methods to find the

Figure 5. Force control loops: a) constrained force control loop; b) standard force control loop

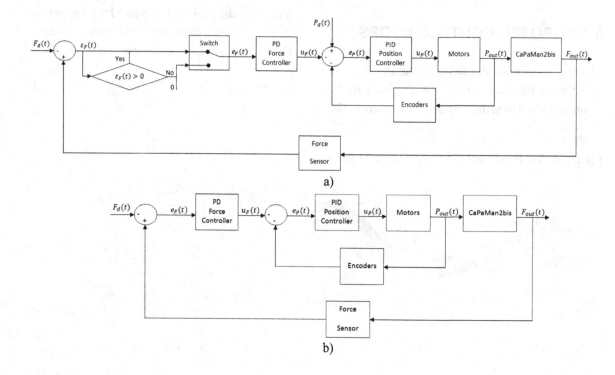

control parameters is not possible. The design of the control parameters have to be done using a manual and Ziegler-Nichols methods.

4. PROGRAMMING AND IMPLEMENTATION OF THE CONTROL LOOPS

The proposed control loops for the CaPaMan2bis have been implemented with LabVIEW system software (National Instruments, 2000a, 2000b). The used of this instrument permits the easy programming and also easy understanding for people that are not familiar with this software. The instrument consists of two parts. The front panel and the block diagram. The front panel gives an interface to the user and permits in general terms, depending on the program, to control the start/stop of the motion, to change the input parameters and to watch some plots and results as the system is working. The block diagram is made of graphical instructions in the form of blocks that permit to move the actuators and acquire data from the sensors. The blocks also read the input parameters given by the user in the front panel and link the plots with the sensors to see how the system is reacting. Some of the blocks are show in Figure 6. In particular, block a is used to initialize the control board with the parameters of MAX software, block b is used to reset the position of and axis to a value or to 0 as default, block c is used to configure and vector space with the axis that is going to used, block d and e are used to storage

the velocity and acceleration of the motion, blocks f is used to smooth the acceleration portions of the motion, block g is used to choose the operation mode of the motion, since this one could be absolute or relative, block h is used to load a target position of all the axis in a vector space, block i is used to start the motion with the lasted store parameters, block j is used to read the position of all the axis in a vector space, block k is used to handler the motion error and sends a message of the type of error that occurred, block m is used to acquired and external analogical signal for example an analogical sensor signal.

4.1. Virtual Instrument for Constrained Control Loop

The front panel of the software implemented for the Constrained Control Loop is presented in Figure 7. The virtual instrument is compound of three phases, showed in Figure 8. The block diagram using LABVIEW is made up of a Sequence Structure that allows the execution of the three independents phases that are called Frames. The block diagram of the virtual instrument for the Constrained Control Loop is showed in Appendix A. Running the instrument in the first phase (frame cero) the legs of the CaPaMan2bis will rotate -30 degrees, reaching with this the higher point for the drill tool. In the second phase (frame one) the drilling tool will drill and the CaPaMan2bis will go down towards the desired position which is set in the front panel. If the force acquired by the sensor is higher than the desired the force control will start

Figure 6. a) Initialize controller; b) reset axis position; c) configure vector space; d) load velocity in RPM; e) load acceleration in RPM/sec; f) load s-curve time; g) set operation mode; h) load vector space position; i) start motion; j) read vector space position; k) motion error handler; l) stop motion; m) read ADC channel

a)   b)   c)   d)   e)   f)   g)   h)   i)   j)   k)   l)   m)

to act until this force is diminished and less than the desired one, this way constraining the force to be less than a value. When the CaPaMan2bis reach the desired position the second phase will end and the in the third phase (frame two) will return to the higher point and the process will end.

The selection of the parameters is achieved using the front panel controls. We have two numerical controls where we can change the desired position of the three legs expressed in degrees and the desired force expressed in Newton. Also the front panel has one boolean indicator and two boolean controls. The first one is true when the CaPaMan2bis reach the higher position to indicate that the system is ready to start drilling, one of the boolean controls is used to start the movement of the CaPaMan2bis to drill and the other one is used to stop the process at any time. The front panel is also compound of four numerical indicator, that show us the current position of each leg and the current force, and eight diagrams that show us the position in degrees of each leg, the force applied to the force sensor, the true force error $e_F(t)$, the force control $u_F(t)$, the force error $\varepsilon_F(t)$ and the force error percentage define by the following equation:

$$e_F\%(t) = \frac{e_F(t)}{F_d(t)} x100 \qquad (5)$$

4.2. Virtual Instrument for Standard Control Loop

The front panel presented for the Standard Control Loop is the same one presented for the Constrained Control Loop in Figure 7. This is because between them the only matter that changes is the time in which the control is acting. We will see some changing in the second frame of the block diagram of the two control loops. Running the instrument, as before, the legs of the CaPaMan2bis will rotate -30 degrees, reaching with this the higher point for the drill tool. Later the drilling tool will drill and the CaPaMan2bis will go down towards the desired position which is set in the front panel. The force control will start to act since the beginning of the process, always trying to reach the desired position as the Standard control loops work. When the CaPaMan2bis reach the desired position it will return to the higher point and the process will end performing all the phases just as the constrained control loop.

Figure 7. Zoom view of the front panel of the constrained and standard control loops

Figure 8. Phases of the constrained and standard control loops

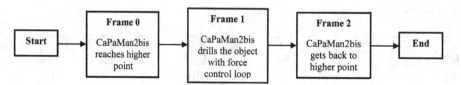

Figure 9. 1) CaPaMan2bis; 2) UMI 7764 NI accessory and servoamplifiers; 3) power supply 24 volt; 4) load cell; 5) power supply 5 V

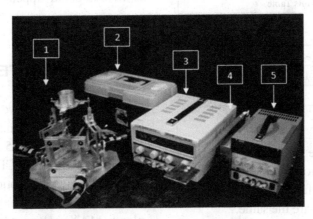

Figure 10. A scheme of the experimental setup in Figure 9

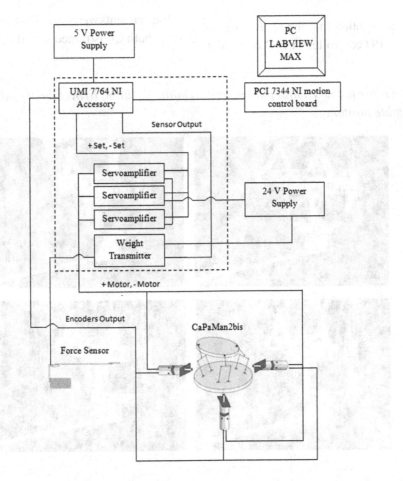

Table 2. Ziegler-Nichols table

Ziegler-Nichols Table			
Control Type	Kp	Ki	Kd
P	0.50Kc	-	-
PI	0.45Kc	1.2Kp/Pc	-
PID	0.60Kc	2Kp/Pc	KpPc/8

The description of the front panel it has been already explained. It is worth to mention that for this case the display of the force error $\varepsilon_F(t)$ and the true force error $e_F(t)$, are the same.

The phases of this virtual instrument are the same as for the previous one, for the same reason explained before; we can see the three phases in Figure 8. The block diagram of the virtual instrument for the Standard Control Loop is showed in Appendix B.

It is worth to mention that the front panel in Figure 7 is for a PD controller. A PID controller

was also developed adding the integral part and is more in detail Appendix C.

5. RESULTS OF EXPERIMENTAL TESTS

Experimental Tests have been carried out using the prototype CaPaMan2bis, a Load Cell, three MAXON motors RE25, three MAXON servo amplifiers 4-Q-DC, a PCI 7344 NI motion control board, an Universal Motion Unit (UMI) 7764 NI accessory, the Measurement & Automation Explorer (MAX) software from NI and LABVIEW system. These elements are shown in Figure 9 and a schematic display is showed in Figure 10.

5.1. PD Controller

Experiments were carried out using a PD controller for both Constrained and Standard control loops.

Figure 11. Procedure to find control parameters: a) initial position; b) overshoot position; c) oscillation position; d) stable position

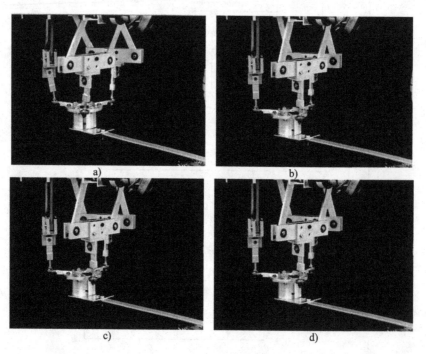

Figure 12. Test result for constant desired force with constrained control loop with PID controller

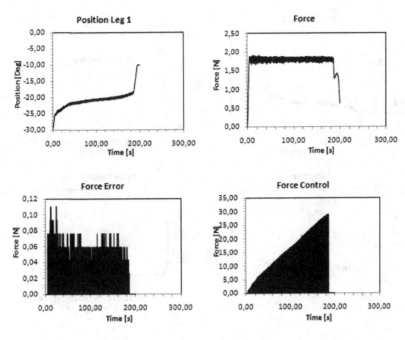

Figure 13. Test result for variable desired force with constrained control loop with PID controller

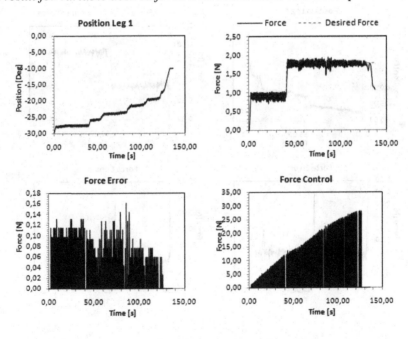

The calibration of the parameters for both control loops was carried out using a manual tuning that is achieved thought the following steps:

- Set Kd value to zero.
- Increase Kp until the output of the loop oscillates.

Figure 14. Test result for constant desired force with standard control loop with PID controller

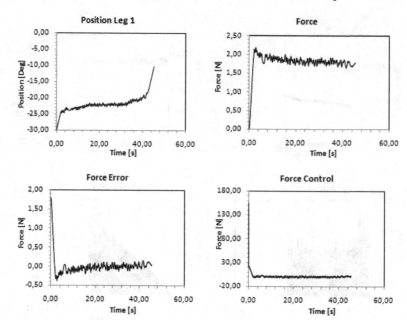

Figure 15. Test result for variable desired force with standard control loop with PID controller

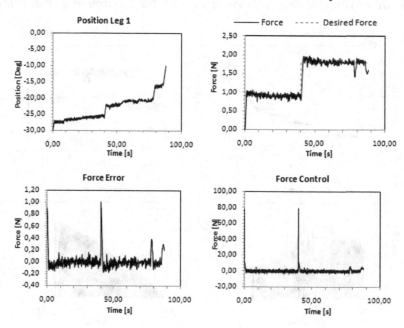

- Set Kp approximately half of the previous value.
- Increase Kd until the output of the loop is acceptably quick to react to and external disturbance.

- Decrease Kd if there is excessive respond and overshooting.

The calibration of the parameters could not be done with the CaPaMan2bis drilling because it means to spend lots of material and could result on the braking of some or several parts of the CaPaMan2bis or burning the motors, this parameters were found putting the force sensor directly on the movable plate of the CaPaMan2bis as it is reported in Figure 11.

5.2. PID Controller

A PID controller was also applied to both Constrained and Standard control loops. The calibration for this controller was made using the Ziegler–Nichols method (Dorf & Bishop, 2005). In order to found the control parameters using the method mentioned we follow this steps:

- Set Kd and Ki values to zero.
- Increase Kp until the output of the loop oscillates. This is the critical Gain Kc, together with the oscillation period Pc are used to set the PID gains as Table 2 indicates.

5.3. Drilling Tests

Several tests were carried out for both control loops with both PD and PID controllers. Tests with constant desired force and variable desired force where applied with different desired forces for better validation of the control loops. Wax and wood were the materials used to make the tests. For each control loop one test is showed where we can see good results for the implementation of the control loops and design of the controller parameters.

5.3.1. Constrained Control Loop

Figure 12 shows that the force does not stabilize for matters of the drilling dynamic, backslash and vibration on the force sensor. Nevertheless it oscillates with a small error margin near the desired force of 1.8 N and the oscillations get smaller as the time passes. Also the tests shown

a quick response since the settling time is small. In Figure 13 it can be observe that for a variable desired force the control implementation shows also a quick respond and good accuracy. For a higher force the oscillations are small giving good expectation when the object to drill requires a higher force.

5.3.2. Standard Control Loop

For the Standard Control Loop it can be observe in Figure 14 that the force also does not stabilize. It can be observe that the overshoot is higher compared with the previous case but the process ends faster. Also error margin is small and the oscillations are near the desired force of 1.8 N. The responds is also quick with a small settling time. For a variable force Figure 15 it can be observe that also the control implementation has a quick respond and good accuracy.

CONCLUSION

In this paper two control algorithms are proposed to obtain a force regulation in the operation of the robot CaPaMan2bis. The implementation of the algorithms has been made by using LabVIEW programming environment and suitable tests have been carried out successfully at LARM. Experimental tests have been carried out for validating the proposed algorithms. Results of the experimental tests show the effectiveness and high speed of the control loops. However, the Constrained Control Loop with PID controller results to be the best at matters of keeping the force controlled, with a lower overshoot, less offset and less settling time. Moreover the control algorithms have been tried out with a variable desired force, obtaining satisfactory results for both control loops. Going farther CaPaMan2bis has been successfully tested even under fast dynamic conditions moved by a SCARA robot to be position over the object to drill.

REFERENCES

Adept Inc. (2009). *Industrial robots.* Retrieved from http://www.adept.com/products/robots

Aguirre, G., Acevedo, M., Carbone, G., & Ottaviano, E. (2003). Kinematic and dynamic analysis of a 3 DOF parallel manipulator by symbolic formulations. In Ambrosio, J. A. C. (Ed.), *Multibody dynamics*. Lisbon, Portugal.

Bonev, I. A. (2009). *Delta parallel robot - the story of success.* Retrieved from http://www.parallemic.org/Reviews/Review002.html

Briones-Leon, A., Carbone, G., & Ceccarelli, M. (2009). Control de posición y fuerza del robot capaman 2 bis en tareas de barrenado. In *Proceedings of the 7th Congreso Internacional en Innovación y Desarrollo Tecnológico*, Cuernavaca, Mexico.

Carbone, G., & Ceccarelli, M. (2005a). A serial-parallel robotic architecture for surgical tasks. *Robotica, 23*, 345–354. doi:10.1017/S0263574704000967.

Carbone, G., & Ceccarelli, M. (2005b). Numerical and experimental analysis of the stiffness performances of parallel manipulators. In *Proceedings of the 2nd International Colloquium Collaborative Research Centre*, Braunschweig, Germany.

Carbone, G., Ceccarelli, M., & Cimpoeru, I. (2005). Experimental tests with a macro-milli robotic system. In *Proceedings of the 14th International Workshop on Robotics in Alpe-Adria-Danube Region*, Bucharest, Romania.

Carbone, G., Ceccarelli, M., Ottaviano, E., Checcacci, D., Frisoli, A., Avizzano, C., & Bergamasco, M. (2003). A study of feasibility for a macro-milli serial-parallel robotic manipulator for surgery operated by a 3 DOFS haptic device. In *Proceedings of the 12th International Workshop on Robotics in Alpe-Adria-Danube Region*, Cassino, Italy.

Ceccarelli, M. (1997). A new 3 DOF spatial parallel mechanism. *Mechanism and Machine Theory, 32*(8), 895–902. doi:10.1016/S0094-114X(97)00019-0.

Ceccarelli, M. (1997). Displacement analysis of a Turin platform parallel manipulator. *Advanced Robotics, 11*, 17–31. doi:10.1163/156855397X00029.

Ceccarelli, M. (1998). A stiffness analysis for CaPaMan (Cassino parallel manipulator). In *Proceedings of the Conference on New Machine Concepts for Handling and Manufacturing Devices on the Basis of Parallel Structures*, Braunschweig, Germany (VDI 1427, pp. 67-80).

Ceccarelli, M., & Decio, P. (1999). A dynamic analysis of Cassino parallel manipulator in natural coordinates. In *Proceedings of the International Workshop on Parallel Machines*, Milano, Italy (pp. 87-92).

Ceccarelli, M., Decio, P., & Jimenez, J. (2002). Dynamic performance of CaPaMan by numerical simulations. *Mechanism and Machine Theory, 37*, 241–266. doi:10.1016/S0094-114X(01)00079-9.

Clavel, R. (1988). Delta, a fast robot with parallel geometry. In *Proceedings of the 18th International Symposium on Industrial Robots*, Lausanne, Switzerland (pp. 91-100).

Dorf, R., & Bishop, R. (2005). *Sistemas de control moderno*. Upper Saddle River, NJ: Pearson/Prentice Hall.

Erlbacher, E. (2000). *Force control basics*. Dallas, TX: Push Corp..

Gosselin, C. (1988). *Kinematics analysis optimization and programming of parallel robot manipulators*. Unpublished doctoral dissertation, McGill University, Montreal, QC, Canada.

Hernández-Martínez, E., Ceccarelli, M., Carbone, G., & López-Cajún, C. (2008). Simulación de un manipulador paralelo espacial de 3 grados de libertad. In *Proceedings of the 5th Congreso Bolivariano de Ingeniería Mecánica, II Congreso Binacional de Ingeniería Mecánica*, Cúcuta, Colombia.

Iannone, S., Carbone, G., & Ceccarelli, M. (2008). Regulation and control of LARM Hand III. In *Proceedings of the International Symposium on Multibody Systems and Mechatronics*, San Juan, Puerto Rico.

LAUMAS. (2009). *Datasheet células de carga para plataformas 150 x 150 mm*. Retrieved from http://www.laumas.com/celletutte_es.htm

Mendes, J., & Ceccarelli, M. (2001). A closed-form formulation for the inverse dynamics of Cassino parallel manipulator. *Multibody System Dynamics, 5*, 185–210. doi:10.1023/A:1009845926734.

Merlet, J. (1988). Force-feedback control of parallel manipulators. In. *Proceedings of the IEEE International Conference on Robotics and Automation, 3*, 1484–1489.

Merlet, J. P. (2006). *Parallel robots*. Dordrecht, The Netherlands: Springer-Verlag.

Merlet, J. P. (2009). *Personal website*. Retrieved from http://www-sop.inria.fr/members/Jean-Pierre.Merlet/merlet_eng.html

National Instruments. (2000a). *LabVIEW user manual*. Austin, TX: National Instruments Corporation.

National Instruments. (2000b). *LabVIEW measurement manual*. Austin, TX: National Instruments Corporation.

Nava, N., Carbone, G., & Ceccarelli, M. (2006). CaPaMan2bis as trunk module in CALUMA (Cassino low-cost hUMAnoid robot). In *Proceedings of the 2nd IEEE International Conference on Robotics, Automation and Mechatronics*, Bangkok, Thailand.

Ottaviano, E., & Ceccarelli, M. (2002). Optimal design of CaPaMan (Cassino parallel manipulator) with a specified orientation workspace. *Robotica, 20*, 159–166. doi:10.1017/S026357470100385X.

Ottaviano, E., & Ceccarelli, M. (2006). An application of a 3-DOF parallel manipulator for earthquake simulations. *IEEE Transactions on Mechatronics, 11*(2), 240–246. doi:10.1109/TMECH.2006.871103.

Stewart, D. (1965-66). A platform with 6 degrees of freedom. *Proceedings - Institution of Mechanical Engineers, 180*(15), 371–386. doi:10.1243/PIME_PROC_1965_180_029_02.

Villegas, E. (2009). *Development of a force control for CaPaMan2bis as a terminal tool for drilling*. Cassino, Italy: University of Cassino, Laboratory of Robotics and Mechatronics.

Wenger, P., & Chablat, D. (2002). Design of a three-axis isotropic parallel manipulator for machining applications: The Orthoglide. In *Proceedings of the Workshop on Fundamental Issues and Future Research Directions for Parallel Mechanisms and Manipulators*, Montreal, QC, Canada.

Wolf, A., Ottaviano, E., Shoham, M., & Ceccarelli, M. (2004). Application of line geometry and linear complex approximation to singularity analysis of the 3-DOF CaPaMan parallel manipulator. *Mechanism and Machine Theory, 39*, 75–95. doi:10.1016/S0094-114X(03)00105-8.

Yiu, K. (2002). *Geometry, dynamics and control of parallel manipulators*. Unpublished doctoral dissertation, Hong Kong University of Science and Technology, Hong Kong.

Zabalsa, I., & Ros, J. (2007). Aplicaciones actuales de los robots paralelos. In *Proceedings of the 8° Congreso Iberoamericano de Ingeniería Mecánica*, Navarra, Spain.

APPENDIX A

Virtual Instrument for Constrained Control Loop

In Figure 16 we can see a part of the first Sequence Frame (Frame number zero). We found one while loop (Figure 16.a) and one case structure (Figure 16.b). The while loop is used to position all the legs of the CaPaMan2bis in -30 degrees and also to acquire the weight of the object about to drill to calibrate the sensor and subtract this amount so the force before drilling is zero. The inputs of the while loop that can be observe in 1 as small squares on the left in Figure 16.a are: the number of the motion control board installed, the initialization to each axis to cero and the error signal. The outputs on the right in 2 are the number the motion control board, the vector space with the three axis, the error signal, a Boolean value indicating the reach of -30 degrees for each leg that will be the inputs of the case structure in Figure 13.c. Also a voltage is measure with the Read ADC block in the sensor that symbolizes the weight of the object to drill and send to Frame One in 3. The case loop is used to halt the motion when we reach the desired position and finish the frame to go to the next one. This halt of the motors is done by the Stop block when the input of the Case Structure is true. Inside the while loop it is read the velocity, ac-

Figure 16. Frame zero of the virtual instrument for the constrained and standard control loops: a) while loop; b) case structure

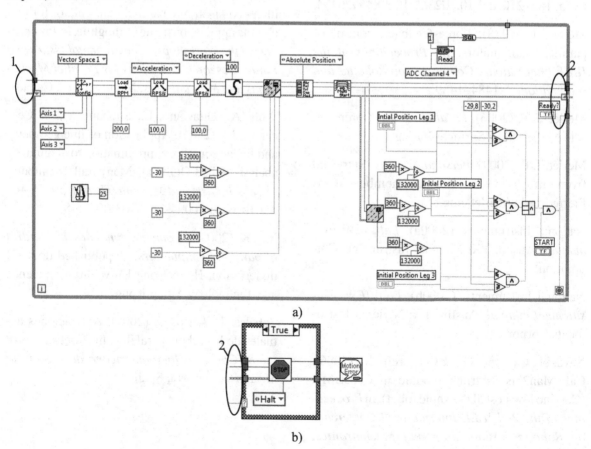

a)

b)

Figure 17. Frame one of the virtual instrument for the constrained control loop—while loop

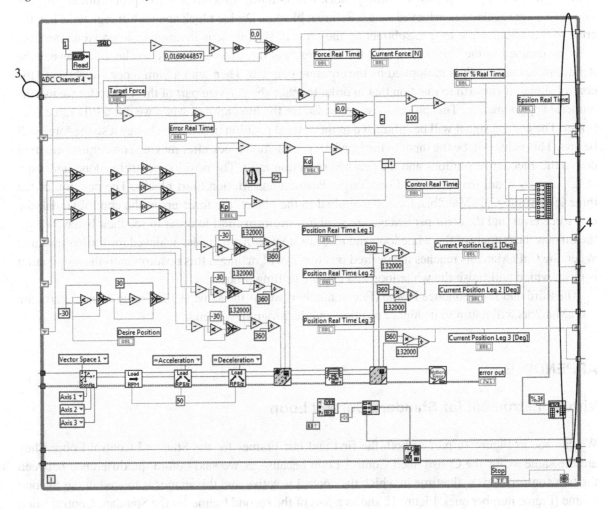

celeration and deceleration of the motors and the desired position and also it is check that this position have been reached, which will stop the motors if is fulfill.

A part of the second Sequence Frame (Frame number one) is showed in Figure 17. The force control loop is found in this frame. By the Read ADC block the voltage given by a certain force applied on the object is measured and subtract by the offset given by the weight of the object to drill, passed to this frame from frame zero as we can see in 3. This operation is achieved with the block Subtract. We also multiply the precious result by a number that is the scope of the sensor and this way we transform the current given by the sensor in force, operation achieved by the Multiply block. This force is read by a waveform chart to show its behavior as the time passed. It also subtracts the target force to determine when the constrained control should act. To compared the target force with the current force it is used the Less Than block that returns true if the current force is less than the target force and false otherwise. When the current force applied to the object is less than the target force using the Selector block it is send a zero value as the error and also the CaPaMan2bis will follow the movement to the desired position, but when the current force is higher than the target force the error will be the difference between this two. This error will be the input of the force control, in this case PD force controller. The proportional regulator is achieved with a Multiply block which multiply the error with a proportional gain Kp. The derivative regulator is achieved with a Subtract block, a Divide block and a Shift register. Then the error of the actual while loop is subtract by the error of the previous iteration. The previous iteration value is obtained by the Shift register. So the difference of the errors is divided by the time between the iterations and this result is multiplied by the constant gain Kd. Then with a Sum block we obtained the control signal. It is worth to mention that in order to make the derivate part of this controller we use the numeric approximation. This process was done also for the integral part when we use a PID force controller. The control signal will be subtracted to the current position in order to change it so the force will be less. This value will be the input to the Load Target Position block after making some transformation due to the encoder revolutions and the gearhed reduction ratio. The position control is done internally in the motion board thanks to the Load Target Position and Motion Start blocks. The position of the three legs of the CaPaMan2bis, the force applied to the object, the force error, the control signal, the true force error and the error percentage are send to waveform chart in order to see their behavior over the time as the outputs of the while loop and also are storage in a text file, achieved for further use in 4. When the CaPaMan2bis reaches the desired position it will maintain this position until the stop button is push, which will finish the while loop and end the drilling process.

The third and last Sequence Frame (Frame number two) is the same as the Frame number zero, the CaPaMan2bis will return to its higher position and the program will end.

APPENDIX B

Virtual Instrument for Standard Control Loop

We can see in Figure 16 part of both the first and last Frames for the Standard Control Loop. They are the same as for the Constrained Control Loop because as we said before the difference between the two control loops is the time in which the control is active and this matter is treated in the Second Frame (Frame number one). Figure 18 shows a part of the second Frame for the Standard Control Loop the force control is found. The difference between this Frame and the one for the Constrained Control Loop resides in the evaluation of the condition current force less than target force. By this manner this Standard Control Loop actuates at all times and its goal is to take the force to the desired one instead of constraining it. By this we observe that the error is always the difference between the target force and the current force. Also the control signal subtracts the current position and the process ends when it reaches the desired position or the stop button is press, as we can see in the bottom part.

APPENDIX C

Virtual Instrument for Implementation of PID Controller

To implement the integral part some parts were add to where the proportional and derivative regulators are, in Figure 17 and Figure 18. These parts are reported in Figure 19. The derivative and proportional parts were already explained previously. For the integral part we use the Built Array, Insert into Array,

Figure 18. Frame one of the virtual instrument for the standard control loop—while loop

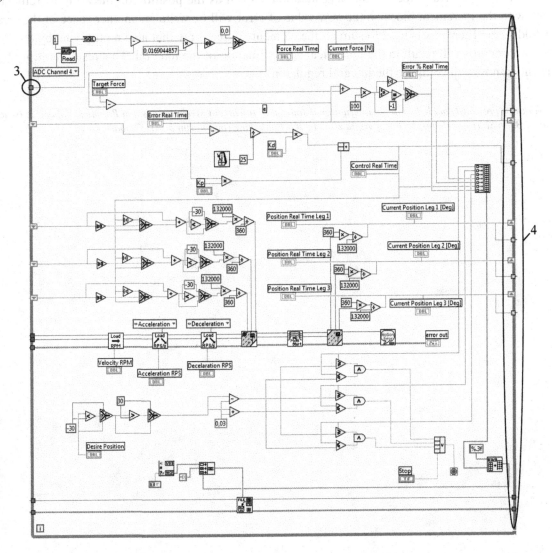

Figure 19. Implementation of the integral part

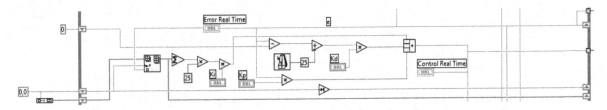

Add Array Elements and two Shift registers. Built Array is use to build and array of one element that will be zero, being the input of the Insert into Array block as well as the position into the array and the value of the element that will be the error in the current iteration. The obtained array it is send as the input for the Built Array block in the next iteration as well as the position on the array incremented by one with the Increment block so the new value of error will be write on another space of the array. The Add Array Elements is used to sum all the elements in the array that also multiplying by the time between iterations will result in the numerical approximation of the integral. This result multiply by a constant gain Ki will compose the integral regulator.

This work was previously published in the International Journal of Intelligent Mechatronics and Robotics, Volume 1, Issue 4, edited by Bijan Shirinzadeh, pp. 1-18, copyright 2011 by IGI Publishing (an imprint of IGI Global).

Chapter 16
High Performance Control of Stewart Platform Manipulator Using Sliding Mode Control with Synchronization Error

Dereje Shiferaw
Indian Institute of Technology Roorkee, India & Graphic Era University, India

Anamika Jain
Graphic Era University, India

R. Mitra
Indian Institute of Technology Roorkee, India

ABSTRACT

This paper presents the design and analysis of a high performance robust controller for the Stewart platform manipulator. The controller is a variable structure controller that uses a linear sliding surface which is designed to drive both tracking and synchronization errors to zero. In the controller the model based equivalent control part of the sliding mode controller is computed in task space and the discontinuous switching controller part is computed in joint space and hence it is a hybrid of the two approaches. The hybrid implementation helps to reduce computation time and to achieve high performance in task space without the need to measure or estimate 6DOF task space positions. Effect of actuator friction, backlash and parameter variation due to loading have been studied and simulation results confirmed that the controller is robust and achieves better tracking accuracy than other types of sliding mode controllers and simple PID controller.

DOI: 10.4018/978-1-4666-3634-7.ch016

INTRODUCTION

The Stewart platform manipulator is a 6DOF parallel manipulator having a fixed base and a moveable platform connected together by six extensible legs, see Figure 1. It has advantages of high precision positioning capacity, high structural rigidity, and strong carrying capacity (Fichter, Kerr, & Rees-Jones, 2009; Guo & Li, 2006; Li & Salcudean, 1997; Merlet, 2006; Dasgupta & Mruthyunthaya, 2000). Potential areas of application include flight and other motion simulators, light weight and high precision machining, data driven manufacturing, dexterous surgical robots and active vibration control systems for large space structures (Merlet, 2006; Sirouspour & Salcudean, 2001; Su & Duan, 2000). The control of this highly coupled nonlinear dynamic system has been a hot research issue in the last few decades. The control of this manipulator can be formulated in either joint space in terms of the length of the legs or task space in terms of Cartesian position and orientation of the moveable platform (Ghorbel, Chételat, Gunawardana, & Longchamp, 2000; Kima, Chobe, & Lee, 2005; Ting, Chen, & Wang, 1999). Each of these approaches has its own advantages and disadvantages. In the joint space approach, the controller is a collection of single input single output (SISO) systems implemented using local information on each actuator length only and the coupling between legs is ignored or is considered as a disturbance (Zhao, Li, & Gao, 2008). The advantage of this approach is that local information required for feedback is obtained easily using simple sensors. This helps to easily implement the individual SISO controllers in parallel resulting in the execution of control algorithm at a reasonably fast speed (Zhao, Li, & Gao, 2008). However, joint space implementation has certain drawbacks. The first one is due to the inherent nonlinear relationship between length of legs and end effector pose, a disturbance in legs may lead to a big error in end effector pose while leg error is small (Paccot,

Andreff, Martinet, & Khalil, 2006). The second problem of joint space SISO implementation is lack of synchronization in the individual control loops. The lack of synchronization leads to a large coupling error which results in lesser accuracy. Moreover, the accumulation of coupling errors generates excessive force and may damage the manipulator itself (Zhao, Li, & Gao 2008). Due to these, joint space approach alone cannot achieve high performance (Kima, Chobe, & Lee, 2005; Zhao, Li, & Gao 2008; Davliakos & Papadopoulos, 2008; Sirouspour & Salcudean, 2001).

To solve these problems, many researchers have proposed joint space feedback control methods improved by using feedforward loops. While Su, Duan, Zheng, Zhang, Chen, and Mi (2004) used a tracking differentiator as a feedforward controller to improve joint space PID controller performance, Zhao, Li, and Gao (2008) designed an adaptive controller having feedback and feedforward parts to compensate for synchronization error of joint space control. However, the above schemes cannot achieve high performance, especially at high speeds due to the fact that the linear PD or PID controllers cannot compensate leg dynamics.

On the other hand, in task space approach, the system is treated as multiple input multiple output (MIMO) system and the coupling between legs is visible. Therefore, task space scheme has the potential to give a superior 6DOF tracking performance under various system uncertainties (Kima, Chobe, & Lee, 2005). Nevertheless this approach needs 6DOF end effector pose, which is very costly if sensors are to be used for measurement or is cumbersome if estimation algorithms are to be used (Kima, Chobe, & Lee, 2005; Nguyen, Zhou, & Antraz, 1997).

From the given discussion, it is clear that none of the approaches is without disadvantage. This shows that employing a hybrid control approach which utilize the advantages of both approaches and minimize their disadvantages would be a

Figure 1. Generalized Stewart platform manipulator with reference frames attached

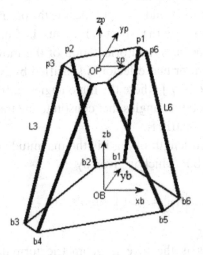

better solution. Accordingly, some authors have proposed model based control methods which utilize advantages of task space approach with joint space. A first proposal of this sort is that of Beji, Abichou, and Pascal (1998). In their paper, they proposed a control method for an electrically driven manipulator where the desired force is calculated in task space using desired position and velocity signals and the leg dynamics compensation is achieved by a force convergent principle. Similarly Fu and Yu (2006) proposed a hybrid control algorithm where the platform dynamics is calculated in task space and the leg dynamics is compensated in joint space. In that paper, the led dynamics is calculated using Newton Euler method. However, the method needs task space positions and orientations and the authors did not specify the method used to get the pose of the end effector. Similarly Lee, Song, Choi, and Hong (2003) used inverse dynamics control with approximate dynamics as a feedforward controller and they employed a linear H-infinity controller in the feedback path to compensate for the approximation error. The last two proposals are promising since they are robust controllers and have the potential to achieve high performance due

to the model based feedforward part. Nevertheless, the complexity of the H-infinity controller and the need for task space position and orientation feedback make the controllers less suitable and a relatively easier, robust and high performance controller is very much required. This paper addresses such a solution.

The control algorithm proposed in this paper is a sliding mode controller which uses a sliding surface which is designed to drive both tracking and synchronization errors to zero. In the controller the model based equivalent control part is computed in task space and the discontinuous switching controller part is computed in joint space. The controller has similar structure to the last three because it combines task space feedforward with joint space feedback. But the controller presented here has two basic differences. The first difference is the feedforward controller part and the second is the nonlinear controller used in joint space. The feedforward controller used in our proposal is an equivalent control signal of a sliding mode controller (SMC) and has a better control performance. It is clear that sliding mode control (SMC) is a robust controller design method, which is effective in controlling systems with significant uncertainties including parameter variations, unmodeled dynamics and external disturbances (Hung, Gao, & Hung, 1993; Dereje & Mitra, 2010). It has been proposed for the control of various systems including Stewart platform manipulator (Kim & Lee, 1998; Iqbal & Bhatti, 2008; Negash & Mitra, 2010). The other difference of the present controller to the hybrid controller proposed by Beji, Abichou, and Pascal (1998) Fu and Yu (2006) Lee, Song, Choi, and Hong (2003) is unlike theirs in our proposed controller, a discontinuous switching controller is used in joint space to compensate the uncertainty in the manipulator leg dynamics. This discontionous controller uses a newly proposed linear sliding surface which is designed to drive both tracking error and synchronization error to zero. One other most important difference is in our proposed controller, though the model based part is computed in task space, it does not need measurement or estimation of task space positions and orientations. In the equivalent controller, the dynamics of the manipulator is calculated using desired trajectory parameters and the mismatch due to the use of desired values instead of actual values is compensated by the joint space discontinuous part. This means the controller avoids the need for task space position estimation or measurements. Hence the controller is able to achieve high performance and low cost implementation at the same time.

The presentation begins with dynamic modeling of the 6DOF manipulator. Then design of the controller and stability analysis follow. In the paper, effect of actuator friction and other uncertainties is thoroughly discussed. The paper is organized as follows: the next section discusses dynamic modeling, while the next section gives the design of the controller and robustness and stability analysis are given in. We devote the next section to simulation result and discussion. Lastly conclusion follows.

MODELING OF THE MANIPULATOR

Kinematic Modeling

For geometric and kinematic modeling, the following conventions are used. The centers of the universal and spherical joints are denoted by B_i (i =1, 2 ... 6) and P_i (i = 1, 2... 6) respectively. Reference frames F_b and F_p are attached to the base and the platform as shown in Figure 1. The position vector of the center of universal joints B_i in frame F_b is b_i and the position vector of the

center of spherical joints P_i in frame F_p is p_i. Let r= [r_x, r_y, r_z] be the position of the origin O_p with respect to O_b and also let R denote the orientation of frame F_p with respect to F_b. Thus the Cartesian space position and orientation of the moveable platform or end effector is specified by X= [r_x, r_y, r_z, α, β, γ] where the three angles α, β, γ are three rotation angles that constitute the transformation matrix R.

Then length of leg i is the magnitude of the vector B_iP_i which is given by

$$\left| B_iP_i \right| = \left\| Rp_i + r - b_i \right\| = q_i \tag{1}$$

This is the inverse geometric formula that gives the length of each leg for a given desired position and orientation of the end effector. The direct geometric model which gives the position r= [r_x r_y r_z] and orientation angles α, β, γ for a given measured value of q_i, i=1, 2... 6 is nonlinear and is solved using numerical methods.

The inverse kinematic model gives the velocity of the active joint $\dot{q}$ for a given end effector linear and angular velocity and is given as

$$\dot{X} = J^{-1}\dot{q} \tag{2}$$

where J is the Jacobian matrix of the platform with respect to the base frame.

Dynamic Modeling

The dynamic modeling of Stewart platform manipulator has been extensively studied by many researchers (Guo & Li, 2006; Ghobakhloo, Eghtesad, & Azadi, 2006). The various methods used are Newton Euler, Lagrangean and the principle

Box 1.

$$\left(M_1\left(X\right) + M_2\left(X\right) \right)\ddot{X} + \left(C_1\left(X,\dot{X}\right) + C_2\left(X,\dot{X}\right) \right)\dot{X} + G_1\left(X\right) + G_2\left(X\right) = J^{-T}\left(\tau - f_f\right) \tag{3}$$

of virtual work. Using the Lagrangean method, dynamic model of the manipulator in task space is given as Equation 3 in Box 1, where X=[r_x, r_y, r_z, α, β, γ] is Cartesian space position and orientation of the moveable platform or end effector, $M_1(.)$, $C_1(.,.)$ and $G_1(.)$ are inertia matrix, coriolis/centrifugal coefficient matrix and gravitational torque of the platform respectively. $M_2(.)$, $C_2(.,.)$ and $G_2(.)$ are corresponding parameters of the legs, J is manipulator Jacobian, τ is joint space actuator torque and f_f is actuator friction. The following assumptions are taken.

Assumption1: The inertia matrices are positive definite and nonsingular.
Assumption 2: The manipulator Jacobian matrix J is assumed to be nonsingular over the whole works space of the manipulator.
Assumption 3: The uncertainties in the inertia, Coriolis and centrifugal and gravitational matrices are bounded and can be given as nominal and deviation as:

$$M_1 = M_{1N} + \Delta M_1$$

$$M_2 = M_{2N} + \Delta M_2$$

$$C_1 = C_{1N} + \Delta C_1$$

$$C_2 = C_{2N} + \Delta C_2$$

$$G_1 = G_{1N} + \Delta G$$

$$G_2 = G_{2N} + \Delta G$$

The above uncertainties are parametric uncertainties and they can be compensated by using high gain feedback if the motion of the manipulator is slow. At high speed, the uncertainties become highly nonlinear and compensating by using high gain feedback is not enough. Another source of

uncertainty which has to be considered in robust controller design is actuator friction, which limits the magnitude of gain to be used and degrades tracking performance. In Stewart manipulator, various methods have been used to model and compensate effect of actuator friction including, replacing rolling joints by flexure mounts (Beji, Abichou, & Pascal, 1998), use of Fridland Park friction estimator (Kima, Chobe, & Lee, 2005) and so on. In this paper, the effect of friction is compensated using sliding mode control and the gain of the switching function is used to compensate for it.

DESIGN OF THE CONTROLLER

In designing a controller for Stewart platform manipulator, two different approaches are possible. The first one is to convert the desired task space position, velocity and acceleration of the platform center to desired joint leg lengths and close the loop by using measured leg lengths as feedback. This approach is known as joint space. In this approach, the individual leg measurements and desired values are taken separately and control is single input single output (SISO). The advantage of this approach is that local information required for feedback is obtained easily using simple sensors. Moreover, the individual SISO controllers can be easily implemented in parallel and hence the control algorithm can be executed reasonably fast. A sliding mode controller of this sort is shown in Figure 2a. In the block diagram, the six dimensional desired positions, velocity and accelerations $xd, \dot{x}d, \ddot{x}d$ are converted to desired leg lengths and velocities ($qd, \dot{q}d$) using the inverse kinematics (IK) block. Then the joint length error e and velocity error $\dot{e}$ are calculated using the measured length q and velocity $\dot{q}$ which is obtained by differentiating and filtering q. Then the discontinuous control torque part is obtained using the sliding surface, switching function $f_s(.)$ and the sliding gain K. The equivalent torque part

Table 1. Geometric specifications of Stewart platform

Joint positions						
Base	$\frac{\pi}{6}$	$\frac{\pi}{2}$	$\frac{5\pi}{6}$	$\frac{7\pi}{6}$	$\frac{3\pi}{2}$	$\frac{11\pi}{6}$
Platform	$\frac{\pi}{12}$	$\frac{7\pi}{12}$	$\frac{3\pi}{4}$	$\frac{5\pi}{4}$	$\frac{17\pi}{12}$	$\frac{23\pi}{12}$

Base radius 0.8m

Platform radius 0.5m

Mass of platform 32kg

Mass of upper leg 4kg

Mass of lower leg 4kg

Initial Height 1.5m

Platform Inertia $I_{xx}=2$, $I_{yy}=2$ and $I_{zz}=4$

Leg Inertia upper $I_{xx}=0.75$, $I_{yy}=0.75$, $I_{zz}=0.018$

Leg Inertia lower $I_{xx}=0.03$, $I_{yy}=0.03$, $I_{zz}=0.002$

CG of upper leg 0.75m from top

CG of lower leg 0.15m from base

is obtained from the joint space dynamic parameters. This approach has the advantage that it does not need evaluation of the forward kinematics. However lack of synchronization puts a limit to its performance and hence this approach cannot give high performance. Furthermore, the difficulty of computing dynamic parameters needed for the equivalent controller part makes it difficult for real time implementation.

In the second approach, the desired task space positions, velocities and accelerations are not converted to desired leg lengths rather they are used directly while feedback is obtained by taking measured or estimated task space position, velocity and acceleration. A task space sliding mode controller is shown in Figure 2b. As shown in the block diagram, the forward kinematics takes in measured leg lengths and gives estimated task

space positions. Then the estimated positions are differentiated and filtered to obtain estimated task space velocities. Then the desired task space positions and velocities are compared with their corresponding estimated values and the error signals are used to form sliding surface. The switching control part is formed using switching function $f_s(.)$ and gain K. An equivalent force component is also calculated in task space using dynamic parameters. Then the sum of the two control signal parts are transformed to joint space using the Jacobian matrix. The sliding mode controller in this approach is therefore a multiple input multiple output system (MIMO). Computation of the dynamic parameters is relatively simpler in this approach and it has the potential to achieve high performance as it avoids synchronization error. However the complexity incurred in getting the task space feedback signals has delayed its practical application.

Our proposed hybrid controller combines the advantages of the above two approaches. The structure of the controller is as shown in Figure 2c. The feed forward part of the controller which takes in the task space acceleration, velocity and position computes the equivalent control force. This is similar to the implementation in Figure 2b with the difference that in the task space sliding mode controller of Figure 2b, estimated acceleration, velocity and positions are used for the computation of the dynamic parameters while in the present case desired values are utilized. This avoids the need for forward kinematics estimation. However, there will be some disturbance torque due to the difference between actual and estimated values and it is assumed that the effect of this disturbance torque will be compensated by the joint space switching controller. In the second part of the controller, a discontinuous switching controller is implemented in joint space. The block named as IK is the inverse kinematics computation needed to convert the task space desired positions, velocities into desired leg elongations and velocities. This is similar to the joint space sliding

Figure 2. a) Block diagram representation of joint space sliding mode controller; b) block diagram representation of task space sliding mode controller; c) block diagram representation of the hybrid control structure

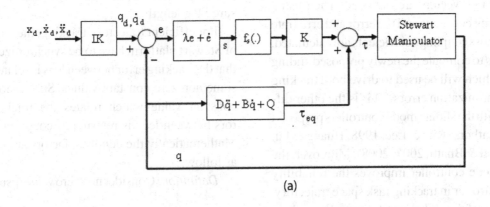

(a)

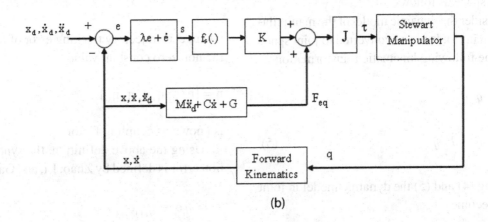

(b)

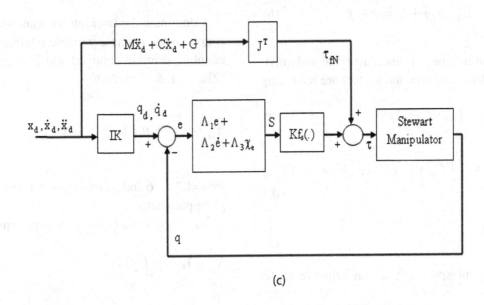

(c)

mode controller shown in Figure 2a. Then using the measured leg lengths and velocities obtained by differentiating the measured leg lengths, error and error rate vectors are obtained. Then using the coupling constants and the error and error rate vectors, cross coupling coefficients are calculated. This helps to compute the newly proposed sliding surface which will be used to drive both tracking and synchronization errors. This is the other difference with the sliding mode controllers proposed by other authors (Kim & Lee, 1998; Huang & Fu, 2005; Iqbal & Bhatti, 2007, 2008). Moreover, the sliding mode controller improves the reliability of the controller in tracking task space trajectory. The design of the sliding mode controller is done in joint space as follows.

Consider the dynamic model of the manipulator given in (3). It can be converted to joint space using the following kinematic transformations.

$$\dot{x} = J^{-1}\dot{q} \tag{4}$$

$$\ddot{x} = J^{-1}\ddot{q} + \dot{J}^{-1}\dot{q} \tag{5}$$

Using (4) and (5) the dynamic model in joint space becomes

$$D\big(q\big)\ddot{q} + B\big(q,\dot{q}\big)\dot{q} + Q = \tau - f_f \tag{6}$$

where the inertia, coriolis centrifugal and gravitational torque matrix and vectors are related by

$$D\big(q\big) = J^T\big(M_1 + M_2\big)J^{-T} \tag{7}$$

$$B\big(q,\dot{q}\big) = J^T\big(\big(C_1 + C_2\big)J^{-T} + \big(M_1 + M_2\big)\dot{J}^{-T}\big) \tag{8}$$

$$Q = J^T G \tag{9}$$

The joint space leg length error vector is given by

$$e = q_d - q. \tag{10}$$

where q_d is the desired leg length and q is measured leg length.

For efficient trajectory tracking, the actuators of Stewart platform have to be synchronized such that the tracking error between legs is related with some non zero constant values. Such a non-zero constant value which relates the tracking errors between legs is referred as coupling factor. Mathematically the coupling factors are defined as follows.

Definition: Consider non zero value η_i such that

$$\eta_1 e_1 = \eta_2 e_2 = \eta_3 e_3 = \eta_4 e_4 = \eta_5 e_5 = \eta_6 e_6 \tag{11}$$

where e_i is trajectory tracking error of leg i, then the non zero constant value

$$\eta = [\,\eta_1\ \eta_2\ \eta_3\ \eta_4\ \eta_5\ \eta_6\,]$$

is known as coupling factor

Using the above definition, the synchronization error is defined by Zhao, Li, and Gao (2008)

$$\varepsilon_i = \eta_i e_i - \eta_{i+1} e_{i+1} \tag{12}$$

Combining the tracking error and synchronization errors, a single variable referred as cross coupling error is obtained and it is defined as (Zhao, Li, & Gao, 2008)

$$\chi_{ei} = \eta_i e_i + \mu \int_0^t \big(\varepsilon_i - \varepsilon_{i-1}\big) dw \tag{13}$$

for i=1,2,… 6 and i-1=6 when i=1 and μ is coupling parameter

In vector form, (13) can be represented as

$$\chi_e = Ie + \mu \int f\big(e\big) \tag{14}$$

where χ_e is vector of the cross coupling errors, I is vector of the coupling factors η, f is a vector function which shows the difference between synchronization errors of successive legs. Taking the derivative of both sides of (14), the dynamic of the cross coupling error is given as

$$\dot{\chi}_e = I\dot{e} + \mu f(e) \tag{15}$$

Then using the joint space tracking error and rewriting the dynamic Equation (6) in state space form and then combining it with dynamics of the synchronization error given in (15), the new dynamic of the Stewart platform manipulator with three state variables at each leg is given as

$$\dot{y}_1 = y_2$$
$$\dot{y}_2 = \ddot{q}_d - D(q)^{-1}\left(\tau - f_f - B(q,\dot{q}) - Q(q)\right) \tag{16}$$
$$\dot{y}_3 = Iy_2 + \mu f(y_1)$$

where the states y_1 and y_2 are the tracking error e and its derivative $\dot{e}$ and y_3 is the cross coupling error χ_e.

In sliding mode control, there are two basic steps and the first step is the design of a stable sliding surface which is formed from the states of a system to be controlled and the second step is the design of control law which will drive the system states towards the stable sliding surface. The sliding surface has to be designed such that when the states are on the sliding surface, the system has to be insensitive to external disturbances and the states should move towards an equilibrium point. In the present case, the system dynamics (16) is formulated in terms of tracking error and synchronization error. If a new sliding surface containing the three states is proposed so that when the system is on that sliding surface, both synchronization error and tracking error asymptotically move to zero, then we can drive both synchronization error and tracking error to zero by using an appropriate control law. Hence, the following new sliding surface, which is used to constrain tracking error and synchronization error, is defined for joint space sliding mode control

$$S = \Lambda_1 y_1 + \Lambda_2 y_2 + \Lambda_3 y_3 \tag{17}$$

The matrices Λ_1, Λ_2 and Λ_3 are 6x6 and without loss of generality, Λ_2 is assumed to be identity matrix. From (17) it is clear that, when s=0, the sliding surface is a plane in 3D which passes through the origin and if a control signal is designed properly, both the tracking error and synchronization error can be driven to zero.

Taking the derivative of S and equating it to zero, the equivalent controller shown in Box 2 is obtained.

Substituting (17) into (15) and assuming that the dynamic parameters are computed exactly, the sliding dynamics of the system becomes

$$\dot{y}_1 = y_2$$
$$\dot{y}_2 = -\Lambda_1 y_2 - \Lambda_3\left(Iy_1 + \mu f(y_1)\right) - D(q)^{-1} f_f$$
$$\dot{y}_3 = Iy_2 + \mu f(y_1) \tag{19}$$

If friction is neglected, the sliding dynamics is a second order ordinary differential equation and both the tracking error and the synchronization error can be made to zero and asymptotic stability

Box 2.

$$\tau = D(q)\left(\ddot{q}_d + \Lambda_1 y_2 + \Lambda_3\left(Iy_2 + \mu f(y_1)\right)\right) + B(q,\dot{q})\dot{q} + Q(q) \tag{18}$$

can be achieved. But in practical systems, friction is a critical factor and it cannot be neglected. Hence the sliding dynamics cannot achieve asymptotic stability but the tracking error and synchronization errors can be made to be bounded with in a tolerable limit. Another point to be noted is, the parameters of the equivalent controller (18) are in joint space and it has been reported by various researchers that computation of these model parameters in joint space is difficult and needs complex coordinate transformations (Ting, Chen, & Wang, 1999). Using the transformations given in (7-9), the task space equivalent of these model parameters can be obtained and the equivalent control signal can be calculated in task space as shown. Using (7),(8) and (9), (18) can be rewritten as Equation 20 in Box 3.

This can be rewritten again as

$$\tau = J^T \tau_f + J^T \left(M_1 + M_2 \right) J^{-T} \left(\Lambda_1 y_2 + \Lambda_3 \left(I y_2 + \mu f\left(y_1\right) \right) \right) \tag{21}$$

where

$$\tau_f = \left(M_1 + M_2 \right) \left(J^{-T} \ddot{q}_d + \dot{J}^{-T} \dot{q} \right) + \left(C_1 + C_2 \right) J^{-T} \dot{q} + G_1 + G_2 \tag{22}$$

If $\dot{q}$ is replaced by $\dot{q}_d$, τf can be rewritten as

$$\tau_f = \left(M_1 + M_2 \right) \ddot{X}_d + \left(C_1 + C_2 \right) \dot{X}_d + G_1 + G_2 \tag{23}$$

and it can be easily calculated in task space without the need for estimating of the pose of the end effector. This is one of the advantages of the proposed controller. It gets rid of the need for the estimation or measurement of the forward kinematics which is complex if estimation is to be used or costly if measurement is to be done. Replacing $\dot{q}$ by $\dot{q}_d$ has certain effects on the sliding dynamics and the effect of the change and presence of other uncertainties is analyzed in the next section.

ROBUST CONTROLLER ANALYSIS

The uncertainties in the Gough-Stewart manipulator include parameter variation, actuator friction and backlash. The uncertainties are assumed to be bounded and parameter uncertainties are assumed to be additive so that the parameters are given as nominal and deviation. Using assumption 3, the uncertain dynamic model of (1) can be rewritten as Equation 24 in Box 4, Where d is parametric uncertainty given by

$$d\left(X\right) = \left(\Delta M_1 + \Delta M_2 \right) \ddot{X} + \left(\Delta C_1 + \Delta C_2 \right) \dot{X} + \Delta G_1 + \Delta G_2 \tag{25}$$

Box 3.

$$\tau = J^T \left(M_1 + M_2 \right) J^{-T} \left(\ddot{q}_d + \Lambda_1 y_2 + \Lambda_3 \left(I y_2 + \mu f\left(y_1\right) \right) \right) + J^T \left(\left(C_1 + C_2 \right) J^{-T} + \left(M_1 + M_2 \right) \dot{J}^{-T} \right) \dot{q} + J^T G \tag{20}$$

Box 4.

$$\left(M_{1N}\left(X\right) + M_{2N}\left(X\right) \right) \ddot{X} + \left(C_{1N}\left(X, \dot{X}\right) + C_{2N}\left(X, \dot{X}\right) \right) \dot{X} + G_{1N}\left(X\right) + G_{2N}\left(X\right) = J^{-T} \left(\tau - f_f \right) - d(X) \tag{24}$$

Table 2. No load tracking performance of the three controllers

Controller Parameters	Translational Motion Errors	Rotational Motion Errors	Remarks
PID	$e_x=e_y=4\%*$ $e_z=0.4\%*$	roll=1% pitch=1% yaw=1.2×10^{-3}m	yaw angle error is absolute error
Simple SMC $\Lambda_1=1000$ $\Lambda_2=1$ $\Lambda_3=0$ K=6000	$e_x=e_y=0.5\%*$ $e_z=0.05\%*$	roll=0.133% pitch=0.133% yaw=1.2×10^{-3}m	yaw angle error is absolute error
New SMC $\Lambda_1=250$ $\Lambda_2=1$ $\Lambda_3=400$ K=6000	$e_x=e_y=0.25\%*$ $e_z=0.05\%*$	roll=0.067% pitch=0.067% yaw=2×10^{-4}m	yaw angle error is absolute error

* the errors are expressed as percentage of the maximum displacements in the respective directions

Table 3. Tracking error performance of the three controllers when platform is carrying payload of 200Kg and actuator friction is considered

Controller Parameters	Translational Motion Errors	Rotational Motion Errors	Remarks
PID	$e_x=e_y=25\%*$ $e_z=2\%*$	roll=6% pitch=6% yaw=5×10^{-3}m	Yaw angle error is absolute error
Simple SMC $\Lambda_1=1000$ $\Lambda_2=1$ $\Lambda_3=0$ K=6000	$e_x=e_y=5\%*$ $e_z=0.267\%*$	roll=1% pitch=1% yaw=1×10^{-3}m	Yaw angle error is absolute error
New SMC $\Lambda_1=250$ $\Lambda_2=1$ $\Lambda_3=400$ K=6000	$e_x=3.5\%*$ $e_y=4.5\%*$ $e_z=0.133\%*$	roll=3% pitch=3% yaw=5×10^{-4}m	Yaw angle error is absolute error

* the errors are expressed as percentage of the maximum displacements in the respective directions

Given the assumptions 1-3, control law (26) (see Box 5) is proposed to stabilize the uncertain dynamic system (16), i.e., the controller is able to drive the system towards the sliding surface in a finite time and will keep the system states on the sliding surface.

With φ being a small boundary value around the sliding surface and τ_{fN} is the task space equivalent controller calculated using nominal parameters. To analyze the stability of the system under the given controller, the control signal has to be rewritten in joint space form since the errors are in joint space. Using the conversion matrices given in (9), the control signal can be written as Equation 27 in Box 6.

Taking the Lyapunov function

$$V = \frac{1}{2}s^T s \qquad (28)$$

Taking the derivative of V

$$\dot{V} = S\dot{S} \qquad (29)$$

Box 5.

$$\tau = J^{T}\tau_{fN} + J^{T}\left(M_{1N} + M_{2N}\right)J^{-T}\left(\Lambda_{1}y_{2} + \Lambda_{3}\left(Iy_{2} + \mu f\left(y_{1}\right)\right)\right) + J^{T}\left(Kf_{s}\left(s\right)\right) \tag{26}$$

where $f_s(s)$ is switching control signal given by

$$f_{s}\left(s\right) = \begin{cases} +1 \text{ for } s \geq \varphi \\ s \text{ for } -\varphi < s < \varphi \\ -1 \text{ for } s \leq \varphi \end{cases}$$

Box 6.

$$\tau = D_{N}\left(q\right)\ddot{q}_{d} + B_{N}\left(q,\dot{q}\right)\dot{q}_{d} + Q_{N}\left(q\right) + D_{N}\left(q\right)\left(\Lambda_{1}y_{2} + \Lambda_{3}\left(Iy_{1} + \mu f\left(y_{1}\right)\right)\right) + Kf_{s}\left(s\right) \tag{27}$$

and $\dot{S}$ is calculated from (17) as

$$\dot{S} = \Lambda_{1}\dot{y}_{1} + \Lambda_{2}\dot{y}_{2} + \Lambda_{3}\dot{y}_{3} \tag{30}$$

and substituting for the state derivatives from (19)

$$\dot{S} = \Lambda_{1}y_{2} - \Lambda_{2}\left(\ddot{q}_{d} - \ddot{q}\right) + \Lambda_{3}\left(Iy_{2} + \mu f\left(y_{1}\right)\right) \tag{31}$$

Substituting for $\ddot{q}$ from (4) into (28) using the result into (26) (see Box 7).

Then substituting the control signal from (23) (see Box 8).

In (31) the matrix D is the uncertain inertia matrix which can be written as nominal and deviation, as

$$D = D_{N} + \Delta D,$$

and using the Sherman-Morrison formula, the inverse of D can be written as

$$D^{-1} = \left(D_{N} + \Delta D\right)^{-1} = D_{N}^{-1} - \frac{1}{1+g}D_{N}^{-1}\Delta D D_{N}^{-1} \tag{35}$$

where g is tr $(D_{N}\Delta D)^{-1}$

Then, using (32), (31) can be rewritten as Equation 36 in Box 9.

Hence the system can be stable and the sliding surface is attractive if the gain of the switching function is selected as

$$K \geq \left\|\Delta V\right\| + \left\|\Delta Q\right\| + \left\|f_{f}\right\| + \left\|h\right\| \tag{37}$$

Box 7.

$$\dot{V} = S\left(\Lambda_{2}\left(\ddot{q}_{d} - D^{-1}\left(\tau - f_{f} - B - Q\right)\right) + \Lambda_{1}y_{2} + \Lambda_{3}\left(Iy_{2} + \mu f\left(y_{1}\right)\right)\right) \tag{32}$$

Box 8.

$$\dot{V} = S\left(\begin{array}{l} \ddot{q}_d - D^{-1}\left(D_N(q)\ddot{q}_d + V_N(q,\dot{q})\dot{q}_d + Q_N(q)\right) + D^{-1}\left(D_N(q)\left(\Lambda_1 y_2 + \Lambda_3\left(Iy_2 + \mu f(y_1)\right)\right)\right) + \\ K_f(S) - f_f - V - Q + \Lambda_1 y_2 + \Lambda_3\left(Iy_2 + \mu f(y_1)\right) \end{array}\right) \tag{33}$$

$$\dot{V} = S\left(\left(\ddot{q}_d + \Lambda_1 y_2 + \Lambda_3\left(Iy_2 + \mu f(y_1)\right)\right)\left(I - D^{-1}D_N(q)\right) + D^{-1}\left(\Delta V + \Delta Q\right) + D^{-1}\left(K_f(S) - f_f\right)\right) \tag{34}$$

$$h = D\left(\ddot{q}_d + \Lambda_1 y_2 + \Lambda_3\left(Iy_2 + \mu f(y_1)\right)\right)\left(-\frac{1}{1+g}D_N^{-1}\Delta D\right) \tag{38}$$

The term h is due to the parameter uncertainty in the inertia matrix, which is mainly due to position of end effector and payload. To avoid the need for excessively high value of gain, which may excite high frequency oscillation or vibration of the platform, friction torque can be separately compensated

SIMULATION RESULTS AND DISCUSSION

For the simulation study of the performance of the controller, a typical 6-6 geometry Gough-Stewart platform with the geometric parameters given in Table 1 (Beji, Abichou, & Pascal, 1998) is implemented using SimMechanics tool box of MATLAB. A friction model containing viscous friction and coulomb friction, which is used in most robotic controllers, is included into the

SimMechanics model to simulate the effect of actuator friction,. The model equation is given as

$$f_f = k_v \dot{q} + k_c \, sign(\dot{q}) \tag{39}$$

and also a random disturbance torque is added to simulate other external disturbance effects. Actuator dynamics is neglected assuming that electric motors with current/torque control are to be used. It is to be reminded that the practical performance of sliding mode controller is also affected by back lash effects. Therefore, a nonlinear backlash simulator is connected in series with the joint actuator to test the chattering that may occur when the sliding mode controller is implemented digitally. To test for robustness against load variation, the controller is allowed to work with payload variations from no load to 200Kg.

The trajectory considered contains translational motions of heave, surge and sway including rotational motions of roll and yaw. The most important thing to be considered in the trajectory is its speed. It is a fast trajectory where the platform moves heave at (400mm/1.98Hz), surge and sway at (150mm/5.9Hz). The rotational motions

Box 9.

$$\dot{V} = -DS\left(D\left(\ddot{q}_d + \Lambda_1 y_2 + \Lambda_3\left(Iy_2 + \mu f(X_1)\right)\right)\left(\frac{1}{1+g}D_N^{-1}\Delta D\right) + \Delta V + \Delta Q + K_f(S) - f_f\right) \tag{36}$$

Figure 3. X direction task space tracking error of PID, simple SMC and new SMC

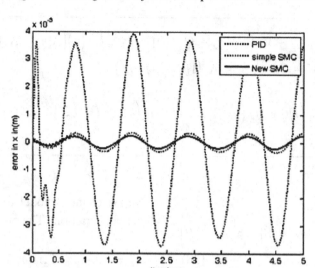

Figure 4. Y direction task space tracking error of PID, simple SMC and new SMC

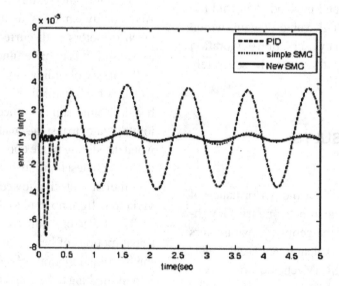

are also fast with rate of roll and yaw motions at (10°/2.7Hz, 10°/2.3Hz) and the maximum desired acceleration is 10g. The trajectories are designed to achieve zero velocity and acceleration initially and are as follows.

$$x(t) = 0.10\left\{1 - \exp\left(-3\pi t/10\right)\right\}\cos(1.88\pi t), m$$

$$y(t) = 0.10\left\{1 - \exp\left(-3\pi t/10\right)\right\}\sin(1.88\pi t), m$$

$$z(t) = 1.563 + \frac{0.15}{1+0.9t}\sin\left\{0.2\pi t\left(\frac{0.1+5.9t}{10.5}\right) + \frac{\pi}{24}\right\}, m$$

$$\alpha(t) = 0.15\left\{1 - \exp(-\pi t)\right\}\sin(0.86\pi t), rad$$

Figure 5. Z direction task space tracking error of PID, simple SMC and new SMC

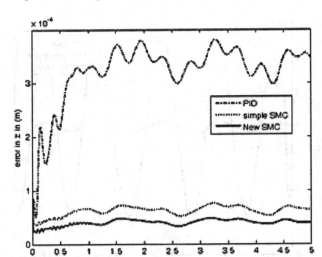

Figure 6. Roll angle tracking error of PID, simple SMC and new SMC

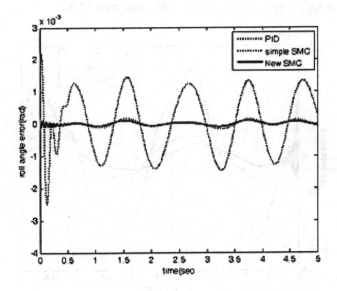

$$\beta(t) = 0, \deg$$

$$\gamma(t) = 0.15 \left\{ 1 - \exp(-\pi t) \right\} \sin(0.74\pi t), rad$$

To compare the performance of the controller with existing controllers, a standard PID controller, with its parameters tuned using genetic algorithm, is implemented. Moreover, a simple sliding mode controller, which has been suggested by many authors and contains joint error and error rate, is also implemented and compared with the new controller. The simulation results for the different cases are shown in Figure 3 through 18. The parameters used in the controllers are

Sliding surface parameters (Λ's) are taken as

$$\Lambda_1 = \text{diag}(250\ 250\ 250\ 250\ 250\ 250)$$

Figure 7. Pitch angle tracking error of PID, simple SMC and new SMC

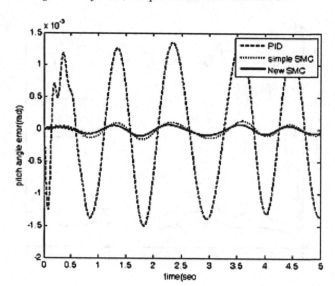

Figure 8. Yaw angle tracking error of PID, simple SMC and new SMC

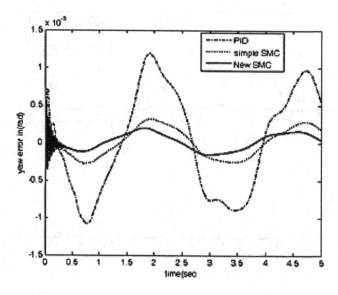

Λ_2=diag(1 1 1 1 1 1)

Λ_3=diag(400 400 400 400 400 400)

The coupling coefficients η_i are all taken to be unity.

As given in (37), to determine the maximum value of K, the uncertainty bounds have to be known. For the present case, the mass of the platform is 32Kg and this is assumed as the nominal. Then the payload can vary from 0 to 200Kg and hence uncertainty on mass and Inertia can be determined. The coriolis and centrifugal uncertainty is determined from the maximum velocity and acceleration. In our case the maximum velocity in x and y direction is 591mm/sec and maximum

Figure 9. X direction task space tracking error of PID, simple SMC and new SMC with 200Kg payload

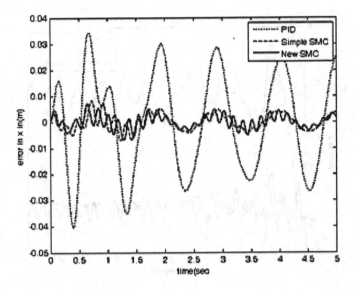

Figure 10. Y direction task space tracking error of PID, simple SMC and new SMC with 200Kg payload

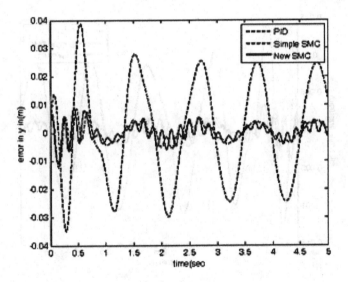

acceleration is assumed to be 10g. For gravitational uncertainty, the upward motion range is from 1.3m to 1.8m and hence using the mass variation gravitational torque variation will be determined.

Accordingly using (37)

K ≥ 200kg*1.8m*9.8m/sec²
+200Kgm*10*9.8m/sec²

hence

K=diag(6000 6000 6000 6000 6000 6000).

The PID controller parameters obtained using GA tuning are:

$$kp = \begin{bmatrix} 7.25 & 2.54 & 3.45 & 7.46 & 2.54 & 3.15 \end{bmatrix} x10^5$$

Figure 11. Z direction task space tracking error of PID, simple SMC and new SMC with 200Kg payload

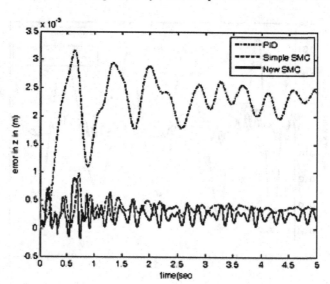

Figure 12. Roll angle tracking error of PID, simple SMC and new SMC with 200Kg payload

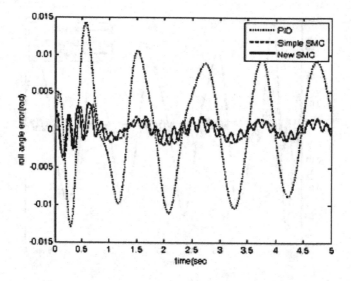

$$kd = \begin{bmatrix} 9.12 & 9.35 & 8.45 & 7.88 & 6.87 & 5.15 \end{bmatrix} x10^4$$

$$Ki = \begin{bmatrix} 1.023 & 0.611 & 1.45 & 0.92 & 0.86 & 0.77 \end{bmatrix} x10^3$$

The performance of the controllers for no load case is shown in Figures 3 through 8 while Figures 9 through 14 show the performance of the controllers when the platform is carrying full load of 200Kg and friction is considered. The friction force considered is given by (39) with k_v=50 and k_c=20 and a disturbance torque taken from a uniform random distribution with maximum value of 50NM is also added to each leg. To see the performance of the controllers in task space, their performance is compared in task space by

Figure 13. Pitch angle tracking error of PID, simple SMC and new SMC with 200Kg payload

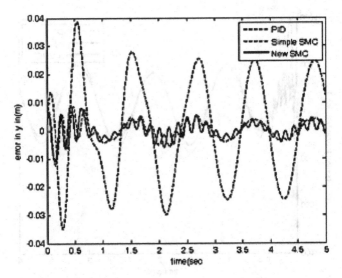

Figure 14. Yaw angle tracking error of PID, simple SMC and new SMC with 200Kg payload

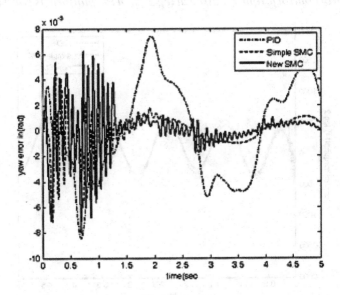

using a numerical forward kinematic estimation algorithm which is not used for control but only for comparison of outputs. The controller parameters used and summary of tracking errors in percentage of the maximum translation and rotations of the manipulator are given in Table 2 and Table 3 for no load and full load cases respectively. As can be seen from the table and the figures, the new sliding mode controller, which employs a new sliding surface containing synchronization error, is performing better than simple sliding mode controller. In the simple sliding mode controller, the tracking errors in x, y and z directions increased by 10, 10 and 5.34 percent respectively when the platform is fully loaded and friction is considered but in the new sliding mode controller,

Figure 15. Comparison of control torque required at leg 1when the platform is carrying payload of 200Kg

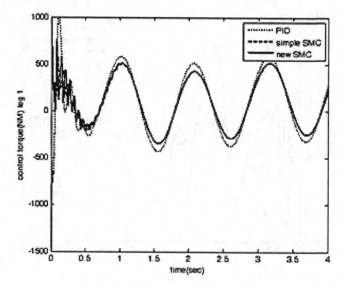

Figure 16. Comparison of control torque required at leg 2when the platform is carrying payload of 200Kg

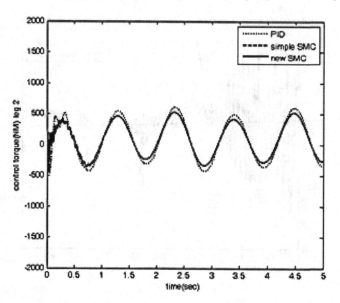

the corresponding values are only 8,6 and 2.66. Generally, in both no load and full load cases, the new sliding mode controller is more than 10% better in tracking error performance. The control effort of the three controllers is given in Figures 14 through 20 and the figures reveal that the control signals are smooth and practically realizable. The signals for the simple sliding mode control and the new sliding mode control are almost the same and have slight oscillations at the beginning. This is because; both of them make use of the desired acceleration which is very high, 10g.

Figure 17. Comparison of control torque required at leg 3when the platform is carrying payload of 200Kg

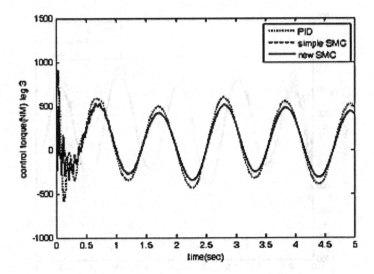

Figure 18. Comparison of control torque required at leg 4when the platform is carrying payload of 200Kg

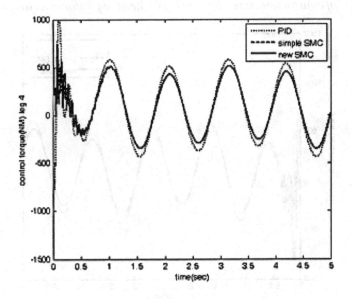

CONCLUSION

The design and simulation study of a sliding mode controller that can drive both tracking error and synchronization asymptotically to zero is discussed. The controller is a hybrid of task space and joint space because the equivalent control signal is obtained from task space in a feedforward manner while the switching controller is computed in joint space. The controller has two main advantages. The first one is it avoids the use of the estimation or measurement of the forward kinematics which is very difficult or costly and the other advantage is it can easily be implemented

Figure 19. Comparison of control torque required at leg 5when the platform is carrying payload of 200Kg

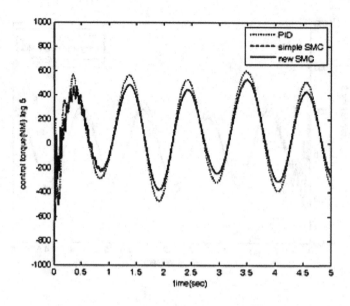

Figure 20. Comparison of control torque required at leg 6when the platform is carrying payload of 200Kg

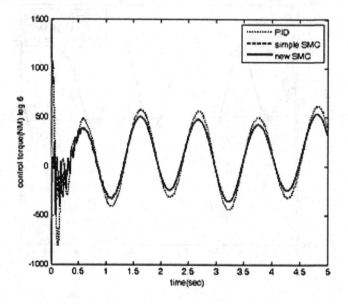

in hardware for real time application. Moreover, the controller considers synchronization error and hence results in a more precise and stable task space trajectory tracking and is safer for the operation of manipulator. The simulation results show that the controller performs better than simple sliding mode controller. The proposed control law needs to be experimentally verified and the new sliding surface can be improved by using neural network and fuzzy logic systems.

REFERENCES

Beji, L., Abichou, A., & Pascal, M. (1998, May). Tracking control of a parallel robot in the task space. In *Proceedings of the IEEE International Conference on Robotics & Automation*, Leuven, Belgium (pp. 2309-2304).

Chin, I.-H., & Li, C.-F. (2005, August 28-31). Smooth sliding mode tracking control of the Stewart platform. In *Proceedings of the IEEE Conference on Control Applications*, Toronto, ON, Canada (pp. 43-48).

Dasgupta, B., & Mruthyunthaya, T. S. (2000). The Stewart platform manipulator: a review. *Mechanism and Machine Theory*, *35*(1), 15–40. doi:10.1016/S0094-114X(99)00006-3.

Davliakos, I., & Papadopoulos, E. (2008). Model-based control of a 6 DOF electro hydraulic Stewart-Gough platform. *Journal of Mechanism and Machine Theory*, *43*(11), 1385–1400. doi:10.1016/j.mechmachtheory.2007.12.002.

Dereje, S., & Mitra, R. (2010). Fuzzy logic tuned PID controller for Stewart platform manipulator. In *Proceedings of the International Conference on Computer Applications in Electrical Engineering and Recent Advances*, Roorkee, India (pp. 272-257).

Dereje, S., & Mitra, R. (2010). Neuro-fuzzy sliding mode control: design and stability analysis. *International Journal of Computational Intelligence Studies*, *1*(3), 242–255. doi:10.1504/IJCISTUDIES.2010.034888.

Dongya, Z., Shaoyuan, L., & Feng, G. (2008). Fully adaptive feedforward feedback synchronized tracking control for Gough-Stewart platform systems. *International Journal of Control. Automation and Systems*, *6*(5), 689–701.

Fichter, E. F., Kerr, D. R., & Jones, J. R. (2009). The Gough–Stewart platform parallel manipulator: a retrospective appreciation. *Proceedings of the Institution of Mechanical Engineers. Part C, Journal of Mechanical Engineering Science*, *223*(1). doi:10.1243/09544062JMES1137.

Flavien, P., Nicolas, A., Philippe, M., & Khalil, W. (2006). Vision based computed torque controller for parallel robots. In *Proceedings of the IEEE 32 Annual Conference on Industrial Electronics* (pp. 3851-3856).

Ghobakhloo, A., Eghtesad, M., & Azadi, M. (2006). Position control of Gough-Stewart-Gough platform using inverse dynamics with full dynamics. In *Proceedings of the IEEE 9th International Workshop on Advanced Motion Control* (pp. 50-55).

Ghorbel, F. H., Chételat, O., Gunawardana, R., & Longchamp, R. (2000). Modeling and set point control of closed chain mechanisms, theory and experiment. *IEEE Transactions on Control Systems Technology*, *8*(5), 801–815. doi:10.1109/87.865853.

Guo, H. B., & Li, H. R. (2006). Dynamic analysis and simulation of a six degree of freedom Stewart platform manipulator. *Proceedings of the Institution of Mechanical Engineers. Part C, Journal of Mechanical Engineering Science*, *220*(1), 61–72. doi:10.1243/095440605X32075.

Hag, S. K., Young, M. C., & Kyo, I. L. (2005). Robust nonlinear task space control for 6 DOF parallel manipulator. *Automatica*, *41*, 1591–1600. doi:10.1016/j.automatica.2005.04.014.

Hung, J. Y., Gao, W., & Hung, J. C. (1993). Variable structure control: A survey. *IEEE Transactions on Industrial Electronics*, *40*(1), 2–19. doi:10.1109/41.184817.

Iqbal, S., & Bhatti, A. I. (2007). Robust sliding-mode controller design for a Stewart platform. In *Proceedings of the International Bhurban Conference on Applied Sciences & Technology*, Islamabad, Pakistan (pp. 155-160).

Iqbal, S., & Bhatti, A. I. (2008). Dynamic analysis and robust control design for Gough-Stewart platform with moving payloads. In *Proceedings of the 17th World Congress the International Federation of Automatic Control*, Seoul, Korea (pp. 5324-5329).

Li, D., & Salcudean, S. E. (1997, April). Modeling, simulation, and control of a hydraulic Stewart platform. In *Proceedings of the IEEE International Conference on Robotics and Automation*, Albuquerque, NM (pp. 3360-3366).

Merlet, J. P. (2006). *Parallel robots* (2nd ed.). Sophia-Antipolis, France: Springer.

Nag, I.-K., & Chong, W.-L. (1998). High speed tracking control of Gough-Stewart platform manipulator via enhanced sliding model controller. In *Proceedings of the International Conference on Robotics and Automation*, Leuven, Belgium (pp. 2716-2721).

Negash, D. S., & Mitra, R. (2010). Integral sliding mode control for trajectory tracking control of Stewart platform manipulator. In *Proceedings of the IEEE-ICIIS International Conference on Industrial and Information Systems*, Surathkal, India (pp. 650-654).

Niguyen, C. C., Zhen, L. Z., & Sami, S. A. (1997). Efficient computation of forward kinematics and Jacobian matrix of a Gough-Stewart platform based manipulator. In. *Proceedings of the IEEE International Conference on Robotics and Automation*, *1*, 869–874.

Se-Han, L., Jae-Bok, S., Woo-Chun, C., & Daehie, H. (2003). Position control of a Stewart platform using inverse dynamics control with approximate dynamics. *Mechatronics*, *13*, 605–619. doi:10.1016/S0957-4158(02)00033-8.

Shaowen, F., & Yu, Y. (2006). Non-linear robust control with partial inverse dynamic compensation for a Stewart platform manipulator. *International Journal of Modelling. Identification and Control*, *1*(1), 44–51. doi:10.1504/IJMIC.2006.008647.

Sirouspour, M. R., & Salcudean, S. E. (2001). Nonlinear control of hydraulic robots. *IEEE Transactions on Robotics and Automation*, *17*(2), 173–192. doi:10.1109/70.928562.

Su, Y. X., & Duan, B. Y. (2000). The application of the Stewart platform in large spherical radio telescopes. *Journal of Robotic Systems*, *17*(7), 375–383. doi:10.1002/1097-4563(200007)17:7<375::AID-ROB3>3.0.CO;2-7.

Su, Y. X., Duan, B. Y., Zheng, C. H., Zhang, Y. F., Chen, G. D., & Mi, J. W. (2004). Disturbance-rejection high-precision motion control of a Stewart platform. *IEEE Transactions on Control Systems Technology*, *12*(3). doi:10.1109/TCST.2004.824315.

Yung, T., Yu-Shin, C., & Shih-Ming, W. (1999). Task space control algorithm for Gough-Stewart platform. In *Proceedings of the 38th Conference on Decision and Control*, Phoenix, AZ (pp. 3857-3862).

This work was previously published in the International Journal of Intelligent Mechatronics and Robotics, Volume 1, Issue 4, edited by Bijan Shirinzadeh, pp. 19-43, copyright 2011 by IGI Publishing (an imprint of IGI Global).

Chapter 17
Kinematics and Dynamics Modeling of a New 4–DOF Cable–Driven Parallel Manipulator

Hamoon Hadian
Isfahan University of Technology, Iran

Yasser Amooshahi
Isfahan University of Technology, Iran

Abbas Fattah
University of Delaware, USA

ABSTRACT

This paper addresses the kinematics and dynamics modeling of a 4-DOF cable-driven parallel manipulator with new architecture and a typical Computed Torque Method (CTM) controller is developed for dynamic model in SimMechanics. The novelty of kinematic architecture and the closed loop formulation is presented. The workspace model of mechanism's dynamic is obtained in an efficient and compact form by means of natural orthogonal complement (NOC) method which leads to the elimination of the nonworking kinematic-constraint wrenches and also to the derivation of the minimum number of equations. To verify the dynamic model and analyze the dynamical properties of novel 4-DOF cable-driven parallel manipulator, a typical CTM control scheme in joint-space is designed for dynamic model in SimMechanics.

DOI: 10.4018/978-1-4666-3634-7.ch017

INTRODUCTION

Cable-driven parallel manipulator is a special class of parallel manipulator in which the moving platform is driven by cables, instead of rigid links. In recent years, these manipulators have been researched extensively (Ebert-Uphoff & Voglewede, 2004; Kawamura, Kino, & Won, 2000; Albus, Bostelman, & Dagalakis, 1993; Mustafa, Yang, Yeo, Lin & Chen, 2008; Homma, Fukuda, Sugawara, Nagata, & Usuba, 2003; Takemura, Enomoto, Tanaka, Denou, Kobayashi, & Tadokoro, 2005). Because of their advantages to provide a light-weight structure, low inertial properties, and large reachable workspace. A cable robot has a light-weight structure with low moving mass because the actuators are always mounted onto the base and the driving cables have negligible masses. Large workspace is achievable for a cable robot since the cables can be wound onto drums to provide infinite length, unlike the rigid cables with fixed lengths. These advantages make the cable manipulator a promising candidate for applications requiring high speed, high acceleration, and high payload but with moderate stiffness and accuracy (Ming & Higuchi, 1994; Ebert-Uphoff & Voglewede, 2004; Kawamura, Kino, & Won, 2000; Albus, Bostelman, & Dagalakis, 1993; Mustafa, Yang, Yeo, Lin & Chen, 2008). Other advantages of these robots include their scalability, adaptability, and safety. As results, cable robots have been employed for long-range position measurement devices (Takemura, Enomoto, Tanaka, Denou, Kobayashi, & Tadokoro, 2005), service robots (Takahashi & Tsubouchi, 2000; Mustafa, Yang, Yeo, Lin, & Chen, 2008), and rehabilitation systems (Homma, Fukuda, Sugawara, Nagata, & Usuba, 2003). It is noted that the unilateral driving property of cables makes the well-developed modeling and analysis methods for conventional rigid-cable parallel manipulators not applicable to cable robots.

There is much prior work in kinematic, static, and dynamic analysis of cable robotic systems. Analysis, simulations and detail hardware implementation of cable-driven robots provided in (Williams & Gallina, 2001). Dynamic analysis of cable array robotic cranes is presented in Shiang, Cannon, and Gorman (1999) in the case of rigid cables and in Shiang, Cannon, and Gorman (2000) for flexible cables. The governing equations of motion for the parallel manipulators could be derived by Newton-Euler method (Fattah & Kasaei, 2000), Lagrange formula (Guo & Li, 2006) and screw theory (Kong & Gosselin, 2005). Other work in design and control of fully constrained cable driven robots includes the WARP (Maeda, Tadokoro, Takamori, Hiller, & Verhoeven, 1999) and FALCON (Kawamura, Choe, Tanaka, & Pandian, 1995) systems. Prior art in trajectory control of under-constrained cable robots is somewhat limited. The authors of Yamamoto, Yanai, and Mohri (2004) employ inverse dynamics and feedforward and feedback control method to provide trajectory control of an incompletely constrained wrench-type cable robot with mobile actuators. Control is achieved through a PD controller and a pre-compensator. The authors of Alp and Agrawal (2002), Basar and Agrawal (2002), and Oh and Agrawal (2005) provide simulation and experimental results of two closed-loop asymptotic control mechanisms based on Lyapunov design techniques and feedback linearization respectively. Moreover, isotropic design of spatial cable robots, particularly a 6-6 cable-suspended parallel robot is studied in Hadian and Fattah (2008). Dexterity analysis of a 3-DOF cable-driven parallel manipulator with a new architecture based on distribution of tension among cables as well as robot stiffness is conducted in Hadian and Fattah (2009).

However, to the best knowledge of the authors, the proposed novel cable-driven manipulator in this paper has not been presented elsewhere. In this paper we address the kinematics and dynamics modeling of a cable-driven parallel manipulator with a new architecture supposed to be used for the moving mechanism of a flight simulator. The merits of this type of architecture, as compared with conventional type like Stewart platform, are independent and low degrees of freedom (DOF), easy to control, advantage of manufacturing, low cost, chance to increase DOF by changing the type of joints. This parallel manipulator is composed of a moving platform (MP), a base platform (BP), four cables, and a passive leg as depicted in Figure 1. The closest architecture to this new 4DOF cable robot is presented in Hadian and Fattah (2011). The 3DOF cable robot is proposed in Hadian and Fattah (2011) consists of a MP and a BP that are connected to each other by means of three cables and one passive leg. According to the cable numbers classification, that robot is under-constrained, but it is redundant architecturally due to the presence of the passive leg, the MP is located on the top of BP and controlled by extending or retracting the cables around the pulleys that are mounted on BP. The passive leg connects the centers of both platforms to each other in which it is attached to the MP by a universal joint and fixed to the BP. The system has 3DOF with three motors and pulleys that are mounted on the BP. Hence the MP has three independent motions: translational along the z-axis (heave), rotation about the y-axis (pitch) and rotation about the x-axis (roll) (Hadian & Fattah, 2011). The main advantage of this new 4DOF cable parallel manipulator at hand is one more DOF, i.e., Yaw rotation about z-axis over that 3DOF cable robot. It may be noted that this current new robot is able to perform all maneuvers of the classic 6DOF Stewart platform by low degrees of freedom, low cost and easy to control.

The main contribution of this paper is investigation of kinematics and dynamics modeling of a 4 degree-of-freedom cable-driven manipulator with new architecture. The main power of this novel mechanism would be the production capability of all six degree-of-freedom with low cost, easy to control, low degrees of freedom, advantages of manufacturing, reconfigurability, low inertial properties, large workspace and thus deployable for flight simulators applications. The remainder of this paper is as follows: first, hardware overview of the robot at hand including architecture design, kinematic equations, and Jacobian matrix is fully described. Afterwards, the dynamic equation of this novel parallel cable mechanism is derived using NOC method. Then, a dynamic model in SimMechanics is presented. Afterwards, a typical CTM control scheme in joint-space is developed to verify the dynamic model and analyze the dynamical properties of novel 4-DOF cable-driven parallel manipulator. Moreover, kinematics, dynamics and control simulation results of a computation example of 4-DOF cable-driven parallel manipulator is discussed. Eventually, the paper based on the achievement results is concluded.

KINEMATICS

In this section, the kinematics of a cable manipulator with a new architecture is studied. This cable robot consists of a MP and a BP that are connected to each other by means of four cables and one passive leg, as shown in Figure 1. For

Figure 1. Schematic design of the robot at hand (left) and kinematics of the passive leg (right)

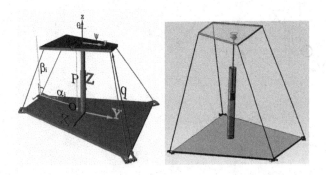

any cable-driven manipulator, an equivalent rigid link counterpart can be found by replacing each cable by a rigid link and ball-and-socket joints at the ends. If the cable has a variable length, then a cylindrical element should be used to represent the cable in the rigid link manipulator.

The central leg is an element that produces a force between the base platform and the moving platform in order to keep all the cables in tension. The central leg can be an active element which generates a desired force. It can be also a passive element such as a cylinder energized by compressed air or a compressive spring designed properly to provide the sufficient force required to maintain tension in the cables. As shown in Figure 1, the direction of the central leg force, pointing towards the moving platform. In this paper, the central leg is considered as a passive leg that is connected to centroid of MP by spherical joint and is fixed to BP. The DOF for the system at hand is obtained to be 4 using Chebyshev-Grübler-Kutzbach formula. Thus four cable pulleys and actuators are used to drive four cables with idle prismatic joint for the central leg. As shown in Figures 1 and 2, by extending or retracting the cables around the pulleys the passive leg acts as an idle prismatic joint moves such as a cylinder along z-axis and thus provides the heave motion of the MP while

all the cables remain in tension. This new design with the central leg and four cables has four motors and pulleys that are mounted on the BP. Hence it provides four independent DOF for the MP, namely, heave h, vertical displacement of MP along z-axis; pitch ψ, rotation of MP about y-axis; roll φ, rotation of MP about x-axis, and yaw θ, rotation of MP about z-axis as shown in Figure 1.

The independent DOF for MP is necessary for moving mechanisms in flight simulators in which any desired and independent motion for the MP should be available. This is a main advantage of this new type of manipulator as compared to its counterpart with other DOF. According to the classification of cable robots, this robot is under-constrained, but it is redundant architecturally due to the presence of the passive leg.

It is noted that the MP is located on the top of BP and controlled by extending or retracting the cables around the pulleys that are mounted on BP. The passive leg connects the centers of both platforms to each other in which it is attached to the MP by a spherical joint and fixed to the BP. The frames XYZ and xyz are inertial and moving frames attached to the BP and MP at points O and O_m, respectively. Moreover, position vector of O_m with respect to O is defined by p and the rotation matrix of MP with respect to reference

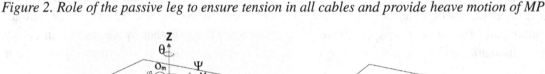

Figure 2. Role of the passive leg to ensure tension in all cables and provide heave motion of MP

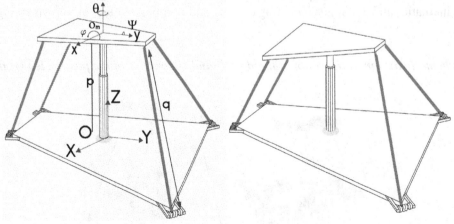

frame by R. The rotation matrix R can be readily determined by having the roll and pitch angle of MP. The ith cable is connected to MP by A_i and to BP by B_i position vector of A_i with respect to O_m is shown by $^m a_i$ and position vector of B_i with respect to O is depicted by b_i

$$^m \mathbf{a}_1 = \left[r_m \cos(\frac{\phi_m}{2}) \quad -r_m \sin(\frac{\phi_m}{2}) \quad 0 \right]^T \quad (1)$$

$$^m \mathbf{a}_2 = \left[r_m \cos(\frac{\phi_m}{2}) \quad r_m \sin(\frac{\phi_m}{2}) \quad 0 \right]^T \quad (2)$$

$$^m \mathbf{a}_3 = \mathbf{Q}_1 \, ^m \mathbf{a}_1 \quad (3)$$

$$^m \mathbf{a}_4 = \mathbf{Q}_2 \, ^m \mathbf{a}_2 \quad (4)$$

$$\mathbf{b}_1 = \left[r_b \cos(\frac{\phi_b}{2}) \quad -r_b \sin(\frac{\phi_b}{2}) \quad 0 \right]^T \quad (5)$$

$$\mathbf{b}_2 = \left[r_b \cos(\frac{\phi_b}{2}) \quad r_b \sin(\frac{\phi_b}{2}) \quad 0 \right]^T \quad (6)$$

$$\mathbf{b}_3 = \mathbf{Q}_2 \mathbf{b}_1 \quad (7)$$

$$\mathbf{b}_4 = \mathbf{Q}_1 \mathbf{b}_2$$

where ϕ_m and ϕ_b are the separation angles of the MP and BP, respectively, and also r_m and r_b are the radii of the circumferential circles of the MP and BP, respectively, as shown in Figure 3. Moreover, Q_1 and Q_2 are rotation matrices about the z-axis through angles 150° and 210°, respectively, as

$$\mathbf{Q}_1 = \begin{bmatrix} \cos(5\pi/6) & -\sin(5\pi/6) & 0 \\ \sin(5\pi/6) & \cos(5\pi/6) & 0 \\ 0 & 0 & 1 \end{bmatrix} \quad (8)$$

$$\mathbf{Q}_2 = \begin{bmatrix} \cos(7\pi/6) & -\sin(7\pi/6) & 0 \\ \sin(7\pi/6) & \cos(7\pi/6) & 0 \\ 0 & 0 & 1 \end{bmatrix} \quad (9)$$

The inverse position kinematics is defined as follows: Given P and R, determine the motion of actuators of the cables, i.e., q_i and e_i. The vector $\mathbf{l}_i = q_i \mathbf{e}_i$ can be written as

$$\mathbf{l}_i = q_i \mathbf{e}_i = \mathbf{p} + \mathbf{a}_i - \mathbf{b}_i$$

For

$$i = 1,2,3,4 \quad (10)$$

The length of ith leg can be expressed as

Figure 3. Geometric shapes of MP (left) and BP (right)

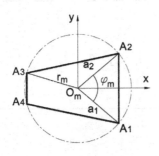 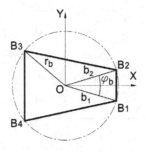

$$q_i^2 = \mathbf{l_i}^T \mathbf{l_i} = (\mathbf{p} + \mathbf{a}_i - \mathbf{b}_i)^T (\mathbf{p} + \mathbf{a}_i - \mathbf{b}_i) \quad (11)$$

Having $\mathbf{b}_i$, $\mathbf{p}$ and $\mathbf{a}_i = R\ ^E\mathbf{a}_i$, length q_i can be easily determined from Equation (11). Moreover, the unit vector $\mathbf{e}_i$ can be obtained from Equation (10) as

$$\mathbf{e}_i = \frac{\mathbf{l_i}}{q_i} = \frac{\mathbf{p} + \mathbf{a}_i - \mathbf{b}_i}{q_i} \quad (12)$$

In other words, the unit vector $\mathbf{e}_i$ can be written in terms of angles α_i and β_i as

$$\mathbf{e}_i = \begin{bmatrix} \sin\beta_i \cos\alpha_i & \sin\beta_i \sin\alpha_i & \cos\beta_i \end{bmatrix} \quad (13)$$

As shown in Figure 1, α_i and β_i the angles of cables with coordinate axes are passive joint variables that are determimed from Equation 13 in term of $\mathbf{e}_i$ for i=1, 2, 3, 4.

The inverse velocity kinematics for the manipulator at hand can be defined as follows: Given the linear velocity of point O_m of MP, i.e., $\dot{\mathbf{p}}$ and $\dot{\mathbf{R}}$, determine the velocity of each cable, namely, $\dot{q}_i$ and $\dot{e}_i$. This can be done by differentiating both sides of Equation 11 with respect to

Table 1. Parameters of the 4-DOF cable robot

Parameters	Value
Radius of base platform, r_b (m)	2
Radius of moving platform, r_e (m)	1
Minimal/Maximal stroke of passive leg (m)	0/1
Initial length of passive leg (m) Minimal/Maximal rotation angles, φ, ψ, θ (deg)	1 -30/30
Mass of moving platform and payloads, m(Kg)	100
Moment of inertia I_{xx}, I_{yy}, I_{zz} (Kg.m²)	120

time, thus obtaining Equation 14 in Box 1, where $\mathbf{b}_i$ is constant and thus $\dot{\mathbf{b}}_i = \mathbf{0}$. Moreover, defining ω as the angular velocity of MP with respect to reference frame and expanding Equation (14), one can obtain

$$\left[\dot{\mathbf{a}}_i \times (\mathbf{p} + \mathbf{a}_i - \mathbf{b}_i) \right]^T \omega + (\mathbf{p} + \mathbf{a}_i - \mathbf{b}_i)^T \dot{\mathbf{p}} = q_i \dot{q}_i \quad (15)$$

Equation (16) can be written in matrix form upon substituting $i=1,2,3,4$ as follows

$$\mathbf{A} \mathbf{t}_{MP} = \mathbf{B} \dot{\mathbf{q}} \quad (16)$$

Box 1.

$$(\dot{\mathbf{p}} + \dot{\mathbf{a}}_i - \dot{\mathbf{b}}_i)^T (\mathbf{p} + \mathbf{a}_i - \mathbf{b}_i) + (\mathbf{p} + \mathbf{a}_i - \mathbf{b}_i)^T (\dot{\mathbf{p}} + \dot{\mathbf{a}}_i - \dot{\mathbf{b}}_i) = 2 q_i \dot{q}_i \quad (14)$$

Box 2.

$$\mathbf{A} = \begin{bmatrix} (\mathbf{a}_i \times \mathbf{l}_i)^T & \mathbf{l}_i^T \end{bmatrix}, \quad \mathbf{B} = \begin{bmatrix} q_1 & 0 & 0 & 0 \\ 0 & q_2 & 0 & 0 \\ 0 & 0 & q_3 & 0 \\ 0 & 0 & 0 & q_4 \end{bmatrix}, \quad \dot{\mathbf{q}} = \begin{bmatrix} \dot{q}_1 & \dot{q}_2 & \dot{q}_3 & \dot{q}_4 \end{bmatrix}^T \quad (17)$$

where $\mathbf{t}_{MP} = \begin{bmatrix} \omega^T & \dot{\mathbf{p}}^T \end{bmatrix}^T$ is a 6-dimensional twist vector of MP comprising of ω and the linear velocity of point O_m in reference frame. Moreover, A is a 4×6 matrix, B is a 4×4 diagonal matrix, and $\dot{\mathbf{q}}$ is a 4-dimensional vector of the cables velocities defined, respectively, as Equation 17 in Box 2.

The angular velocity ω of MP can be determined by writing $\dot{\mathbf{R}}$ as $\Omega = \dot{R}^T R$ where Ω is the cross-product matrix of ω. The vector ω can be readily determined from the off diagonal arrays of the skew-symmetric matrix Ω. Next, from Equation (16), $\dot{\mathbf{q}}$ can be written as

$$\dot{\mathbf{q}} = \mathbf{B}^{-1}\mathbf{A}\mathbf{t}_{MP} \qquad (18)$$

Here, the inverse of B can be readily calculated because B is a diagonal matrix.

Thereafter, $\dot{\mathbf{e}}_i$ can be determined by differentiating both sides of Equation (12) with respect to time, namely,

$$\dot{\mathbf{e}}_i = \frac{(\dot{\mathbf{p}} + \dot{\mathbf{R}}\mathbf{a}_i - \dot{q}_i\mathbf{e}_i)}{q_i} \qquad (19)$$

The time rate of changes for the angles α_i and β_i, i.e., $\dot{\alpha}_i$ and $\dot{\beta}_i$ are determined upon differentiating both sides of Equation (13) with respect to time.

Finally, the inverse acceleration kinematics are solved to obtain the cable acceleration $\ddot{q}_i$ and angular acceleration of cables with the axes of the reference frame, i.e., $\ddot{\alpha}_i$ and $\ddot{\beta}_i$ for a given linear and angular acceleration of MP. To this end, upon differentiating both sides of Equation (19) with respect to time, thus obtaining $\ddot{\mathbf{q}}$ as

$$\ddot{\mathbf{q}} = \dot{\mathbf{B}}^{-1}\mathbf{A}\mathbf{t}_{MP} + \mathbf{B}^{-1}\dot{\mathbf{A}}\mathbf{t}_{MP} + \mathbf{B}^{-1}\mathbf{A}\dot{\mathbf{t}}_{MP} \qquad (20)$$

Then $\ddot{\mathbf{e}}_i$ can be determined from differentiation of Equation (19) with respect to time. Finally, $\ddot{\alpha}_i$ and $\ddot{\beta}_i$ obtain from twice differentiation of Equation (13) with respect to time.

DYNAMICS

The governing equations of motion of this manipulator can be expressed in terms of nonlinear differential equations by modeling the manipulator as a mechanical system with kinematic loops. The independent governing equations of motion of the system can be determined using NOC method. This method is based on determining the orthogonal complement of the kinematic constraint velocity matrix. The form of this matrix depends on whether the system is being formulated in joint space or in Cartesian space. It has been shown that using the methodology of NOC leads to the elimination of the nonworking kinematic-constraint wrenches and also to the derivation of the minimum number of equations. A basic problem related to dynamics of the manipulator at hand is the study of inverse and forward dynamics. Inverse dynamics has potential applications in obtaining the power of actuators and in controlling the system of the manipulator at hand. To this end, having the motion of MP and using the inverse kinematics of the system, the time history of cables motion and their time derivatives are determined. Thereafter, the actuator forces of the cables can be obtained such that it produces the desired motion of MP.

The governing equations of motion can be determined as follows: First the dynamical equations of motion for each cable of the legs and MP are written using Newton-Euler equations (Fattah & Kasaei, 2000). These equations are expressed in terms of the twist vector of each cable that is a six dimensional vector composed of angular and linear velocities of the cable. Then by assembling all equations of motion of all cables together, the governing equations of the whole system are obtained. The next step is to formulate the NOC matrix N and pre-multiplying the governing equations by N to obtain the minimum number of equations of motion of the system at hand.

Modeling

The Newton-Euler's formula for each cable of the system can be written as (Fattah & Kasaei, 2000; Amooshahi & Hadian, 2009)

$$M_i \dot{t}_i + \Omega_i M_i t_i = W_i^E + W_i^c \qquad (21)$$

where t_i is twist vector of cable i which can be expressed in terms of angular velocity of cable i, i.e., ω_i and linear velocity of center of mass of cable i, i.e., $\dot{c}_i$ as

$$t_i = \begin{bmatrix} \omega_i \\ \dot{c}_i \end{bmatrix} \qquad (22)$$

Moreover, M_i and Ω_i are extended mass matrix and angular velocity matrix that can be defined as

$$M_i = \begin{bmatrix} I_i & 0_{33} \\ 0_{33} & m_i 1_{33} \end{bmatrix} \Omega_i = \begin{bmatrix} \omega_i \times 1_{33} & 0_{33} \\ 0_{33} & 0_{33} \end{bmatrix} \qquad (23)$$

Here I_i is inertia matrix of cable i about its center of mass, 0_{33} and 1_{33} are 3×3 zero and identity matrices, respectively, m_i is the mass of cable i and $\omega_i \times 1_{33}$ is cross product matrix of angular velocity. W_i^E is the external forces and moments as well as actuators forces applied on cable i and W_i^c is nonworking kinematic constraint wrenches.

For any cable-driven parallel manipulator, an equivalent rigid link counterpart can be found by replacing each cable by a rigid link and ball-and-socket joints at the ends. If the cable has a variable length, then a cylindrical element should be used to represent the cable in the rigid link manipulator. This analogy is valid as long as the cable-based manipulator is rigid according to the following definition (Behzadipour & Khajepour, 2006):

Rigidity

A cable-driven parallel manipulator is rigid at a certain pose with respect to a given external load (including dynamic loads) and passive leg force if and only if all cables are in tension. A positive cable force is considered as a tensile force in the cable. Therefore, the governing equations of motion of the whole system are determined by assembling the dynamics of all rigid bodies, represented by Equation (21), thereby obtaining

$$M\dot{t} + \Omega Mt = W^E + W^c \qquad (24)$$

where M is the generalized extended mass matrix, Ω is the generalized angular velocity matrix, t is the generalized twist vector and W is the generalized wrench vector of the system and can be written as

$$\mathbf{M} = diag(\mathbf{M}_1, \mathbf{M}_2, \ldots, \mathbf{M}_r) \qquad (25)$$

$$\Omega = diag(\Omega_1, \Omega_2, \ldots, \Omega_r) \qquad (26)$$

$$\mathbf{t} = \begin{bmatrix} \mathbf{t}_1^\mathbf{T} & \mathbf{t}_2^\mathbf{T} & \cdots & \mathbf{t}_r^\mathbf{T} \end{bmatrix}^T \qquad (27)$$

$$\mathbf{W} = \begin{bmatrix} \mathbf{W}_1^\mathbf{T} & \mathbf{W}_2^\mathbf{T} & \cdots & \mathbf{W}_r^\mathbf{T} \end{bmatrix}^T \qquad (28)$$

Here, r is the number of moving links plus the MP of the system. For an n-degree-of-freedom (DOF) system, the generalized twist vector t can be expressed as a linear transformation of $\dot{q}$, which is an n-dimensional vector of independent generalized speeds, namely, t=N$\dot{q}$, where N is orthogonal complement of kinematic constraint velocity matrix and because of its definition, it was named the natural orthogonal complement. By definition, the power developed by the non-working kinematic constraint wrench W^C vanishes and therefore by pre-multiplying both sides of Equation (24) by N^T and inserting $t = N\dot{q}$, the governing equations of motion of the system can be obtained as

$$N^T M N \ddot{q} + N^T \left(M\dot{N} + \Omega M N \right) \dot{q} = N^T W^E$$
$$W^E = W^M + W^G + W^L + W^D \tag{29}$$

where W^M, W^G, W^L and W^D are actuator wrench, gravity wrench, fixed passive leg wrench and damping wrench, all on MP, respectively. The generalized angular velocity matrix Ω can be determined by kinematic analysis of the manipulator at hand and the generalized mass matrix is readily computed by having the inertia of each cable and MP. The effect of damping forces is neglected in this project. The gravity wrench for each cable is a zero vector except for moving platform that is a vector which all arrays is zero except the last one which is $m_i g$ (m_i is the mass of cable i). As mentioned in kinematic section, the central leg is an element that produces a force between the base platform and the moving platform in order to keep all the cables in tension. In this paper, passive element such as a cylinder energized by compressed air or a compressive spring is designed properly to provide the sufficient force required to maintain tension in the cables. The value of pushing passive leg force on MP is assumed twice the weight of MP in order to ensure the tension condition in all cables. Therefore, the passive leg wrench indicates a vector which all arrays is zero except the last one which is $2mg$. Here m is the mass of MP in kilogram.

Having the geometric and inertia properties of the manipulator at hand and solving its kinematics, it is possible to compute $T = N^T W^M$ from Equation (29). It may be noted that the components of the vector are actuator forces of the four cables because the lengths of four cables have been chosen

as the independent generalized coordinates for the system at hand. The computation of N is the most important part which will be described in next subsection.

Computation of NOC Matrix N

The natural orthogonal complement (NOC) matrix N can be computed symbolically or numerically. Symbolic computation of N for parallel cable manipulators and in general for mechanical systems with kinematic loops is very cumbersome and it is sometimes impossible to express explicit relations in terms of independent generalized speeds. Therefore, numerical computation of N is an alternative method which can be used. Here, N is computed symbolically as follows: With reference to Figure 1, q_i the length of cables are joint space variables; φ roll, ψ pitch, θ yaw and h heave of MP are Cartesian space variables; and α_i and β_i the angles of cables with coordinate axes are passive joint variables. The twist vector of each cable can be expressed in terms of independent generalized speeds $\dot{q} = \left[\dot{q}_1, \dot{q}_2, \dot{q}_3, \dot{q}_4\right]^T$ and passive generalized speeds

$$\eta = \left[\dot{\alpha}_1, \dot{\beta}_1, \dot{\alpha}_2, \dot{\beta}_2, \dot{\alpha}_3, \dot{\beta}_3, \dot{\alpha}_4, \dot{\beta}_4\right]^T.$$

Then expressing the kinematic constraint equations governing the system and in the light of their time derivatives, vector η can be expressed in terms of $\dot{q}$ and thereafter the generalized twist vector of the system can be expressed, in turn, in terms of $\dot{q}$. Finally, having the above relations, matrix N can be derived symbolically.

The twist vector of each cable is written as (Amooshahi & Hadian, 2009)

$$t_{1i} = \begin{bmatrix} \Lambda_i \\ l_{1i}\Gamma_i \end{bmatrix} \begin{bmatrix} \dot{\alpha}_i \\ \dot{\beta}_i \end{bmatrix} \quad i = 1, 2, 3, 4 \tag{30}$$

$$t_{2i} = \begin{bmatrix} 0_{31} \\ e_i \end{bmatrix} \dot{q}_i + \begin{bmatrix} \Lambda_i \\ (q_i - l_{2i})\Gamma_i \end{bmatrix} \begin{bmatrix} \dot{\alpha}_i \\ \dot{\beta}_i \end{bmatrix} \qquad (31)$$
$$i = 1, 2, 3, 4$$

Here t_{1i} and l_{1i} denote the twist vector of ith cable and center of ith cable length where the robot is posed at home position, respectively. Consequently, t_{2i} and l_{2i} denote the twist vector of ith cable and center of ith cable length where the robot moves from its home position.where Λ_i and Γ_i are 3×2 matrices written as

$$\Lambda_i = \begin{bmatrix} 0 & -\sin\alpha_i \\ 0 & -\cos\alpha_i \\ 1 & 0 \end{bmatrix} \qquad i = 1, 2, 3, 4 \qquad (32)$$

$$\Gamma_i = \begin{bmatrix} -\sin\alpha_i \sin\beta_i & \cos\alpha_i \cos\beta_i \\ \cos\alpha_i \sin\beta_i & \sin\alpha_i \cos\beta_i \\ 0 & -\sin\alpha_i \end{bmatrix} \qquad (33)$$
$$i = 1, 2, 3, 4$$

The velocities of four non-collinear points of the MP are expressed in terms of the independent generalized speed $\dot{q}$ and passive joint speeds η.

Upon substitution of twist vectors of each cable from Equations (30) and (31) and MP into Equation (27), one obtains,

$$t = N_1 \dot{q} + N_2 \eta \qquad (34)$$

where N_1 and N_2 are 54×3 and 54×6 matrices. By substituting η into Equation (34) and factoring out $\dot{q}$, one derives,

$$\mathbf{t} = (\mathbf{N}_1 + \mathbf{N}_2\mathbf{J})\dot{\mathbf{q}} \qquad (35)$$

Therefore, the matrix N obtains as

$$\mathbf{N} = \mathbf{N}_1 + \mathbf{N}_2\mathbf{J} \qquad (36)$$

Moreover, $\dot{\mathbf{N}}$ is also computed as

$$\dot{\mathbf{N}} = \dot{\mathbf{N}}_1 + \dot{\mathbf{N}}_2\mathbf{J} + \mathbf{N}_2\dot{\mathbf{J}} \qquad (37)$$

The detail computation of N_1 and N_2 matrices could be found out in (Amooshahi & Hadian, 2009).

Results and Discussions

To verify the dynamic model and analyze the dynamical properties of novel 4-DOF cable-driven parallel manipulator, a typical CTM control scheme in joint-space is designed for dynamic model in SimMechanics. The control block diagram is shown in Figure 4, and a set of configuration parameters for a case of 4-DOF cable-driven parallel manipulator is given in Table 1.

The solver 'ode45', which is based on the fourth-order and fifth-order Runge-Kutta formulas with fixed step (sampling time is 1 or 2 ms), is utilized to solve the nonlinear system of differential equations for the simulation model.

For solving inverse position kinematics, position vector of MP, p, with respect to reference frame is defined by a prescribed cycloid maneuver as well as the rotation angles of MP

$$\mathbf{p} = \begin{bmatrix} 0 & 0 & h(t) \end{bmatrix}^T \qquad (38)$$

$$h(t) = 1 + \frac{2}{2\pi}\left(\frac{2\pi}{T}t - \sin\left(\frac{2\pi}{T}t\right)\right) \ (m) \qquad (39)$$

$$\phi_{(t)} = \frac{\pi}{6}\left(2\frac{t}{T} - \frac{1}{\pi}\sin\left(\frac{2\pi}{T}t\right) - 1\right) \ (rad) \qquad (40)$$

$$\psi_{(t)} = \frac{\pi}{6}\left(2\frac{t}{T} - \frac{1}{\pi}\sin\left(\frac{2\pi}{T}t\right) - 1\right) \ (rad) \qquad (41)$$

$$\theta_{(t)} = \frac{\pi}{6}\left(2\frac{t}{T} - \frac{1}{\pi}\sin\left(\frac{2\pi}{T}t\right) - 1\right) \ (rad) \qquad (42)$$

$$0 \leq t \leq T \quad T = 3$$

where $h(t)$ shows the heave of MP and T is the period of the maneuver in second. The rotation matrix R of MP with respect to reference frame for the motion of heave is the 3×3 identity matrix. However, R can be defined, in turn, for the pitch, roll and yaw motion of MP as

$$\mathbf{R}_{Roll} = \begin{bmatrix} 1 & 0 & 0 \\ 0 & \cos(\varphi_{(t)}) & -\sin(\varphi_{(t)}) \\ 0 & \sin(\varphi_{(t)}) & \cos(\varphi_{(t)}) \end{bmatrix} \quad (43)$$

$$\mathbf{R}_{Pitch} = \begin{bmatrix} \cos(\psi_{(t)}) & 0 & \sin(\psi_{(t)}) \\ 0 & 1 & 0 \\ -\sin(\psi_{(t)}) & 0 & \cos(\psi_{(t)}) \end{bmatrix} \quad (44)$$

$$\mathbf{R}_{Yaw} = \begin{bmatrix} \cos(\theta_{(t)}) & -\sin(\theta_{(t)}) & 0 \\ \sin(\theta_{(t)}) & \cos(\theta_{(t)}) & 0 \\ 0 & 0 & 1 \end{bmatrix} \quad (45)$$

while in the combined motion of roll-pitch-yaw heave(RPYH) for the MP, the rotation matrix is written as $\mathbf{R} = \mathbf{R}_{Yaw}\mathbf{R}_{Roll}\mathbf{R}_{Pitch}$

After defining p and R for any desired motion of MP and computing their time rate of changes, inverse kinematics of the problem is solved for any chosen motion of MP. The inverse kinematics is defined to calculate the desired cable displacements with the desired trajectory which is straightforward and direct under closed form solution for the spatial cable-driven parallel manipulator. The inputs of the controllers are the errors of displacement of cables between the desired q_{di}, and actual leg displacement q_i, meanwhile, the outputs of them are the actuator output force, driving the

Figure 4. Block diagram of computed torque method controller

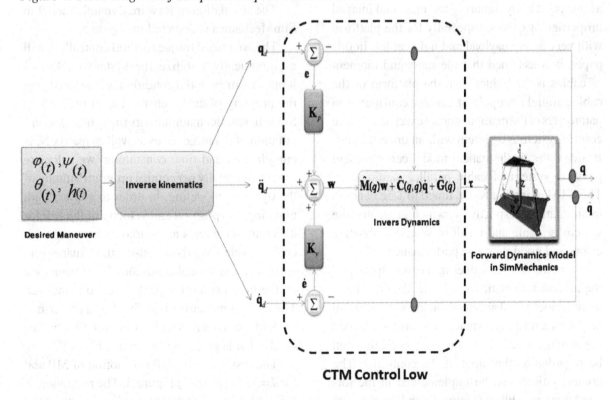

CTM Control Low

Figure 5. Forward dynamic model in SimMechanics (left) and its simulation result (right)

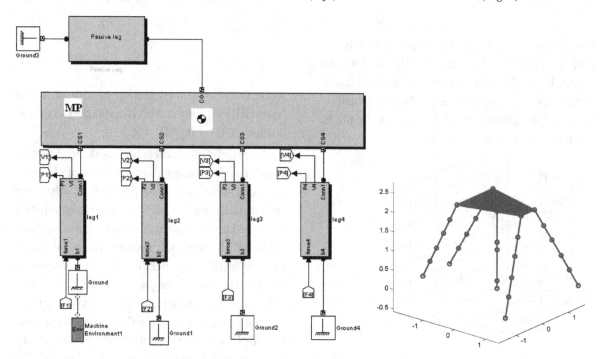

cable parallel manipulator (Figure 4). The 4-DOF cable-driven parallel manipulator may be treated as one rigid body, ignoring the mass and inertial properties of cables, especially for the platform with very heavy payload and light cables. In this paper, it is assumed that the mass and moment of cables is far lighter than the platform of the cable parallel manipulator, and the configuration parameters of the manipulator are exact. Hence, the control system is a system without uncertainties. Besides, the mathematical model can represent physical system of cable parallel manipulator. The CTM gains K_p, K_v are tuned to 1.5e3 and 30 in simulation, respectively, which guarantee the system is stable and result in small steady-state errors and good dynamic performance.

From the desired values in the workspace and the inverse kinematic model, the ideal displacement, velocity and acceleration in joint space can be got as a reference values. The errors between the actual condition and the ideal condition can be regarded as the input of the controller. The compensations can be implemented in the forward dynamics block (Figure 4), and the accurate

outputs can be calculated by the precise dynamic model.

The simulation of forward dynamic model in SimMechanics is depicted in Figure 5.

The computed torque method controllers will asymptotically stabilize the system to desired inputs, from an initial condition, while satisfying the property of cable tension for all time $t > 0$. Since in parallel manipulators the symbolic computation of dynamic terms as well as matrix N, is complicated and time consuming, we calculate them numerically per control program running of the dynamic modeling. In order to address path planning, the position and velocity of the MP for different maneuvers as desired inputs is considered. Path planning should ensure those maneuvers and coincide the final condition of first maneuver on the initial condition of the second maneuver. The desired maneuver is defined by a prescribed cycloid maneuver as well as the rotation angles of MP that is given in Equations (39)-(42).

The results for the RPYH motion of MP and for T=3 s is shown in Figure 6. The responses of the cable robot in SimMechanics, i.e., the time

Figure 6. Time history of length (a), speed (b), and acceleration (c) of the cables

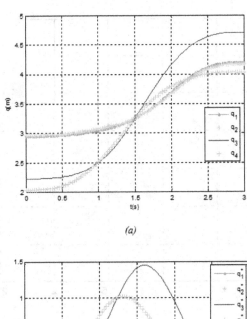

(a)

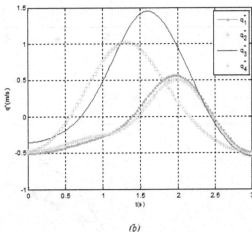

(b)

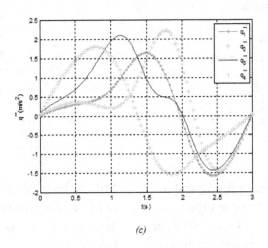

(c)

Figure 7. Time history of forces in which cables exerted on MP

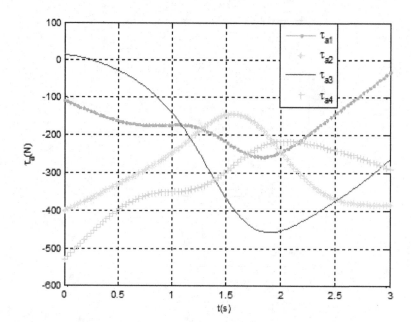

Figure 8. Desired cable motion and their online motions

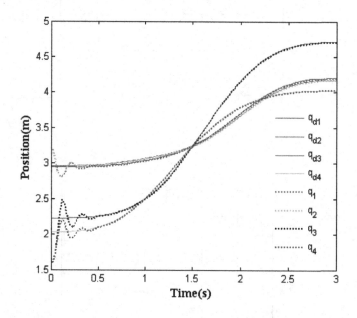

history of cables length, linear velocity and acceleration of the cables are shown in Figure 6.

The forces of the actuators of each cable for specific maneuvers of MP, i.e., roll, pitch, yaw and heave are determined using the result of previous section. The results show that these forces depend highly on the kind of motion of

MP, the mass of MP, position of center of mass of MP, and moment of inertia of the MP. The time history of mentioned cable forces are depicted in Figure 7.

It may be noted that Figure 7 shows the forces that cables apply on the MP and thus the negative value of these forces would be the cable

Figure 9. Error of cables position

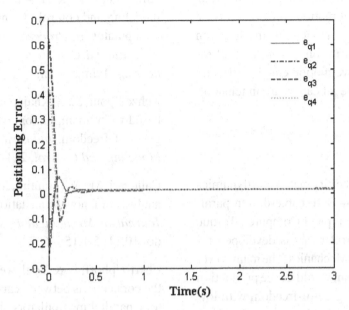

Figure 10. Simulation of the robot maneuver

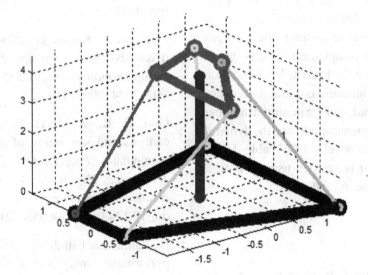

tension. Thereby, it is inferred from Figure 7, all cables remain in tension when the robot performs the given maneuver. As shown in Figure 7, there is only one exception for third cable that indicates positive value for cable force apply on MP to 0.2 second. Obviously, such errors can be occurred due to the complex calculation of those forces and thus it would be negligible compared with the period of 3 second.

Simulation results of the desired motion of four cables q_{di} as well as online tracking motion of those cables q_i are shown in Figure 8. As shown in Figure 8, the controller asymptotically tracks the cables motion to the desired motion of all cables. Also, the simulation results of positioning error of four cables are depicted in Figure 9. It is observed from this graph that computed torque method controller asymptotically stabilized the system to the desired motion of cables and thus positioning error of all motion converge to zero. Figure 10 shows the robot that performs the given maneuver at the specific configuration of all rotation angles and heave motion equal to 30° and 2m, respectively while all cables are in tension.

CONCLUSION

This paper addresses the kinematics and dynamics modeling of a novel 4-DOF cable-driven parallel manipulator and a typical Computed Torque Method (CTM) controller that is developed for dynamic model in SimMechanics. The main power of this novel mechanism would be the production capability of all six degree-of-freedom with low cost, easy to control, low degrees of freedom, advantages of manufacturing, large workspace and thus deployable for flight simulators applications. The workspace model of mechanism's dynamic is obtained in an efficient and compact form by means of natural orthogonal complement (NOC) method. The methodology of NOC leads to the elimination of the nonworking kinematic-constraint wrenches and also to the derivation of the minimum number of equations. The dynamic model and the dynamical properties of the novel 4-DOF cable-driven parallel manipulator is verified using a typical CTM control scheme in joint-space.

REFERENCES

Albus, J. S., Bostelman, R. V., & Dagalakis, N. (1993). The NIST ROBOCRANE. *Journal of Robotic Systems*, *10*(5), 709–724. doi:10.1002/rob.4620100509.

Alp, A. B., & Agrawal, S. K. (2002). Cable suspended robots: Feedback controllers with positive inputs. In *Proceedings of the American Control Conference*, Anchorage, AK (pp. 815-820).

Amooshahi, Y., & Hadian, H. (2009). Dynamic modeling and control of a novel 4-DOF parallel manipulator. In *Proceedings of the 18th Annual International Conference on Mechanical Engineering*, Tehran, Iran.

Behzadipour, S., & Khajepour, A. (2006). Cable-based robot manipulators with translational degrees of freedom. *Industrial Robotics: Theory, Modeling and Control*, 211-236.

Callegari, M., & Tarantini, M. (2003). Kinematic analysis of a novel translational platform. *ASME Journal of Mechanical Design*, *125*, 308–315. doi:10.1115/1.1563637.

Ebert-Uphoff, I., & Voglewede, P. A. (2004). On the connections between cable-driven manipulators, parallel manipulators and grasping. In *Proceedings of the IEEE International Conference on Robotics & Automation*, New Orleans, LA (pp. 4521-4526).

Fattah, A., & Kasaei, G. (2000). Kinematics and dynamics of parallel manipulator with a new architecture. *Robotica*, *18*(29), 535–543. doi:10.1017/S026357470000271X.

Guo, H. B., & Li, H. R. (2006). Dynamics analysis and simulation of a Stewart platform manipulator. *Proceedings of the Institution of Mechanical Engineers. Part C, Journal of Mechanical Engineering Science*, *220*(1), 61–72. doi:10.1243/095440605X32075.

Hadian, H., & Fattah, A. (2008). Best kinematic performance analysis of a 6-6 cable-suspended parallel robot. In *Proceedings of the IEEE/ASME International Conference on Mechatronic and Embedded Systems and Applications*, Beijing, China (pp. 510-515).

Hadian, H., & Fattah, A. (2009). On the study of dexterity measure for a novel 3-DOF cable-driven parallel manipulator. In *Proceedings of the 17th Annual International Conference on Mechanical Engineering*, Tehran, Iran.

Homma, K., Fukuda, O., Sugawara, J., Nagata, Y., & Usuba, M. A. (2003). Wire-driven leg rehabilitation system: Development of a 4-dof experimental system. In *Proceedings of the IEEE/ASME International Conference on Advanced Intelligent Mechatronics*, Kobe, Japan (pp. 908-913).

Kawamura, S., Choe, W., Tanaka, S., & Pandian, S. (1995). Development of an ultrahigh speed robot falcon using wire drive system. In *Proceedings of the IEEE International Conference on Robotics and Automation* (pp. 215-220).

Kawamura, S., Kino, H., & Won, C. (2000). High-speed manipulation by using a parallel wire-driven robots. *Robotica*, *18*, 13–21. doi:10.1017/S0263574799002477.

Kong, X., & Gosselin, C. (2005). Type synthesis of 4-dof sp-equivalent parallel manipulators: A virtual-chain approach. In *Proceedings of the CK International Workshop on Computational Kinematics*.

Maeda, K., Tadokoro, S., Takamori, T., Hiller, M., & Verhoeven, R. (1999). On design of a redundant wire-driven parallel robot warp manipulator. In *Proceedings of the International Conference on Robotics and Automation* (pp. 895-900).

Ming, A., & Higuchi, T. (1994). Study on multiple degree-of-freedom positioning mechanism using wires (part 1) - concept, design and control. *International Journal of the Japanese Society for Precision Engineering*, *28*, 131–138.

Mustafa, S. K., Yang, G., Yeo, S. H., Lin, W., & Chen, I. M. (2008). Self-calibration of a biologically inspired 7 DOF cable-driven robotic arm. *IEEE/ASME Transactions on Mechatronics*, *13*, 66–75. doi:10.1109/TMECH.2007.915024.

Oh, S., & Agrawal, S. (2005). Cable suspended planar robots with redundant cables: Controllers with positive inputs. *IEEE Transactions on Robotics*, *21*(3), 457–465. doi:10.1109/TRO.2004.838029.

Shiang, W., Cannon, D., & Gorman, J. (1999). Dynamic analysis of the cable array robotic crane. In *Proceedings of the IEEE International Conference on Robotics and Automation*.

Shiang, W., Cannon, D., & Gorman, J. (2000). Optimal force distribution applied to a robotic crane with flexible cables. In *Proceedings of the IEEE International Conference on Robotics and Automation*.

Takahashi, Y., & Tsubouchi, O. (2000). Tension control of wire suspended mechanism and application to bathroom cleaning robot. In *Proceedings of the 39th SICE Annual Conference*, Iizuka, Japan (pp. 143-147).

Takemura, F., Enomoto, M., Tanaka, T., Denou, K., Kobayashi, Y., & Tadokoro, S. (2005). Development of the balloon-cable driven robot for information collection from sky and proposal of the search strategy at a major disaster. In *Proceedings of the IEEE/ASME International Conference on Advanced Intelligent Mechatronics*, Monterey, CA (pp. 658-663).

Williams, R. L., Gallina, P., & Rossi, A. (2001). Planar cable-direct-driven robots, part ii: Dynamics and control. In *Proceedings of the ASME Design Technical Conference*.

Yamamoto, M., Yanai, N., & Mohri, A. (2004). Trajectory control of incompletely restrained parallel-wire-suspended mechanism based on inverse dynamics. *IEEE Transactions on Robotics*, *20*(5), 840–850. doi:10.1109/TRO.2004.829501.

This work was previously published in the International Journal of Intelligent Mechatronics and Robotics, Volume 1, Issue 4, edited by Bijan Shirinzadeh, pp. 44-60, copyright 2011 by IGI Publishing (an imprint of IGI Global).

Chapter 18
Kinematic Isotropic Configuration of Spatial Cable–Driven Parallel Robots

Hamoon Hadian
Isfahan University of Technology, Iran

Abbas Fattah
University of Delaware, USA

ABSTRACT

In this paper, the authors study the kinematic isotropic configuration of spatial cable-driven parallel robots by means of four different methods, namely, (i) symbolic method, (ii) geometric workspace, (iii) numerical workspace and global tension index (GTI), and (iv) numerical approach. The authors apply the mentioned techniques to two types of spatial cable-driven parallel manipulators to obtain their isotropic postures. These are a 6-6 cable-suspended parallel robot and a novel restricted three-degree-of-freedom cable-driven parallel robot. Eventually, the results of isotropic conditions of both cable robots are compared to show their applications.

INTRODUCTION

Cable robots are typically of a kinematic structure similar to parallel manipulators. The key difference, however, is that while the legs of a parallel manipulator impose bidirectional constraints, the cables of a cable robot impose unidirectional constraints, since a cable can only pull, not push. This difference makes it impossible to transfer many of the more advanced analysis tools used

for parallel manipulators to cable robots. Instead, many tools from grasping are more suitable, since the fingers of a grasp are also unidirectional (each finger can only push, not pull). This mathematical connection has been pointed out by several researchers (Voglewede & Ebert-Uphoff, 2005).

However, Cable robots are a kind of parallel manipulators that consists of a MP connected in parallel to a base by light weight cables and driven actuators that enable controlled release of cables.

DOI: 10.4018/978-1-4666-3634-7.ch018

The motors and pulleys are mounted on the base platform and positioned in the extremities of the robot workspace. The motors can control the moving platform by extending or retracting the cables around the pulleys. Cable robots have several advantages in comparison with conventional parallel manipulators: 1) low inertial properties and high payload-to-weight ratio due to few moving parts, 2) potentially large workspace, since the cables may support a wide range of motion of MP, 3) reconfigurability because of remote location of motors and controls, 4) rapid deployability for their simple components, 5) economical construction and maintenance.

Due to these characteristics, cable robots are ideal for many applications, such as locomotion interface using two cable-driven parallel mechanisms (Perreault & Gosselin, 2007), handling of hazardous materials and disaster search and rescue efforts (Bosscher, Williams, & Tummino, 2005), a balloon cable-driven robot for information collection from sky and search strategy at a major disaster (Takemura et al., 2005), ultra-high-speed cable robot for pick-and-place applications (Dekker, Khajepour, & Behzadipour, 2006), radiotelescope application by using workspace optimization of a very large cable-driven mechanism (Bouchard & Gosselin, 2007), gait rehabilitation that deploys a fully-constrained cable robot consists of eight cables (string-man) (Surdilovic & Bernhardt, 2004), and a 5 degrees-of-freedom wire-based robot (NeRe-Bot), designed for the treatment of patients with stroke-related paralyzed or paretic upper limb during the acute phase (Rosati, Gallina, Masiero, & Rossi, 2005).

There are many research works in optimal design and synthesis of rigid-link mechanisms and cable manipulators (Alici & Shirinzadeh, 2004; Li & Xu, 2006). Isotropic design of two types of spatial parallel manipulators: a three-degrees-of-freedom and the Stewart-Gough platform has been studied (Fattah & Ghasemi, 2002), a geometrical approach for the study and design of cable parallel robots are presented in (Behzadipour & Khajepour, 2004), best design for planar cable-direct-driven robot and haptic interfaces with one degree of actuation redundancy has been studied in Williams and Gallina (2002), complete kinematic and manipulability analyses for a planar 4 wire driven 3-DOF mechanism are presented in Gallina and Rosati (2002), analysis of the best kinematic performance for a 6-6 cable-suspended parallel robot is conducted in Hadian and Fattah (2008). Recently, there are some important investigations on the workspace of cable robots (Brau, Gosselin, & Lallemand, 2005; Gouttefarde & Gosselin, 2006; Bosscher & Riechel, 2006; Diao & Ma, 2007; Bruckmann, Mikelsons, Hiller, & Schramm, 2007; Gouttefarde, Merlet, & Daney, 2006; Verhoeven & Hiller, 2002). However, to the best knowledge of the authors, there is not any research work on the isotropic posture of cable robots. The kinematic isotropic configuration deals with Jacobian matrix which relates the input and output velocity of a system, i.e., a manipulator with kinematic isotropic configuration has the best kinematic performance and does not have any singularity configuration in its entire workspace. Isotropicity of a robotic manipulator is related to condition number of its Jacobian matrix, which can be obtained using singular values.

The key contribution of this paper is investigation of kinematic isotropic posture of cable robots. This can be done using the following four techniques: a) symbolic computation to obtain the isotropic and cable tension conditions, b) geometric workspace to determine the range of motion of MP without singularity, c) numerical workspace and GTI as objective functions to optimize the design parameters of robot, and d) numerical scheme to calculate the isotropic configuration of cable robots. Although first two techniques have already reported in the literature (Fattah & Ghasemi, 2002; Bosscher, Riechel, & Ebert-Uphoff, 2006), however in this research work, we intend to apply these methods to two types of spatial cable manipulators, i.e., a novel restricted three-degrees-of-freedom cable-driven parallel robot and a 6-6 cable-suspended parallel robot to investigate their isotropic configurations. The remainder of this paper is as follows: In first section, we provide the kinematic modeling of the two foregoing robots. Next, we present four different methods for isotropic posture of cable robots in. We apply these methods to two aforementioned cable robots. Finally, we draw some key conclusions based on the achievement in this research work in last section.

KINEMATIC MODELING

In this section, we present the kinematics modeling of two spatial cable robots, i.e., their Jacobian matrices are derived and respective design parameters of these robots are introduced. Also, the classification of cable robots based on the relation between degree-of-freedom of the cable robot and number of cables, is taken into account. Cable robots can be classified as fully-constrained or under-constrained (Verhoeven, Hiller, & Tadokoro, 1998). In fully-constrained case, the pose of the MP can be completely determined by the current lengths of cables. In these manipulators, there is always at least one extra cable in addition to the number of DOF of the cable robot. In contrast, under-constrained cable robots use fewer

Figure 1. A general 6-6 cable-suspended parallel robot

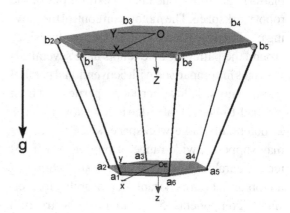

or equal cables than number of DOF and thus the pose of the MP is not completely determined by the lengths of the cables. Instead, most of these manipulators rely on the presence of gravity to determine the pose of the MP.

6-6 Cable-Suspended Parallel Robot

Consider 6-6 cable suspended parallel robot shown in Figure 1. Regarding the foregoing classification, the 6-6 cable suspended parallel robot is taken into account as under-constrained cable robot because it has 6 DOF and 6 cables (actuators). The base vertices of the base platform (BP) of the manipulator, i.e., $b_1, \ldots, b_6$ are located within the same plane $Z = 0$, as shown in Figure 2. The

Box 1.

$$
\begin{aligned}
{}^{E}\mathbf{a}_1 &= \begin{bmatrix} r_e \cos(\alpha) & r_e \sin(\alpha) & 0 \end{bmatrix}^T & {}^{o}\mathbf{b}_1 &= \begin{bmatrix} r_b \cos(\beta) & r_b \sin(\beta) & 0 \end{bmatrix}^T \\
{}^{E}\mathbf{a}_2 &= \begin{bmatrix} r_e \cos(\alpha) & -r_e \sin(\alpha) & 0 \end{bmatrix}^T & {}^{o}\mathbf{b}_2 &= \begin{bmatrix} r_b \cos(\beta) & -r_b \sin(\beta) & 0 \end{bmatrix}^T \\
{}^{E}\mathbf{a}_3 &= \mathbf{Q}_1 \, {}^{E}\mathbf{a}_1 & {}^{o}\mathbf{b}_3 &= \mathbf{Q}_1 \, {}^{o}\mathbf{b}_1 \\
{}^{E}\mathbf{a}_4 &= \mathbf{Q}_1 \, {}^{E}\mathbf{a}_2 & {}^{o}\mathbf{b}_4 &= \mathbf{Q}_1 \, {}^{o}\mathbf{b}_2 \\
{}^{E}\mathbf{a}_5 &= \mathbf{Q}_2 \, {}^{E}\mathbf{a}_1 & {}^{o}\mathbf{b}_5 &= \mathbf{Q}_2 \, {}^{o}\mathbf{b}_1 \\
{}^{E}\mathbf{a}_6 &= \mathbf{Q}_2 \, {}^{E}\mathbf{a}_2 & {}^{o}\mathbf{b}_6 &= \mathbf{Q}_2 \, {}^{o}\mathbf{b}_2
\end{aligned}
\tag{1}
$$

Box 2.

$$\mathbf{Q}_1 = \begin{bmatrix} \cos(2\pi/3) & -\sin(2\pi/3) & 0 \\ \sin(2\pi/3) & \cos(2\pi/3) & 0 \\ 0 & 0 & 1 \end{bmatrix} \quad \mathbf{Q}_2 = \begin{bmatrix} \cos(4\pi/3) & -\sin(4\pi/3) & 0 \\ \sin(4\pi/3) & \cos(4\pi/3) & 0 \\ 0 & 0 & 1 \end{bmatrix} \qquad (2)$$

points are placed at radial distance r_b from the origin of the inertial reference frame, i.e., O, the center of base platform (BP).

The moving platform (MP) similarly has a set of vertices $a_1, \ldots, a_6$ located at a distance r_e from the origin of the moving coordinate frame, i.e., O_E, the center of mass of the MP. These points are located on the $z = 0$. The position vectors of vertices of the base platform with respect to frame O, $\mathbf{b}_i = \overrightarrow{Ob_i} (i = 1, \ldots, 6)$ and the position vectors of vertices of the moving platform with respect to frame O_E, $\mathbf{a}_i = \overrightarrow{O_E a_i} (i = 1, \ldots, 6)$ are written as Equation 1 in Box 1.

Moreover, Q_1 and Q_2 are rotation matrices about the Z-axis through angles 120° and 240°, respectively, as Equation 2 in Box 2, where β and α are the separation angles of the BP and MP which are shown in Figure 2. The position from point b_i to point a_i, the cable vector, can be expressed with respect to frame O as

$$^o\mathbf{1}_i = {}^o\mathbf{p}_E + {}^o\mathbf{R}_E{}^E\mathbf{a}_i - {}^o\mathbf{b}_i, \quad i = 1, \ldots, 6, \qquad (3)$$

where $^o\mathbf{p}_E$ represents the position vector of point O_E with respect to O. $^o\mathbf{R}_E$ is the rotation matrix of MP with respect to BP using a fixed axis rotation sequence of ψ, θ and ϕ about X, Y and Z axes, respectively. The magnitude of each $^o\mathbf{1}_i$ vector is

$$l_i^2 = \left\| {}^o\mathbf{1}_i \right\|^2 = {}^o\mathbf{1}_i^{T\,o}\mathbf{1}_i. \qquad (4)$$

Using the time derivative of the kinematic constraint equations, the relation between the cable velocity vector $\dot{\mathbf{q}}$ and the twist of MP, t can be related by the Jacobian matrix as follows:

$$\dot{\mathbf{q}} = \mathbf{J}\mathbf{t} \qquad (5)$$

where

$$\dot{\mathbf{q}} = [\dot{l}_1 \ \dot{l}_2 \ \cdots \ \dot{l}_6]^T \qquad (6)$$

$$\mathbf{t} = \begin{bmatrix} \dot{\mathbf{p}}_E^T & \omega_E^T \end{bmatrix}^T. \qquad (7)$$

Figure 2. MP and BP shapes of the 6-6 cable-suspended parallel robot

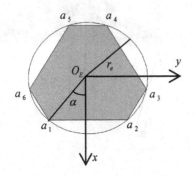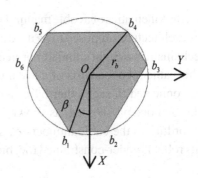

Here, ω_E is the angular velocity vector of the MP with respect to frame O. The Jacobian matrix of the cable robot is written such that its ith row is

$$\mathbf{j} = \left[\left(\mathbf{e}_i \right)^T \quad \left({}^o\mathbf{a}_i \times \mathbf{e}_i \right)^T \right], \quad i = 1, \dots, 6. \qquad (8)$$

where

$$\mathbf{e}_i = \frac{{}^o\mathbf{l}_i}{l_i}, \quad i = 1, \dots, 6. \qquad (9)$$

Also, we can obtain the kinematic relation governing the input and output velocities of the system by another formula as

$$\mathbf{A}\mathbf{t} = \mathbf{B}\dot{\mathbf{q}} \qquad (10)$$

where

$$\mathbf{A} = \left[{}^o\mathbf{l}_i{}^T \quad \left({}^o\mathbf{a}_i \times {}^o\mathbf{l}_i \right)^T \right], \quad i = 1, \cdots, 6 \qquad (11)$$

$$\mathbf{B} = diag \left(l_1, l_2, \cdots, l_6 \right) \qquad (12)$$

Here, A and B are the forward and inverse Jacobian matrices, respectively. Also, we have

$$\mathbf{J} = \mathbf{B}^{-1}\mathbf{A}. \qquad (13)$$

Novel 3-DOF Cable-Driven Parallel Robot

In this section, the kinematics of a cable manipulator with a new architecture supposed to be used as a moving mechanism in a flight simulator project is studied. This cable robot consists of a MP and a BP that are connected to each other by means of three cables and one passive leg, as shown in Figure 3. According to the cable numbers classification, this robot is under-constrained too, but

it is redundant architecturally due to the presence of the passive leg. It is noted that the MP is located on the top of BP and controlled by extending or retracting the cables around the pulleys that are mounted on BP. The passive leg (OO_E) connects the centers of both platforms to each other in which it is attached to the MP by a universal joint and fixed to the BP. The frames XYZ and xyz are inertial and moving frames attached to the BP and MP at points O and O_E, respectively.

The ith cable is connected to MP by $\mathbf{A}_i$ and to BP by $\mathbf{B}_i$, position vector of $\mathbf{A}_i$ with respect to O_E is shown by $\mathbf{a}_i$ and position vector of $\mathbf{B}_i$ with respect to O is depicted by $\mathbf{b}_i$

$$\mathbf{a}_1 = \begin{bmatrix} 0 & r_e & 0 \end{bmatrix}^T \qquad \mathbf{b}_1 = \begin{bmatrix} 0 & r_b & 0 \end{bmatrix}^T$$
$$\mathbf{a}_2 = \mathbf{Q}_1\mathbf{a}_1 \qquad \qquad \mathbf{b}_2 = \mathbf{Q}_1\mathbf{b}_1$$
$$\mathbf{a}_3 = \mathbf{Q}_2\mathbf{a}_1 \qquad \qquad \mathbf{b}_3 = \mathbf{Q}_2\mathbf{b}_1$$

$$(14)$$

Here r_e and r_b are the radii of the circumferential circles of the MP and BP, as shown in Figure 4. It is assumed that both the MP and BP are equilateral triangles.

Moreover, position vector of O_E with respect to O is defined by ${}^o\mathbf{p}_E$ and the rotation matrix of MP with respect to reference frame is given by ${}^o\mathbf{R}_E$. The rotation matrix can be readily deter-

Figure 3. The novel 3-DOF cable-driven parallel robot

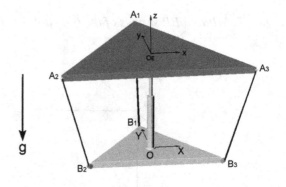

Figure 4. MP and BP shapes of 3-DOF cable-driven parallel robot

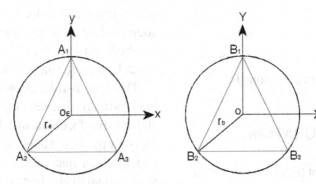

mined by having the roll and pitch angle of MP. The system has 3-DOF with three motors and pulleys that are mounted on the BP. Hence the MP has three independent motions: translational along the Z-axis, h (heave), rotation about the y-axis, θ (pitch) and rotation about the x-axis, φ (roll). The kinematics of this robot is exactly similar to its rigid leg equivalent (Fattah & Kasaei, 2000).

Upon obtaining the kinematic constraint equations and the time differentiation of the equations thus obtained, we can obtain the relation between input or cable velocity, $\dot{\mathbf{q}}$, and the output or Cartesian velocity, $\mathbf{t}$, as follows

$$\mathbf{A}\mathbf{t} = \mathbf{B}\dot{\mathbf{q}} \qquad (15)$$

where

$$\dot{\mathbf{q}} = \begin{bmatrix} \dot{l}_1 & \dot{l}_2 & \dot{l}_3 \end{bmatrix}^T \qquad (16)$$

$$\mathbf{t} = \begin{bmatrix} {}^o\dot{\mathbf{p}}_E^T & {}^o\omega_E^T \end{bmatrix}^T. \qquad (17)$$

Here, l_1, l_2 and l_3 are the lengths of the cables, also ${}^o\dot{\mathbf{p}}_E$ and ${}^o\omega_E$ are the linear and angular velocities of MP with respect to O, respectively. Moreover, $\mathbf{A}$ is a 3×6 matrix, $\mathbf{B}$ is a 3×3 diagonal matrix, as

$$\mathbf{A} = \begin{bmatrix} {}^o\mathbf{l}_1^T & ({}^o\mathbf{a}_1 \times {}^o\mathbf{l}_1)^T \\ {}^o\mathbf{l}_2^T & ({}^o\mathbf{a}_2 \times {}^o\mathbf{l}_2)^T \\ {}^o\mathbf{l}_3^T & ({}^o\mathbf{a}_3 \times {}^o\mathbf{l}_3)^T \end{bmatrix}, \quad \mathbf{B} = \begin{bmatrix} l_1 & 0 & 0 \\ 0 & l_2 & 0 \\ 0 & 0 & l_3 \end{bmatrix} \qquad (18)$$

The vector of cable position ${}^o\mathbf{l}_i$ can be written as

$$ {}^o\mathbf{l}_i = l_i\,{}^o\mathbf{e}_i = {}^o\mathbf{p}_E + {}^o\mathbf{R}_E\mathbf{a}_i - {}^o\mathbf{b}_i \qquad (19) $$

Since, the cable-driven robot under study has 3-DOF, it would be useful to consider the twist of MP as

$$\mathbf{t}_N = \begin{bmatrix} \dot{\varphi} & \dot{\theta} & \dot{h} \end{bmatrix}^T \qquad (20)$$

In this case, it may be modified their relative Jacobian matrices. This can be done using the relation of angular velocity of MP with time rate of rotation angles (i.e., time rate of roll and pitch angles). Therefore, Equation (15) can be rewritten as

$$\mathbf{A}_N\mathbf{t}_N = \mathbf{B}_N\dot{\mathbf{q}} \qquad (21)$$

where

$$\mathbf{A}_N = \begin{bmatrix} \mathbf{s}_4 & \cos\phi\,\mathbf{s}_5 + \sin\phi\,\mathbf{s}_6 & \mathbf{s}_3 \end{bmatrix}, \quad \mathbf{B}_N = \mathbf{B}. \tag{22}$$

Here $\mathbf{A}_N$ is 3×3 matrix and $\mathbf{s}_i$ is the ith column of $\mathbf{A}$ matrix defined in Equation (18).

ISOTROPIC CONFIGURATION

In this section, four different techniques to obtain the isotropic configuration of cable manipulators are presented and reliability of these methods is compared to each others.

Symbolic Method

Isotropy of a robotic manipulator is related to the condition number of its Jacobian matrix, which can be calculated as the ratio of the largest and the smallest singular values. The quality of performance of a robot with respect to the force and velocity transmission can be addressed by using the condition number of the Jacobian matrix of the robot. When the condition number approaches 1 the matrix J is said to be well-conditioned and is far from singularities and conversely as the condition number approaches infinity the matrix is said to be ill-conditioned. Clearly, the condition number attains its minimum value of unity for matrices with identical singular values, such matrices map the unit ball into another ball, although of a different size, and are, thus, called *isotropic*. By extension, we shall call manipulators whose Jacobian matrix can attain isotropic values isotropic as well. The condition number of J can be thought of as indicating the distortion of the unit ball in the space of joint-variables. The larger this distortion, the greater the condition number, the worst-conditioned Jacobians being those that are singular. Isotropy of a robotic manipulator is related to the condition number of its Jacobian matrix, which can be calculated as the ratio of the largest and the smallest singular values. The

quality of performance of a robot with respect to the force and velocity transmission can be addressed by using the condition number of the Jacobian matrix of the robot (Fattah, & Ghasemi, 2002; Hadian & Fattah, 2008)

There are two Jacobian matrices relating the joint and Cartesian velocities, namely the forward and inverse Jacobian matrices. The conditions for isotropy should apply to both matrices; these two Jacobian matrices are isotropic if they are proportional to an identity matrix, namely,

$$\mathbf{A}^T\mathbf{A} = \sigma^2\mathbf{I} \tag{23}$$

$$\mathbf{B}^T\mathbf{B} = \tau^2\mathbf{I} \tag{24}$$

where I is an identity matrix and σ and τ are two scalars. In other words, an isotropic matrix has identical singular values with a condition number of one.

Nevertheless, isotropic conditions for cable manipulators can also be determined by multiplying both sides of the input and output velocity equations, i.e., Equation (10) or Equation (15), by $\mathbf{B}^{-1}$, to obtain

$$\dot{\mathbf{q}} = \mathbf{B}^{-1}\mathbf{A}\mathbf{t} = \mathbf{J}\mathbf{t} \tag{25}$$

Hence, the isotopic conditions can be applied to the Jacobian matrix $\mathbf{J}$ instead of both matrices A and B. However, besides the above mentioned conditions we have to satisfy the cable tension conditions in cable manipulators. This can be done by using static equilibrium of cable robot

$$\mathbf{T} = \mathbf{J}^{-T}\mathbf{w} + \mathbf{y}\lambda \tag{26}$$

Here, $\mathbf{T}$ is the tensile forces vector of cables, $\mathbf{w}$ is the applied wrench on the MP from environment and the vector $\mathbf{y}$ defines as the nullspace or kernel of the Jacobian matrix, i.e., $\mathbf{y} \in \ker(\mathbf{J}^T)$, $\lambda \in R$ is an arbitrary scalar. In order to ensure the cable tension, we can determine the conditions either directly using $\mathbf{T} \geq 0$ or apply the condition that the components of $\mathbf{y}$ should have the same sign in the entire workspace (Williams & Gallina, 2002). The algorithm which should be followed to apply the symbolic method could be written as

1. Solve kinematic equations.
2. Derive both Jacobian matrices A and B.
3. Apply isotropic conditions for matrices A and B separately.
4. Obtain the isotropic configurations via the overlap results of isotropic conditions of both matrices.
5. Solve the static equilibrium of cable robot.
6. Determine the cable tension conditions (ranges of motion that cable met tension).
7. Check the obtained isotropic configuration should be within the tension conditions.
8. If step VII is ensured, the isotropic configuration would be determined. Else there is no isotropic configuration.

Geometric Workspace

In general, the workspace of a cable manipulator is defined as a set of all poses (i.e., both positions and orientations) that the MP can physically reach while the force feasibility condition and perhaps some additional constraints are satisfied (Diao & Ma, 2007).

The wrench-feasible workspace (WFW) presents a method for analytically generating the boundaries of the workspace for cable robots. This method uses the available net wrench set, which is the set of all wrenches that a cable robot can apply to its surroundings without violating tension limits in the cables. The geometric properties of this set permit evaluation of the boundaries for cable parallel robots (Bosscher, Riechel, & Ebert-Uphoff, 2006). So, we can determine the motion ranges of MP with singularity-free configuration while the cable tensions are less than the maxi-

mum tensions allowed in the cables. Since the upper tension limits of the cables are assumed to be high, the geometry of the WFW is dominated by the lower boundaries. The boundary equations corresponding to contact between any vertex v and a side of available net wrench set is of the form

$$\det\left[\$_1 \quad \$_2 \quad \cdots \quad \$_{i-1} \quad (\mathbf{v} - \mathbf{w}_g) \right] = 0 \qquad (27)$$

Here, $\$_j$ is the jth column of Jacobian matrix, where $j=1,...,i$ and $\mathbf{v} = \left[f_x, f_y, f_z, m_x, m_y, m_z \right]$ is an arbitrary applied wrench on the MP, and w_g is the gravitational wrench. Detailed derivation of boundary equations of cable robots is presented in (Bosscher, Riechel, & Ebert-Uphoff, 2006). The procedure is given below should be followed to apply the geometric workspace technique

1. Solve kinematic equations.
2. Derive Jacobian matrix J.
3. Evaluate arbitrary applied wrench and gravitational wrench on MP.
4. Calculate boundary equations of workspace (geometric workspace).
5. Plot workspace boundaries.
6. Determine the motion ranges of MP with singularity-free configuration while the cable tensions are less than the maximum tensions allowed in the cables.

Numerical Workspace and GTI

The static equilibrium of the cable suspended parallel robot is used to find the force of each cable. Force and moment balance on the moving platform is

$$\sum F_x = 0, \quad \sum F_y = 0, \quad \sum F_z = mg, \qquad (28)$$

$$\sum M_{x/MP} = 0, \quad \sum M_{y/MP} = 0, \quad \sum M_{z/MP} = 0, \qquad (29)$$

where it can be seen that there are no moments applied to MP and the only external force is the gravitational force. All other external forces and moments are ignored. To relate the external forces represented by Equations (28) and (29) to the forces in the cables, the dual relationship between kinematics and statics can be used as follows:

$$\mathbf{f}_{ext} = \mathbf{J}^T \mathbf{T} \qquad (30)$$

where $\mathbf{f}_{ext}$ is a six-dimensional vector containing the external forces and moments given by

$$\mathbf{f}_{ext} = \begin{bmatrix} 0 & 0 & mg & 0 & 0 & 0 \end{bmatrix}^T. \qquad (31)$$

The first three rows in Equation (30) represent the external forces and similarly the last three rows represent the external moments in the x, y, and z direction, respectively. Now to obtain the equations for the force of each cable, Equation (30) is rearranged in the following manner:

$$\mathbf{T} = \mathbf{J}^{-T} \mathbf{f}_{ext} \qquad (32)$$

The workspace volume for the cable robot is characterized by the set of points where the center of mass of the moving platform can be positioned while all cables are in tensions (Pusey, Fattah, Agrawal, & Messina, 2004). To this end, at each point within the possible workspace, the equation of the force in each cable is used to see if tension is obtainable. The types of workspace that are studied in this work include the constant orientation workspace and the total orientation workspace. The total orientation workspace can be determined as set of locations of point O_E of MP that can be reached with all orientations within a set of defined ranges on the orientation parameters. The workspace volume is determined for different robots for every constant orientation combination of ψ, θ, and φ. This inevitably led to an extensive collection of constant orientation workspace data

where a graphical representation of the ''good'' points with their corresponding conditioning index value can be displayed for each of the robot and orientation configuration. Due to the vast amount of information involved the individual workspace volume shapes for each robot at every constant orientation is not presented.

However, the trends pertaining to certain aspects of the workspace will be discussed. In Section of numerical workspace computations, the data from the constant orientation workspace is combined for the set of discrete orientations to form the total orientation workspace. In order to present the results for the total orientation workspace in a systematic and comprehensive manner the data is organized into a series of subset plots. If an orientation did not produce any workspace points then the point was set to zero for the surface.

A program is created in MATLAB for workspace analysis of cable robots. The program inputs include the connection points of the base and moving platform, the radius of these connection points on both platforms, the given orientation of the moving platform, the desired search volume and incremental step size for the mesh search. The program checks every point in the volume to see whether the cables yield tensions when the MP is positioned at each point of the possible workspace volume.

For cable robots, a tension factor is used as a performance index to evaluate the quality of force closure at a specific configuration (Hadian & Fattah, 2009; Verhoeven & Hiller, 2002). The tension factor is defined as the minimum tension over the maximum tension of the cables. If **t** is the homogenous solution of the cable tensions, then:

$$tf = \frac{\min(\mathbf{t})}{\max(\mathbf{t})} \qquad (33)$$

The tension factor tf, is a measure of the tension condition of the Jacobian matrix. It reflects the relative tension distribution among the cables for a specific platform pose inside the workspace. In order to evaluate the quality of the entire workspace, the global tension index (GTI) is proposed. The tension factor is a local measure because it characterizes the tension distribution at a given posture of the moving platform. By integrating the local *tf* over the workspace, GTI can be obtained as follows:

$$GTI = \frac{\int_w tf\, dw}{\int_w dw} \qquad (34)$$

where *w* denotes the workspace volume of the manipulator. When the GTI approaches zero, one of the cable tensions is close to zero, i.e., the platform is located near the singular configuration. Hence if GTI approaches one, the platform is positioned far from singularity and thus the cable robot is called isotropic. A larger GTI is more favorable because there is a better tension balance among the cables.

According to the finite workspace generation, the GTI value can be computed numerically. Equation (34) can be written separately for both the constant orientation workspace and the total orientation workspace as:

$$GTI = \begin{cases} \dfrac{\sum_{i=1}^{n_p} tf_i}{n_p} & for\ cons\tan t\ orientation \\[3ex] \dfrac{\sum_{i=1}^{n_p}\sum_{j=1}^{n_o} tf_{i,j}}{n_p n_o} & for\ total\ orientation \end{cases} \qquad (35)$$

where tf_i and $tf_{i,j}$ represent the tension factor of Jacobian matrix that is computed from every posture *i* in the constant orientation workspace and from every position *i* at every orientation *j* in the total orientation workspace, respectively. n_p

and n_o are the number of points composing of the workspace and covering the range of possible orientations, respectively.

It may be noted that the numerical workspace and GTI are applied to optimize the design parameters for achievement of best kinematic performance for cable robots. The algorithm should be followed to apply numerical workspace and GTI could be written as

1. Given geometry of the cable manipulator including
    a. Connection points of the BP and MP (a_i and b_i)
    b. Radius of these connection points on both platforms (r_e and r_b)
2. Given orientation of the MP
3. Given desired search volume
4. Incremental step size along all three axis
5. Check every point in the volume
6. If $\mathbf{T} \geq 0$ the current point is in the workspace volume
7. Else go to the next point and repeat step V
8. Compute the GTI at the point within the workspace volume

Numerical Approach

The isotropic conditions for both manipulators are determined numerically by applying conditions on matrix J. In the isotropic posture, the Jacobian matrix J has non-zero identical singular values or it has unit condition number. Therefore, matrices $\mathbf{JJ}^T$ or $\mathbf{J}^T\mathbf{J}$ becomes proportional to an identity matrix. The isotropy condition for matrix J can be written as

$$\mathbf{J}^T\mathbf{J} = \delta^2\mathbf{I} \qquad (36)$$

where δ is a scalar and I is an identity matrix. Upon substitution of J from Equation (13) into Equation (36), one obtains

$$\mathbf{J}^T\mathbf{J} = \mathbf{A}^T(\mathbf{B}^{-1})^2\mathbf{A} = \delta^2\mathbf{I}. \qquad (37)$$

Moreover, the isotropic configuration of cable manipulators is required to have tensile forces in all cables in their entire workspace, namely, $\mathbf{T} \geq \mathbf{0}$.

Now, if the entries of J have different units, the foregoing definition of condition number cannot be applied, for we would face a problem of ordering singular values of different units from largest to smallest. We resolve this inconsistency by defining the characteristic length (L), by which we divide the Jacobian entries that have unit of length, thereby producing a new Jacobian that is dimensionally homogeneous. Therefore, some of columns are divided by a length L. Thus, matrix $\mathbf{A}$ is partitioned into two submatrices as

$$\mathbf{A} = \left[\mathbf{F}_p \quad \frac{1}{L}\mathbf{F}_o\right] \qquad (38)$$

Upon substituting $\mathbf{A}$ from Equation (33) into Equation (32), we can obtain

$$\mathbf{J}^T\mathbf{J} = \begin{bmatrix} \mathbf{F}_p^T(\mathbf{B}^{-1})^2\mathbf{F}_p & \frac{1}{L}\mathbf{F}_p^T(\mathbf{B}^{-1})^2\mathbf{F}_o \\ \frac{1}{L}\mathbf{F}_o^T(\mathbf{B}^{-1})^2\mathbf{F}_p & \frac{1}{L^2}\mathbf{F}_o^T(\mathbf{B}^{-1})^2\mathbf{F}_o \end{bmatrix} = \begin{bmatrix} \delta^2\mathbf{I} & \mathbf{0} \\ \mathbf{0} & \delta^2\mathbf{I} \end{bmatrix}$$

$$(39)$$

Consequently, we have an over-determined nonlinear constrained system that should be solved numerically. Applying the isotropic and cable tension conditions for cable manipulators and solving the over-determined nonlinear constrained system, it is possible to obtain the isotropic configuration of the manipulators using the optimization toolbox in MATLAB. The algorithm below is given to follow the numerical scheme

1. Solve kinematic equations.
2. Derive both Jacobian matrices A and B.
3. Evaluate general Jacobian matrix J.

4. Apply isotropic conditions for general Jacobian matrix J along with Solve the static equilibrium of cable robot $\mathbf{T} \geq \mathbf{0}$.
5. Derive an over-determined nonlinear constrained system of equations.
6. Solve above system of equations numerically by means of *fmincon* which is a function of optimization toolbox in MATLAB.
7. Obtain isotropic configurations of cable robot.
8. Compare to results of symbolic method.

ISOTROPIC POSTURES OF THE MECHANISMS

In this section, we applied the four aforementioned methods to two cable manipulators, considered previously, to investigate their isotropic conditions.

Isotropic Posture of the 6-6 Cable-Suspended Parallel Robot

Symbolic Method

The isotropic conditions of the manipulator with a general translational motion of the MP are studied, using pure symbolic computation on the isotropic conditions for both Jacobian matrices A and B. The isotropy condition for matrix A is determined first. From Equation (11), the elements of the first three columns of matrix A have the units of length and the second three columns have the unit of (length)^2. Hence, a characteristic length L, is introduced to homogenize the elements of Jacobian matrix so that the condition number is non-dimensional. Therefore, the second three columns are divided by characteristic length L and the matrix A is divided into two sub matrices $\mathbf{F}_p$ and $\mathbf{F}_o$ as

$$\mathbf{A} = \left[\mathbf{F}_p \quad \frac{1}{L}\mathbf{F}_o\right] \qquad (40)$$

Here, $\mathbf{F}_p$ and $\mathbf{F}_o$ are 6×3 matrices. Upon substitution of Equation (40) into Equation (23), the isotropic conditions for the Jacobian matrix A are given as

$$\mathbf{F}_p^{\ T}\mathbf{F}_p = \sigma^2 \mathbf{I}_3 \tag{41}$$

$$\frac{1}{L^2}\mathbf{F}_o^{\ T}\mathbf{F}_o = \sigma^2 \mathbf{I}_3 \tag{42}$$

$$\frac{1}{L}\mathbf{F}_p^{\ T}\mathbf{F}_o = \mathbf{0} \tag{43}$$

The traces of Equations (41) and (42) lead us to

$$\sigma^2 = 2 \tag{44}$$

$$L^2 = \frac{1}{6}\sum_{i=1}^{6}\frac{\left\|\mathbf{m}_i\right\|^2}{l_i^2} \tag{45}$$

where l_i is length of the cable i and m_i is written as

$$\mathbf{m}_i = {}^o\mathbf{a}_i \times {}^o\mathbf{l}_i \qquad (i = 1,...,6) \tag{46}$$

If the translational motion of the MP is only considered, the isotropic conditions lead to the following results

$$p_x = 0, \ p_y = 0, \ p_z \in R$$

$$r_e = r_b \cos(\alpha - \beta), L = r_e \tag{47}$$

Here, p_x, p_y and p_z are the components of the position vector ${}^o\mathbf{p}_E$ defined in Equation (3). Moreover, since the MP has only translational motion, we can show that the projection of each leg along the z-axis is identical and equal to each other. This means that the angle of leg i with respect to the vertical axis for all legs is identical,

$\gamma_i = \gamma$ for $i = 1,...,6$. Therefore, the legs length are equal to each other, namely, $l_i = l$ for $i = 1,...,6$. This makes matrix B to have identical elements and hence matrix B is in the isotropic condition. Also, we can optimize the isotropic condition for both $\mathbf{F}_o$ and $\mathbf{F}_p$ matrices while we have a maximum value of GCI using $\gamma = 45°$ (Fattah & Ghasemi, 2002). As an example, consider the geometric parameters of the robot at hand as

$$r_b = 6(m), r_e = 3\sqrt{3}, \alpha = 30°, \beta = 60° \tag{48}$$

Generally, in order to ensure the cable tension, we can determine the conditions either directly using $\mathbf{T} \geq \mathbf{0}$ or the components of $\mathbf{y}$ (nullspace of Jacobian matrix) should have the same sign in the entire workspace. Since the cable robot under study has six actuators and the task space dimension is also six, the dimension of nullspace is zero and thus we have to use directly $\mathbf{T} \geq \mathbf{0}$ to ensure the cable tension. Therefore, the condition of cable tension for this example leads to the following ranges of motion

$$p_z \geq 0, -1.73 \leq p_y \leq 1.73, -1.5 \leq p_x \leq 1.5. \tag{49}$$

It is clear that the isotropic configuration presented in Equation (47) is in above ranges. Therefore, the configuration specified in Equation (47) is the isotropic posture for the translational motion of the 6-6 cable-suspended parallel robot.

Geometric Workspace

Boundary equations corresponding to contact between any vertex v and six sides of available net wrench set are

$$h_1 = \det[\mathbf{s}_1 \ \mathbf{s}_2 \ \mathbf{s}_3 \ \mathbf{s}_4 \ \mathbf{s}_5 \ (\mathbf{v} - \mathbf{w}_g)] = 0 \tag{50.a}$$

$$h_2 = \det\begin{bmatrix} \mathbf{s}_1 & \mathbf{s}_2 & \mathbf{s}_3 & \mathbf{s}_4 & \mathbf{s}_6 & (\mathbf{v} - \mathbf{w}_g) \end{bmatrix} = 0 \quad (50.b)$$

$$h_3 = \det\begin{bmatrix} \mathbf{s}_1 & \mathbf{s}_2 & \mathbf{s}_3 & \mathbf{s}_5 & \mathbf{s}_6 & (\mathbf{v} - \mathbf{w}_g) \end{bmatrix} = 0 \quad (50.c)$$

$$h_4 = \det\begin{bmatrix} \mathbf{s}_1 & \mathbf{s}_2 & \mathbf{s}_4 & \mathbf{s}_5 & \mathbf{s}_6 & (\mathbf{v} - \mathbf{w}_g) \end{bmatrix} = 0 \quad (50.d)$$

$$h_5 = \det\begin{bmatrix} \mathbf{s}_1 & \mathbf{s}_3 & \mathbf{s}_4 & \mathbf{s}_5 & \mathbf{s}_6 & (\mathbf{v} - \mathbf{w}_g) \end{bmatrix} = 0 \quad (50.e)$$

$$h_6 = \det\begin{bmatrix} \mathbf{s}_2 & \mathbf{s}_3 & \mathbf{s}_4 & \mathbf{s}_5 & \mathbf{s}_6 & (\mathbf{v} - \mathbf{w}_g) \end{bmatrix} = 0 \quad (50.f)$$

where $\mathbf{w}_g = \begin{bmatrix} 0, & 0, & 0, & 0, & 0, & mg \end{bmatrix}^T$ and v is chosen zero vector, because we are seeking on geometry interpretation of static workspace.

If the translational motion of MP of the example given in Equation (48) is considered, the WFW boundaries are derived for the 6-6 cable-suspended parallel robot as

$$h_{1,6} = \sqrt{3}p_y + 3 - p_x = 0 \quad (51.a)$$

$$h_{2,3} = 2p_x + 3 = 0 \quad (51.b)$$

$$h_{4,5} = \sqrt{3}p_y - 3 + p_x = 0 \quad (51.c)$$

It may be noted, the lower and upper boundaries are lead to same results. The boundaries workspace is shown in Figure 5. As a physical interpretation for only translational motion, we expect not to have any changes in the workspace in z direction. In this case there is a planar workspace area such that it can extrude in z direction to make the whole workspace volume.

Numerical Workspace and GTI

The constant orientation workspace for the robot at hand with the geometric parameters such that defined in Equation (48) is obtained by means of the created program in MATLAB mentioned in previous section. Once a particular geometry, MP size, and MP translational motion is selected, the program checks every point in the mesh search volume to see whether the cables yield tension when the MP is positioned at each point of the possible workspace volume.

The procedure of computation of the numerical workspace volume and GTI of this manipulator is as follows:

1. Given geometry of the cable manipulator including:
    a. Connection points of the BP and MP (a_i and b_i),
    b. Radius of these connection points on both platforms (r_e and r_b).
2. Given orientation of the MP.
3. Given desired search volume.
4. Incremental step size along all three axis.
5. Check every point in the volume.
6. If $\mathbf{T} \geq \mathbf{0}$ the current point is in the workspace volume.
7. Else go to the next point and repeat step V.
8. Compute the GTI at the point within the workspace volume.

Figure 5. Analytically boundary of workspace for the 6-6 cable-suspended parallel robot

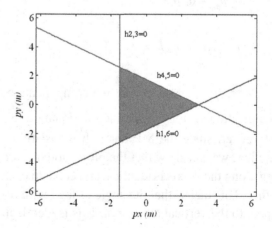

The result of numerical workspace volume for different orientation of the MP is shown in Figure 6. These results can validate the proposed algorithm and may be verified from the results of motion ranges achieved in previous section.

As shown in Figures 7 through 15 along the z direction there is no major change in workspace volume, since only translational motion of MP is considered. Moreover, the ranges of p_x and p_y are also in good agreement with resulted cable tension conditions given in Equation (49).

The size of the separation angle of the MP relative to the BP is varied from the double size to smaller sizes, namely, $0 \leq \gamma = \alpha/\beta \leq 2$. The variation is for four different size of the moving platform that includes $r = r_e/r_b = 0.5, 1, 1.5, 2$. The variation of the GTI is studied for constant position and different orientations of the MP that is limited between $-60° \leq \theta, \psi, \theta \leq 60°$. The results are depicted in Figure 16 for different geometry configuration and sizes of platforms.

From this graphs it is observed the best tension distribution among cables occurs in $r=1$ and $\gamma=0$. This trend was observed to be consistent for any constant position chosen. In attempting a broader range of studies besides those concerning geometric shapes and configurations, different orientations were studied.

Due to the lengthy computational time only four ratios of the MP to the BP and one geometry

Figure 6. Numerical workspace volume of 6-6 cable-suspended parallel robot for $\psi=\varphi=\theta=0$

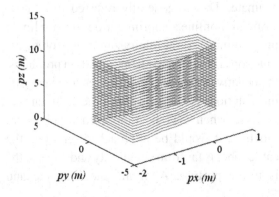

shape were considered. The ratios include $r = 0.5, 1, 1.5, 2$ and the geometry shape is $\gamma=0$. As inferred from Figure 17 the ratio $r=1$ and the configuration of $\theta=\psi=\varphi=0$ yield the best tension distribution among the driving cables.

Numercial Approach

The matrices F_p and F_o of Equation (40) for this manipulator can be written as

$$\mathbf{F}_p = \begin{bmatrix} {}^o\mathbf{l}_1{}^T \\ {}^o\mathbf{l}_2{}^T \\ {}^o\mathbf{l}_3{}^T \\ {}^o\mathbf{l}_4{}^T \\ {}^o\mathbf{l}_5{}^T \\ {}^o\mathbf{l}_6{}^T \end{bmatrix} \quad \mathbf{F}_o = \begin{bmatrix} ({}^o\mathbf{a}_1 \times {}^o\mathbf{l}_1)^T \\ ({}^o\mathbf{a}_2 \times {}^o\mathbf{l}_2)^T \\ ({}^o\mathbf{a}_3 \times {}^o\mathbf{l}_3)^T \\ ({}^o\mathbf{a}_4 \times {}^o\mathbf{l}_4)^T \\ ({}^o\mathbf{a}_5 \times {}^o\mathbf{l}_5)^T \\ ({}^o\mathbf{a}_6 \times {}^o\mathbf{l}_6)^T \end{bmatrix}, \quad (52)$$

The isotropic condition for the 6-6 cable suspended parallel manipulator is written as

$$\mathbf{J}^T\mathbf{J} = \delta^2 \begin{bmatrix} \mathbf{I}_3 & \mathbf{0} \\ \mathbf{0} & \mathbf{I}_3 \end{bmatrix} \quad (53)$$

where

$$\mathbf{J} = \mathbf{B}^{-1}\mathbf{A} \quad (54)$$

Here **A** and **B** are as defined in Equations (11) and (12), where $\mathbf{I}_3$ is a 3×3 identity matrix.

Comparing Equation (52) with Equation (53) and in the light of Equation (39), we can obtain the isotropic conditions for the manipulator as

$$\mathbf{F}_p{}^T\mathbf{B}^{-2}\mathbf{F}_p = \delta^2\mathbf{I}_3 \quad (55)$$

$$\frac{1}{L^2}\mathbf{F}_o{}^T\mathbf{B}^{-2}\mathbf{F}_o = \delta^2\mathbf{I}_3 \quad (56)$$

Figure 7. Top view of workspace volume for $\psi=\varphi=\theta=0$

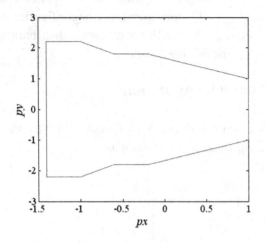

Figure 8. Numerical workspace volume of 6-6 cable-suspended parallel robot for $\psi=\theta=0$, $\varphi=10^{\circ}$

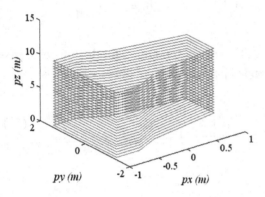

Figure 9. Top view of workspace volume for $\psi=\theta=0$, $\varphi=10^{\circ}$

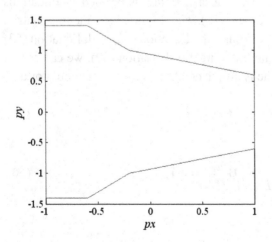

$$\frac{1}{L}\mathbf{F}_p^{\ T}\mathbf{B}^{-2}\mathbf{F}_o = 0 \qquad (57)$$

The traces of Equations (55) and (56) lead us to

$$\delta^2 = 2 \qquad (58)$$

$$L^2 = \frac{1}{6}\sum_{i=1}^{6}\frac{\|\mathbf{m}_i\|^2}{l_i^{2}} \qquad (59)$$

where l_i is length of the cable i and mi is written as

$$\mathbf{m}_i = {}^o\mathbf{a}_i \times {}^o\mathbf{l}_i \qquad (i=1,...,6) \qquad (60)$$

Equations (55), (56), (57) and (59) consist of a set of 22 equations in seven unknowns. We have twelve equations for the first two equations, i.e., Equations (55) and (56), because of the symmetry matrix. There are nine equations for Equation (57) and one equation for Equation (59). The unknowns are the position vector components, i.e., p_x, p_y, p_z; the fixed Euler angles, namely, ψ, φ, θ of the MP, and the characteristic length L. Also, there are six constraint inequalities because of cable tension. Hence, we have an over-determined constrained system that should be solved numerically. In this case, *fmincon* function from optimization toolbox in MATLAB is used to solve this constrained nonlinear system. *fmincon* attempts to find a constrained minimum of a scalar function of several variables starting at an initial estimate. This is generally referred to as constrained nonlinear optimization *or* nonlinear programming. As the minimization of an objective function is common to all optimization problems, the negative value of GTI is chosen to investigate the solution. Negative value of GTI is chosen because when GTI is minimized as objective function, it would be approached one, i.e., the cable robot is far from singularity and close to the isotropic posture. Applying the isotropic and

Figure 10. Numerical workspace volume of 6-6 cable-suspended parallel robot for φ=θ=0, ψ=10°

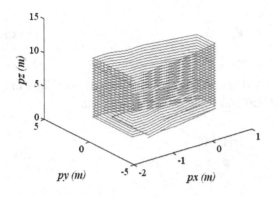

Figure 11. Top view of workspace volume for φ=θ=0, ψ=10°

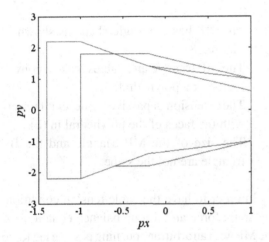

Figure 12. Numerical workspace volume of 6-6 cable-suspended parallel robot for ψ=φ=0, θ=10°

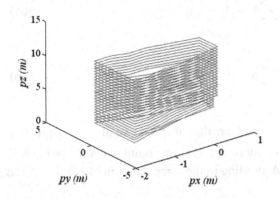

cable tension conditions for numeric example given in Equation (48) and solving the over-determined constrained systems by means of *fmincon* MATLAB function leads to approximate solution for isotropic configuration of the mechanism as

$$p_x = 0, \quad p_y = 0, \quad p_z = 5.9873$$

$$\psi = \varphi = \theta = 0$$

$$L = 6.01, \quad \delta = 1.41. \tag{61}$$

As shown, the numerical results verified the symbolic results obtained in Equation (47).

Isotropic Posture of the Novel 3-DOF Cable-Driven Parallel Robot

Symbolic Method

We determine the isotropy conditions for the Jacobian matrix $\mathbf{A}_N$ defined in Equation (22) first. The characteristic length of the manipulator, i.e., L, is used to homogenize the elements of Jacobian matrix so that the condition number is non-dimensional. Therefore, the first two columns are divided by characteristic length L and the matrix $\mathbf{A}_N$ is divided into two submatrices $\mathbf{F}_o$ and $\mathbf{F}_p$ as

$$\mathbf{A}_N = \begin{bmatrix} \dfrac{1}{L}\mathbf{F}_o & \mathbf{F}_p \end{bmatrix} \tag{62}$$

Upon substitution of Equation (62) into Equation (23), the isotropic condition for Jacobian matrix $\mathbf{A}_N$ implies that

$$\frac{1}{L^2}\mathbf{F}_o{}^T\mathbf{F}_o = \sigma^2\mathbf{I}_2 \tag{63.a}$$

Figure 13. Top view of workspace volume for $\psi=\varphi=0,\ \theta=10^{o}$

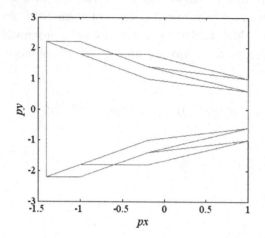

Figure 14. Numerical workspace of 6-6 cable-suspended parallel robot for $\psi=5o,\ \varphi=20o,\ \theta=-10^{o}$

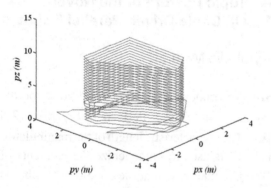

Figure 15. Top view of workspace volume for $\psi=5o,\ \varphi=20o,\ \theta=-10^{o}$

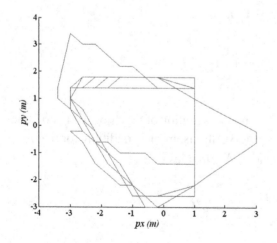

$$\mathbf{F}_p{}^T\mathbf{F}_p = \sigma^2 \tag{63.b}$$

$$\frac{1}{L}\mathbf{F}_o{}^T\mathbf{F}_p = \mathbf{0}. \tag{63.c}$$

Comparing the traces of both sides of Equation (63.a) and Equation (63.b) leads to

$$L^2 = \frac{tr(\mathbf{F}_o{}^T\mathbf{F}_o)}{2(\mathbf{F}_p{}^T\mathbf{F}_p)} \tag{64}$$

In order to prove that all cables are in tension everywhere in its workspace, we use static equilibrium of MP by imposing the following conditions (Behzadipour & Khajepour, 2005)

1. The MP has a triangle shape as shown in Figure 18,
2. The MP triangle and cables make a convex region or a polyhedral,
3. The extension of passive leg never intersects with the faces of the polyhedral in (b),
4. The size of the MP triangle and the BP triangle are not the same.

In order to have the cable tension condition, the static force and torque balance equations of the MP with an arbitrary pushing passive leg force f_i should result in positive cable forces.

The force and torque balance equations are

$$\sum_{i=1}^{3}\mathbf{t}_i = \mathbf{f}_l - m\mathbf{g} \tag{65}$$

$$\sum_{i=1}^{3}\mathbf{a}_i \times \mathbf{t}_i = \mathbf{0} \tag{66}$$

where $\mathbf{t}_i$ are the cable forces and $\mathbf{a}_i$ is the position vector of the attaching point of cable i on the MP. According to the force diagram of Figure 19, the

Figure 16. Variation of GTI versus geometry parameters of 6-6 robot

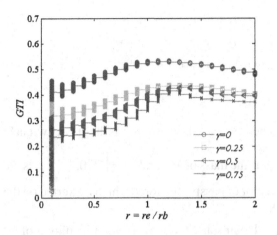

Figure 17. Variation of GTI versus orientation of the MP of 6-6 robot

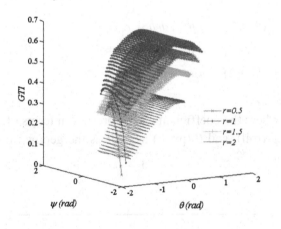

Figure 18. A schematic deign of 3-DOF cable-driven parallel robot

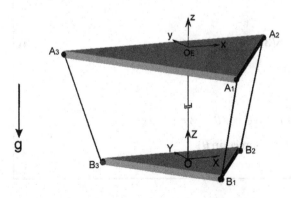

forces t_1, t_2 and t_3 generate a polyhedral. As a result, t_i's can be balanced by f_l as long as f_l is inside the polyhedral. f_l can be easily shown to be inside the polyhedral by considering the conditions b and c.

Here, the passive force is balanced by three positive combined cable forces. Therefore, to ensure the cable tension condition, it is enough to satisfy the torque balance equation. Carrying out the balancing of Equation (65), the range of motion that provide the tension in all cables are obtained as

$$h \geq r_e, \quad -30^o \leq \phi, \theta \leq 30^o \qquad (67)$$

Upon solving Equation (63) symbolically and applying the cable tension condition, the configuration of the MP for the isotropic condition of Jacobian matrix $\mathbf{A}_N$ is determined as

$$\theta = 0, \quad \phi = -\cos^{-1}\left(\frac{r_b}{r_e}\right), \quad h = k_h\sqrt{r_e^2 - r_b^2}. \qquad (68)$$

where

$$k_h \geq \frac{1}{\sqrt{1 - \left(\frac{r_b}{r_e}\right)^2}} \qquad (69)$$

It is inferred from the results that $r_e \geq r_b$, i.e., the size of MP should be larger than BP size. Next, we determine the isotropy condition for matrix **B** using Equation (24). This condition occurs when we have identical diagonal elements for **B**, namely

$$l_1 = l_2 = l_3 \qquad (70)$$

In this case, the roll and pitch angles of the MP are zero. It may be noted that when $\mathbf{A}_N$ is

Figure 19. The polyhedral of force balance on the MP

$$g_1(\theta,\phi,h) = \det\begin{bmatrix} \mathbf{s}_4 & \cos\phi\,\mathbf{s}_5 + \sin\phi\,\mathbf{s}_6 & \mathbf{v} - \mathbf{w}_g \end{bmatrix} = 0$$
$$g_2(\theta,\phi,h) = \det\begin{bmatrix} \mathbf{s}_4 & \mathbf{s}_3 & \mathbf{v} - \mathbf{w}_g \end{bmatrix} = 0$$
$$g_3(\theta,\phi,h) = \det\begin{bmatrix} \cos\phi\,\mathbf{s}_5 + \sin\phi\,\mathbf{s}_6 & \mathbf{s}_3 & \mathbf{v} - \mathbf{w}_g \end{bmatrix} = 0$$

$$(71)$$

where $\mathbf{w}_g = \begin{bmatrix} 0,\ 0,\ -mg \end{bmatrix}^T$ is the gravity wrench, m is the mass of MP and $\mathbf{v} = \begin{bmatrix} 0,\ 0,\ \|\mathbf{f}_l\| \end{bmatrix}^T$ is the vector of passive leg force which is exerted on the MP.

Upon substitution of $\mathbf{s}_i$, the ith column of $\mathbf{A}$, from Equation (18) into Equation (55), the boundary equations are determined in terms of heave, pitch and roll as

isotropic, the other Jacobian matrix $\mathbf{B}$ may not be isotropic and vice versa. So, we study the variation of GTI to optimize design parameter for whole Jacobian matrix of 3-DOF cable robot.

Geometric Workspace

Using Equation (27), the three boundary equations corresponding to contact between the vertex v and sides of available net wrench set are obtained from $\mathbf{A}_N$, the Jacobian matrix defined in Equation (22), as

$$g_1(\theta,\phi,h) = a_1 h^2 + b_1 h + c_1 = 0$$
$$g_2(\theta,\phi,h) = a_2 h^2 + b_2 h + c_2 = 0 \qquad (72)$$
$$g_3(\theta,\phi,h) = a_3 h^2 + b_3 h + c_3 = 0$$

where the coefficients are expressed in terms of φ (roll), and θ (pitch) as well as the geometric

Box 3.

$$a_1 = \sqrt{3}\,r_e^2(f_l + mg)\cos^2\phi\cos\theta$$
$$b_1 = 1/2\,r_e^2 r_b (f_l + mg)\cos\phi\,(\sqrt{3}\sin\phi\cos\theta + 3\sin\theta) \qquad (73)$$
$$c_1 = 1/2\,r_e^2 r_b^2 (f_l + mg)(\sqrt{3}\cos^2\phi\cos\theta - \sqrt{3}\cos\theta + 3\sin\phi\sin\theta)$$
$$a_2 = 3\,r_e(f_l + mg)\cos\phi$$
$$b_2 = 1/2\,r_e(f_l + mg)(3\,r_b\sin\phi + 2\sqrt{3}\,r_e\sin\theta\sin\phi + \sqrt{3}\,r_b\cos\phi\sin\theta) \qquad (74)$$
$$c_2 = 3/2\,r_e^2 r_b (f_l + mg)(\sqrt{3}\sin\phi\cos\phi\sin\theta + \cos^2\phi - 1)$$
$$a_3 = -\sqrt{3}\,r_e(f_l + mg)\cos\phi\cos\theta$$
$$b_3 = -1/2\,r_e(f_l + mg)(2\sqrt{3}\,r_e\cos\phi\sin\phi\cos\theta - \sqrt{3}\,r_b\sin\phi\cos\theta + 3\,r_b\sin\theta) \qquad (75)$$
$$c_3 = -1/2\,r_e^2 r_b (f_l + mg)(\sqrt{3}\cos^2\phi\cos\theta - \sqrt{3}\cos\theta + 3\sin\phi\sin\theta)$$

parameter of robot (see Equations 73 through 75 in Box 3).

As shown in Equation (72), the equations of workspace boundaries are quadratic functions of h (heave) in which, we can determine h as a function of φ and θ analytically. These are six equations that create the same number of surfaces in which indicates the workspace boundaries of the manipulator. The surfaces may be depicted to give the graphical interpretation of the singularity-free workspace boundaries of this cable-driven robot. The workspace volume includes the volume in which consists of overlapping of the workspace boundaries. It may be noted that the numerical values of geometric parameters as well as the applied force of passive leg and the MP's gravity force for this cable-driven parallel robot are considered as

$$r_e = 1.2\,m, \ r_b = 1\,m, \ f_l = \left\| \mathbf{f}_l \right\| = 200\,N, \ mg = 100\,N \tag{76}$$

Upon substitution of numerical values from Equation (76) into Equations (73), (74), and (75), and eventually evaluating the h in terms of θ and φ, the graphical interpretation of workspace boundaries are depicted in Figure 20.

As shown in Figure 20, the geometric workspace volume is generated using the overlapping space of the six boundaries intersections. Note that each boundary is specified by one color. Consequently, the workspace volume of the 3-DOF cable-driven parallel manipulator has a cube shape that is almost constant along Z direction.

Numerical Workspace and GTI

The prescribed search volume for possible workspace volume is limited to the region as

$$-30^\circ \leq \phi, \theta \leq 30^\circ, \ 2(m) \leq h \leq 5(m) \tag{77}$$

Using the same geometric parameters given in Equation (76), A MATLAB program checks every point in the search volume to see whether the cables yield tensions when the MP is positioned at each point of the possible workspace volume. The result is shown in Figure 21.

As shown in Figure 21, the numerical workspace volume of this cable manipulator is in good agreement with the geometric workspace volume presented in Figure 20. This is happened because the ranges of orientation angles of MP in both cases are almost the same and also the trend of both of them versus the h, the heave motion of MP, is constant and similar.

For this novel manipulator, the design parameters that were varied included the different sizes of the MP and BP (ratios of the MP to BP), and size of the fix segment of the passive leg h_f. Therefore the dimensionless design parameters of the current manipulator are considered as $\mu = r_e / r_b$ and $\eta = h_f / r_b$ with the bound areas of $1 \leq \mu \leq 3$, $1 \leq \eta \leq 2$. A similar algorithm to the first robot is used to study the variation of GTI with respect to two dimensionless design parameters. The results are depicted in Figures 22 and 23 for the constant heave motion of the MP while roll and pitch angles are varied between -30° to 30°.

Figure 20. Graphical interpretation of workspace boundaries of the 3-DOF cable-driven parallel robot

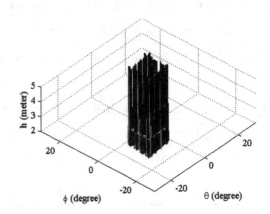

Figure 21. Numerical workspace volume of the 3-DOF cable-driven parallel robot

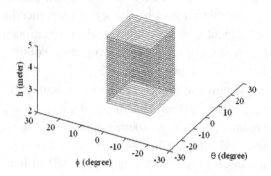

Numerical Approach

The matrices F_o and F_p of Equation (39) for this manipulator can be written as

$$\mathbf{F}_o = \begin{bmatrix} k_1 & k_2 \\ k_3 & k_4 \\ k_5 & k_6 \end{bmatrix} \quad \mathbf{F}_p = \begin{bmatrix} k_7 \\ k_8 \\ k_9 \end{bmatrix}, \tag{78}$$

where the elements of matrices F_o and F_p are defined as Equation 79 in Box 4.

The isotropy condition for matrix J can be written as

$$\mathbf{J}^T \mathbf{J} = \delta^2 \begin{bmatrix} \mathbf{I}_2 & \mathbf{0} \\ \mathbf{0} & 1 \end{bmatrix} \tag{80}$$

where

$$\mathbf{J} = \mathbf{B}_N^{-1} \mathbf{A}_N \tag{81}$$

From the results in Figure 22, it is observed that the GTI value increases slightly as the ratio of the MP to the BP approaches $\mu=2$ value for all values of η, and then gradually decreases as it approaches to $\mu=3$. However, the top line with $\eta=2$ yields the maximum value of GTI. As inferred from Figure 23, the GTI value increases slightly as the MP rotates itself from the orientations pertaining to $\varphi=0$ and $\theta=15^o(0.5^{rad})$ and then gradually decreases as it approaches to $\varphi=\theta=\pm30^o$ from $\varphi=0$ and $\theta=15^o(0.5^{rad})$.

Box 4.

$$k_1 = r_e(r_b \sin\phi + h \cos\phi) \tag{79.a}$$

$$k_2 = 0 \tag{79.b}$$

$$k_3 = -\sqrt{3}/4 \, r_e \, r_b \cos\phi \sin\theta + 1/4 \, r_e \, r_b \sin\phi - \sqrt{3}/2 \, r_e \, h \sin\phi \sin\theta - 1/2 \, r_e \, h \cos\phi \tag{79.c}$$

$$k_4 = 3/4 \, r_e \, r_b \sin\theta - \sqrt{3}/4 \, r_e \, r_b \sin\phi \cos\theta + \sqrt{3}/2 \, r_e \, h \cos\phi \sin\theta \tag{79.d}$$

$$k_5 = \sqrt{3}/4 \, r_e \, r_b \cos\phi \sin\theta + 1/4 \, r_e \, r_b \sin\phi + \sqrt{3}/2 \, r_e \, h \sin\phi \sin\theta - 1/2 \, r_e \, h \cos\phi \tag{79.e}$$

$$k_6 = 3/4 \, r_e \, r_b \sin\theta + \sqrt{3}/4 \, r_e \, r_b \sin\phi \cos\theta - \sqrt{3}/2 \, r_e \, h \cos\phi \cos\theta \tag{79.f}$$

$$k_7 = h + r_e \sin\phi \tag{79.g}$$

$$k_8 = h + \sqrt{3}/2 \, r_e \cos\phi \sin\theta - 1/2 \, r_e \sin\phi \tag{79.h}$$

$$k_9 = h - \sqrt{3}/2 \, r_e \cos\phi \sin\theta - 1/2 \, r_e \sin\phi. \tag{79.i}$$

Figure 22. Variation of GTI versus geometry parameters of 3-DOF robot

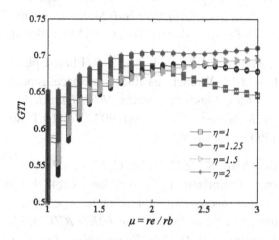

Figure 23. Variation of GTI versus pose of the MP of 3-DOF robot

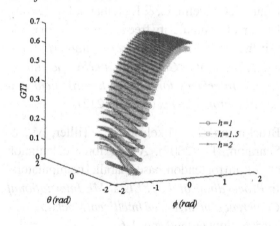

Here $\mathbf{A}_N$ and $\mathbf{B}_N$ are as defined in Equation (22), and $\mathbf{I}_2$ is a 2×2 identity matrix.

Upon substituting Equations (78) into Equation (80) and in the light of Equation (39), we obtain the isotropic conditions for the manipulator as

$$\frac{1}{L^2}\mathbf{F}_o^T\mathbf{B}_N^{-2}\mathbf{F}_o = \sigma^2\mathbf{I}_2 \tag{82}$$

$$\mathbf{F}_p^T\mathbf{B}_N^{-2}\mathbf{F}_p = \sigma^2 \tag{83}$$

$$\frac{1}{L}\mathbf{F}_o^T\mathbf{B}_N^{-2}\mathbf{F}_p = \mathbf{0}. \tag{84}$$

The traces of both sides of Equations(90) and (91) lead to

$$L^2 = \frac{tr(\mathbf{F}_o^T\mathbf{F}_o)}{2(\mathbf{F}_p^T\mathbf{F}_p)} \tag{85}$$

Upon substituting Equations (79) into Equations (82)–(85), we obtain an overdetermined system of equations with seven equations, namely, three equations related to Equation (82) because of the symmetry matrix, one equation for Equation (83) and two equations for Equation (84) in four unknowns, namely, h, ϕ, θ and δ. Also, there are three inequalities because of cable tension. Hence, it is an over-determined constrained system that should be solved numerically. The objective function of the optimization problem is chosen the negative value of GTI. Negative value of GTI is chosen because when GTI is minimized as objective function, it would be approached one, i.e., the cable robot is far from singularity and close to the isotropic postures. Applying the isotropic and cable tension condition for this specific robot and solving the nonlinear optimization problem by means of *fmincon* MATLAB function, the isotropic configuration of the 3-DOF constrained cable-driven parallel robot is given as

$$\theta = 0, \quad \phi = 0, \quad h = 4.75\,m,$$
$$L = 1.7, \quad \delta = 1.2 \tag{86}$$

It may be noted that the manipulator would achieve the best kinematic performance at this configuration. Hence, the manipulator would be in high dexterity through the workspace if the MP configuration is chosen around the resulted posture of the isotropic configuration of the cable robot.

CONCLUSION

The kinematic isotropic posture of two spatial cable robots, a 6-6 cable-suspended parallel robot and a 3-DOF cable-driven parallel manipulator, was investigated to achieve the best kinematic performance for these robots. This objective has been accomplished by introducing four different methods as 1) symbolic computation to obtain the isotropic and cable tension condition symbolically, 2) geometric workspace in which WFW and boundaries workspace were determined analytically and plotted to give the graphical interpretation of singularity-free workspace, 3) numerical workspace and GTI as objective functions to optimize the design parameters of cable robots and 4) numerical approach to evaluate the isotropic configuration of cable robots numerically. The results showed that the best architecture for the 6-6 cable robot is obtained when the MP is of hexagonal and the BP is of equilateral triangular shapes. Also, the ratio of the MP size to BP size was derived for the isotropic posture of the both robots. Upon comparison of the results of both cable manipulators from the dexterity and isotropic viewpoints, it can be concluded that the 3-DOF cable robot can be used where the dexterity and kinematic performance are more important. Nevertheless, the 6-DOF cable robot can be used where the large workspace volume is desirable.

REFERENCES

Alici, G., & Shirinzadeh, B. (2004). Optimum synthesis of a parallel manipulator based on kinematic isotropy and force balancing. *Robotica*, 22(1), 97–108. doi:10.1017/S0263574703005216.

Behzadipour, S., & Khajepour, A. (2004). Design of reduced DOF parallel cable-based robots. *Mechanism and Machine Theory*, 39(10), 1051–1065. doi:10.1016/j.mechmachtheory.2004.05.003.

Behzadipour, S., & Khajepour, A. (2005). A new cable-based parallel robot with three degrees of freedom. *Journal of Multibody System Dynamics*, 13(4), 371–383. doi:10.1007/s11044-005-3985-6.

Bosscher, P., Riechel, A. T., & Ebert-Uphoff, I. (2006). Wrench-feasible workspace generation for cable-driven robots. *IEEE Transactions on Robotics*, 22(5), 890–903. doi:10.1109/TRO.2006.878967.

Bouchard, S., & Gosselin, C. M. (2007). Workspace optimization of a very large cable-driven parallel mechanism for a radiotelescope application. In *Proceedings of the IDETC/CIE ASME International Design Engineering Technical Conferences & Computers and Information in Engineering Conference*, Las Vegas, NV (pp. 1-7).

Brau, E., Gosselin, F., & Lallemand, J. P. (2005). Design of a singularity free architecture for cable driven haptic interfaces. In *Proceedings of the First Joint Eurohaptics Conference and Symposium on Haptic Interfaces for Virtual Environment and Teleoperator Systems* (pp. 208-212).

Bruckmann, T., Mikelsons, L., Hiller, M., & Schramm, D. (2007). A new force calculation algorithm for tendon-based parallel manipulators. In *Proceedings of the IEEE/ASME International Conference on Advanced Intelligent Mechatronics*, Zurich, Switzerland (pp. 1-6).

Dekker, R., Khajepour, A., & Behzadipour, S. (2006). Design and testing of an ultra high-speed cable robot. *International Journal of Robotics and Automation*, 21(1), 25–34. doi:10.2316/Journal.206.2006.1.206-2824.

Diao, X., & Ma, O. (2007). A method of verifying force-closure condition for general cable manipulators with seven cables. *Mechanism and Machine Theory*, 42(12), 1563–1576. doi:10.1016/j.mechmachtheory.2007.06.008.

Fattah, A., & Ghasemi, H. A. M. (2002). Isotropic design of spatial parallel manipulators. *The International Journal of Robotics Research, 21*(9), 811–824. doi:10.1177/02783649020021009842.

Fattah, A., & Kasaei, G. H. (2000). Kinematics and dynamics of a parallel manipulator with a new architecture. *Robotica, 18*(5), 535–543. doi:10.1017/S026357470000271X.

Gallina, P., & Rosati, G. (2002). Manipulability of a planar wire driven haptic device. *Mechanism and Machine Theory, 37*(2), 215–228. doi:10.1016/S0094-114X(01)00076-3.

Gouttefarde, M., & Gosselin, C. M. (2006). Analysis of the wrench-closure workspace of planar parallel cable-driven mechanisms. *IEEE Transactions on Robotics, 22*(3), 434–445. doi:10.1109/TRO.2006.870638.

Gouttefarde, M., Merlet, J.-P., & Daney, D. (2006). Determination of the wrench-closure workspace of 6-DOF parallel cable-driven mechanisms. *Advances in Robot Kinematics*, 315-322.

Hadian, H., & Fattah, A. (2008). Best kinematic performance analysis of a 6-6 Cable-suspended parallel robot. In *Proceedings of the IEEE/ASME International Conference on Mechatronic and Embedded Systems and Applications*, Beijing, China (pp. 510-515).

Hadian, H., & Fattah, A. (2009). On the study of dexterity measure for a novel 3-DOF cable-driven parallel manipulator. In *Proceedings of the 17th Annual International Conference on Mechanical Engineering*, Tehran, Iran (pp. 412-418).

Li, Y., & Xu, Q. (2006). GA-based multi-objective optimal design of a planar 3-DOF cable-driven parallel manipulator. In *Proceedings of the International IEEE Conference on Robotics and Biomimetics*, Kunming, China (pp. 1360-1365).

Perreault, S., & Gosselin, C. (2007). Cable-driven parallel mechanisms: application to a locomotion interface. *Journal of Mechanical Design, 130*(10), 102301–102309. doi:10.1115/1.2965607.

Pusey, J., Fattah, A., Agrawal, S., & Messina, E. (2004). Design and workspace analysis of a 6-6 cable-suspended parallel robot. *Mechanism and Machine Theory, 39*(7), 761–778. doi:10.1016/j.mechmachtheory.2004.02.010.

Rosati, G., Gallina, P., Masiero, S., & Rossi, A. (2005). Design of a new 5 DOF wire-based robot for rehabilitation. In *Proceedings of the IEEE 9th International Conference on Rehabilitation Robotics*, Chicago, IL (pp.430-433).

Surdilovic, D., & Bernhardt, R. (2004). STRING-MAN: a new wire robot for gait rehabilitation. In *Proceedings of the IEEE International Conference on Robotics and Automation*, Berlin, Germany (pp. 2031-2036).

Takemura, F., Enomoto, M., Tanaka, T., Denou, K., Kobayashi, Y., & Tadokoro, S. (2005). Development of the balloon-cable driven robot for information collection from sky and proposal of the search strategy at a major disaster. In *Proceedings of the IEEE/ASME International Conference on Advanced Intelligent Mechatronics*, Monterey, CA (pp. 658-663).

Verhoeven, R., & Hiller, M. (2002). Estimating the controllable workspace of tendon-based Stewart platforms. In *Proceedings of the International Symposium on Advances in Robot Kinematics*, Portoroz, Slovenia (pp. 277-284).

Verhoeven, R., Hiller, M., & Tadokoro, S. (1998). Workspace of tendon-driven Stewart platforms: Basics, classification, details on the planar 2-dof class. In *Proceedings of the International Conference on Motion and Vibration Control* (pp. 871-876).

Voglewede, A. P., & Ebert-Uphoff, I. (2005). Application of the antipodal grasp theorem to cable-driven robots. *IEEE Transactions on Robotics, 21*(4), 713–718. doi:10.1109/TRO.2005.844679.

Williams, R. L. II, & Gallina, P. (2002). Planar cable-direct-driven robots: design for wrench exertion. *Journal of Intelligent & Robotic Systems, 35*, 203–219. doi:10.1023/A:1021158804664.

Chapter 19

Experimental Investigations on the Contour Generation of a Reconfigurable Stewart Platform

G. Satheesh Kumar
Indian Institute of Technology Madras, India

T. Nagarajan
Universiti Teknologi Petronas (UTP), Malaysia

ABSTRACT

Reconfiguration of Stewart platform for varying tasks accentuates the importance for determination of optimum geometry catering to the specified task. The authors in their earlier work (Satheesh et al., 2008) have indicated the non availability of an efficient holistic methodology for determining the optimum geometry. Further, they have proposed a solution using the variable geometry approach through the formulation of dimensionless parameters in combination with generic parameters like configuration and joint vector. The methodology proposed provides an approach to develop a complete set of design tool for any new reconfigurable Stewart platform for two identified applications viz., contour generation and vibration isolation. This paper details the experimental investigations carried out to validate the analytical results obtained on a developed Stewart platform test rig and error analysis is performed for contour generation. The experimental natural frequency of the developed Stewart platform has also been obtained.

1. INTRODUCTION

Serial manipulators have the advantage of sweeping workspaces and dexterous maneuverability like the human arm, but their load capacity is rather poor due to the cantilever structure. Consequently,

from strength considerations, the links become bulky on the one hand, while on the other they tend to bend under heavy load and vibrate at high speed. Though possessing a large workspace, their precision positioning capability is poor. This lead to the development of parallel manipulators as

DOI: 10.4018/978-1-4666-3634-7.ch019

alternative solution and considerable work has been done in this domain (Raghavan, 1993; Tsai, 1999; Simaan *et al.,* 2001, 2003).

The Stewart platform, proposed by Stewart as an aircraft simulator (Stewart, 1965), is a six degree-of-freedom parallel manipulator where the end-effector is attached to a movable plate supported by six linear actuated links in parallel. Typically, the number of limbs is equal to the number of degrees of freedom such that every limb is controlled by one actuator and all the actuators can be mounted at or near the fixed base (Fichter, 1986). Recently, due to the advantages of the Stewart platform as compared to conventional open kinematics chain machines, much effort has been dedicated to the implementation of Stewart platforms in machine tools and robotic applications (Dasgupta *et al.,* 2000). The most important fields where Stewart platform has solved the pertinent, time immemorial problems are Machine tool applications, Precision pointing and Vibration isolation applications.

Researchers have constantly tried Stewart platform in various other fields since the time its potential as a six degree-of-freedom mechanism was identified (for flight simulation). The systems that have been developed and readily available in the market as commercial hexapods, such as VARIAX from Giddings & Lewis, Tornado from Hexel Corp., Geodetic from Geodetic Technology Ltd., are mission specific. The rigidity of the legs of the Stewart platform determines the application area. A stiff hexapod, for example, is used as a rigid interface for active damping and precision pointing applications while a soft hexapod is used, in general, for the purpose of active isolation of vibrations (Hanieh, 2003). In all the cases cited above, there is not much freedom available for the end-user to choose between structural rigidity and dexterity to use the same platform for other applications. In effect there exists a need for reconfigurable designs of robotic platforms to meet the raising standards created by the cur-

rently maturing industrial scenario. Along this line Wavering (1998) identified, among many, the potential directions for future work as the alternative kinematic configurations and improved modularity/configurability.

2. RECONFIGURATION

A modular reconfigurable robotic system is a collection of individual link and joint components that can be assembled into different robot geometries for specific task requirements. However, the machining tolerance and assembly errors at the module interconnections affect the positioning accuracy of the end-effector (Chen & Yang, 1997, 1998; Chen & Burdick, 1998). A generic approach which is better than the modular design approach for the prescribed varying tasks should be identified. The authors in their earlier work identified from the literature (Du Plessis & Snyman, 2006; Lin *et al.,* 2003; Chen, 2000, 2001) that there is a lack of an efficient holistic methodology for determining the optimum geometry for the task of reconfiguration. It was also identified (Leger, 1999; Chen *et al.,* 1999, 2001; Xi, 2001) that there is a need to develop a reconfigurable configuration for at least two applications.

A solution (Satheesh *et al.,* 2008) was presented through the formulation of dimensionless parameters in combination with generic parameters like configuration. This methodology is best suited when the problem is approached from the application perspective. The proposed methodology provided a holistic approach to develop a complete set of design tool for any new reconfigurable Stewart platform for two important applications viz., contour generation and vibration isolation. In order to validate the simulation results obtained earlier, experimental investigations made are presented in this research work, for reconfiguration of Stewart platform with variable geometry approach. In the following section the experimental test rig developed for this purpose is explained. Subsequently the experimental setup and results obtained for both the applications are presented.

3. EXPERIMENTAL TEST-RIG

A Stewart platform (PeilPOD3-3) shown in Figure 1 was developed at IIT Madras to conduct initial study on the parameters affecting the modularity. For the purpose of clarity the parameters chosen for the study, in the preceding work of the authors, is included.

3.1. Parameters for Reconfigurability

1. Configuration
2. **bi:** Position vector of the joints in the moving platform
3. **p:** Position vector of the origin of the moving platform coordinate
4. λ: Angle between the moving platform and the legs
5. γ: Angle between the legs
6. Work volume

6-6 and 3-3 configurations are chosen for this study owing to their common usage in industries for their simplicity in design. 3-3 configuration is in general represented by Figure 2 and 6-6 is

Figure 1. PeilPOD3-3 as test-rig

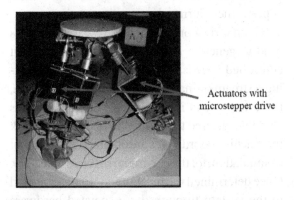

Actuators with
microstepper drive

represented by Figure 3. Figure 2 also represents the parameters λ and γ which are directly dependent on b_i (called as joint vector from now on) and so an explicit study on them is avoided. p is chosen as the fixed parameter (refer to Figure 3 for other parameters). Other parameters do not come under the scope of this research.

3.2. Test-Rig

The developed Stewart platform is a 3-3 configuration with spherical joints at the top and the bottom platform (Ying & Liang, 1994; Nanua & Waldron, 1989; Griffis & Duffy, 1989). From the simulation results it is decided to choose a 3-3 configuration for the experimental test-rig. Other reasons are the ease in changing the b_i parameter during experimentation and the inherent better workspace that 3-3 configuration possess over 6-6 configuration (Huang *et al.*, 1999). The design parameters are chosen such that the platform can operate within the ranges of ± 0.1 m in the X and Y-axes, ± 0.3 m in the Z-axis, $\pm 15°$ in roll and pitch, and $\pm 20°$ in yaw. The disadvantages of relatively small rotating range of the moving platform and complex control required for 6 axes are applicable to this platform also.

The legs connect the top and bottom platforms through spherical joints and a base block (SPS connection). The connection point at the base provides the desired angle of contact between the legs and the platforms. The choice of spherical joints was made to study the effect of passive degrees of freedom (DOF) which is not included under the scope of this research work. Table 1 provides the geometrical parameters and other specifications of the developed Stewart platform, under the nominal conditions. The base block is designed to accommodate all the changes in angles achieved by changing the joint vector. Slots are provided in both the platforms to slide and vary the joint vector to test the platform for different conditions. Provisions are also made in the legs to

Figure 2. Parameters for reconfiguration

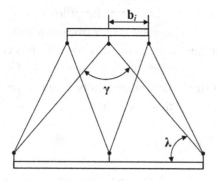

Figure 3. Schematic for 6-6 configuration

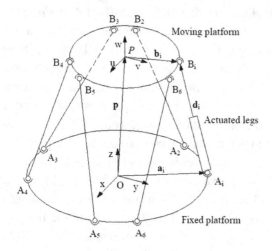

mount displacement and force sensors to measure the leg length variations and forces respectively. The top platform is used to mount accelerometer and other displacement transducers like LVDT for the different sets of experiments.

3.3. Hardware and Path Control

3.3.1. Motion Control

The actuators used are Integrated Microstepping Drive/NEMA 17 linear actuators with micro-stepping drive and onboard electronics. The integrated electronics of the drive eliminates the need to run the motor cabling through the machine, reducing the potential for problems due

to electrical noise. Input voltage requirement for the drive is +12 to +48 VDC and accepts up to 14 resolution settings from 1/2 to 256 microsteps per step providing up to 51,200 steps/rev. The logic inputs are internally limited to allow for a +5VDC power supply. The stepclock pin is driven by the output channels of 8254 timer card. The direction reversal is controlled by 8255 timer for all the six actuators. The enable pin facilitates the independent control for all the six actuators and is controlled by the remaining ports of 8255 timer card. Two SN7407s are used for current control on the signals sent to the direction and enable channels.

Figure 4 provides the schematic representation of connections for the developed experimental setup which includes the timer cards used for driving the actuators. To engage with the onboard electronics of micro-stepper actuators two PCI based 24 channel I/O & Timer interface cards are used.

3.3.2. Path Control

The path control is based upon the geometric path control presented by McCloy and Harris (1986). Both geometric and kinematic methods of path control assume that the manipulator passes through a succession of equilibrium states to achieve coordinated control. Though geometric control involves complex calculations which must be performed in a short time, it is preferred over the kinematic control since only off-line control is performed during the experimentation.

Geometric control in this research work is used to generate motion along desired paths at controlled speeds. This is achieved by calculating a stream of machine coordinates in real time, representative of intermediate world coordinates along the desired trajectory. In this research work the machine coordinates are obtained as the output of simulation, for the maximum stiffness trajectory. Once determined the machine coordinates are fed to the system through the associated hardware

Figure 4. Schematic representation of the experimental setup

controls. The actuators will then drive the end-effector or the centroid of the moving platform to the desired world coordinates through an unpredictable path. By varying the travel/step parameter of the micro stepper motor the resolution of the determined machine coordinates can be altered, up to a minimum possible extent.

4. METHODOLOGY ADOPTED AND DISCUSSIONS

The flowchart presented in Figure 5 represents the complete details of the simulation followed by the procedure of experimentation. The approximated inertia details, used to develop the trajectory tracking algorithm, of PeilPOD3-3 are provided in Table 2. The implementation of the control system to overcome the error is envisaged for further research.

Tracking performance of PeilPOD3-3 is used to study and validate the simulation results obtained for contour generation, which comprehensively authenticates the proposed methodology based on variable-geometry approach. Tracking performance for two trajectories viz., Straightline and Circular trajectories is done and then error analysis is performed on the data obtained. Error analysis is used in advantage to study by comparison, the difference in performance of the system for the maximum stiffness trajectory and other trajectories. Machining error, play at the joints and alignment errors are a few of the reasons identified for the errors obtained on the tracking performance of the system.

The complete setup along with the LVDTs is shown in Figure 6. Two LVDTs of range 20 mm and resolution of 20 microns are mounted at 90° to each other as shown in Figure 5 to obtain the displacement of the system along x-y coordinates. The perpendicularity error of the cube mounted on the moving platform is within 60 microns. Two perpendicular surfaces are mounted on the moving platform over which the LVDTs are allowed to slide to measure the displacements along

Figure 5. Simulation of experiment followed by the procedure

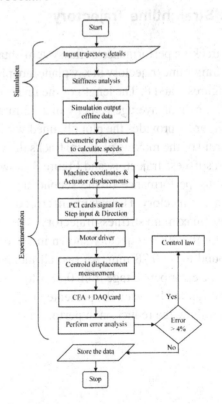

Figure 6. Experimental setup for tracking performance

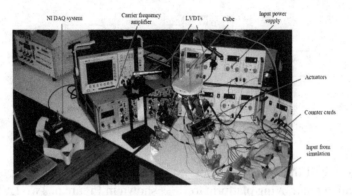

the two perpendicular directions. Table 3 lists all the specifications of the LVDTs used. The data is acquired through the Carrier Frequency Amplifier (CFA) and the LVDT is operated at a voltage of 24V. NI DAQ card (AI-16E-4) is used to acquire the data from the LVDTs for this open-loop control system. It has 16 inputs, with 500 kS/s and 12 bit multifunction I/O capabilities too. NI DAQ signal accessory N114 is used to communicate with PC to process the acquired data.

4.1. Straightline Trajectory

The tracking performance of the system obtained for straightline trajectory in x-y plane, are plotted in Figures 7 and 8. The length of the line tracked is 20 mm at an average velocity of 1.33 mm/s.

Figure 7 provides the plot obtained while the centroid of the moving platform tracks the maximum stiffness trajectory and Figure 8 shows the tracking performance of the second maximum stiffness trajectory. The maximum error obtained for the maximum stiffness trajectory is 0.22 mm and for the other trajectory shown in Figure 8 it is found to be 0.28 mm. An overall increase in average error percentage from 0.88% to 1.57% is also observed between the presented trajectories which show the reduction in performance. A di-

verging trend is observed towards the end of the trajectory for both the cases.

4.2. Circular Trajectory

The diameter of the circular trajectory tracked is 20 mm and the plots obtained while the centroid tracks this trajectory for two cases are provided in Figures 9 and 10. The displacement while tracking the maximum stiffness trajectory is plotted in Figure 9. Figure 10 provides the tracking performance for another circular trajectory with second maximum stiffness. An increase in

Figure 7. Tracking performance for straightline trajectory (maximum stiffness trajectory)

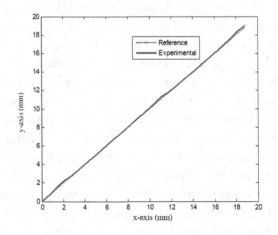

Figure 8. Tracking performance for straightline trajectory (second maximum stiffness)

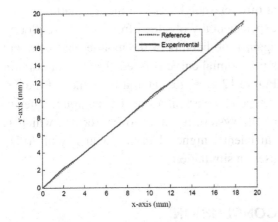

Figure 9. Tracking performance for circular trajectory (maximum stiffness trajectory)

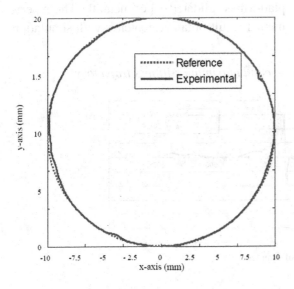

average error percentage from 1.32% to 3.44% is observed for this trajectory over the maximum stiffness trajectory. The maximum error is found to be 0.34 mm for maximum stiffness trajectory and 0.47 mm for another trajectory with second maximum stiffness.

An observation of the error plots for the circular trajectories, shown in Figure 11, shows

repetitive errors at the start of the trajectory. It provides firsthand information on the performance of the platform in support of Figures 9 and 10. The sudden sharp rise in the error, around 355th sample, is attributed to a typical play observed in the platform which is to be solved by arresting the play in the joints. The errors include the noise and non-linearities caused by the mounting that are difficult to eliminate and so calls for the usage of a laser based calibration device. An explicit analysis to ameliorate the causative factors for the errors is avoided as it is not included in the scope of the research.

The results obtained for both circular trajectories shows same pattern of performance deterioration between maximum stiffness trajectory and other trajectory. This is a first step towards showing that configurations which have lower stiffness due to their geometry also suffer from worse performance. Presently experimental measurements are taken on the test rig for other set of experiments that are performed in reiterating the validity of the claim placed by this research work in order to extend it to other areas of research.

Figure 10. Tracking performance for circular trajectory (second maximum stiffness)

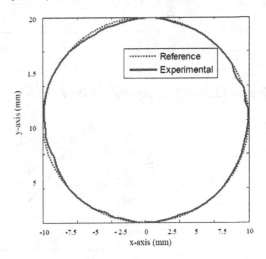

4.3. Natural Frequency

Measurement of natural frequency is a must in our case for the platform to be applied for vibration isolation applications. Experiments were also conducted to quantify the natural frequency of the test-rig in order to establish the frequency at which the simulation is to be performed for vibration isolation application. The procedure adopted is a standard methodology followed for any multibody system. An accelerometer is mounted at the centroid of the moving platform. The platform is held at nominal height and an impulse is produced on the moving platform and the natural frequency is found from the output obtained from the accelerometer. Impulse is generated by means of a small hammer perpendicular to the surface of the top platform and no automated system was used. At the same height the position of the accelerometer

is shifted and natural frequency is recorded for all those positions. The same procedure is repeated at different heights of the mobile platform. The experimentally obtained frequencies are plotted against the heights of the mobile platforms and a polynomial curve is fitted. The lowest point in Figure 12 gives the natural frequency of the developed Stewart platform. The natural frequency of the system is found to be 166 Hz which is sufficiently higher than the frequency of 70 Hz used in simulation.

CONCLUSION

A holistic approach proposed to develop a complete set of design tool for a reconfigurable Stewart platform is validated experimentally. The experimental result obtained matches with simulation

Figure 11. Displacement error for circular trajectory (second maximum stiffness trajectory)

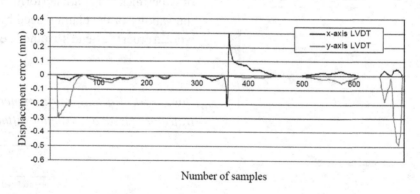

Figure 12. Natural frequency of PeilPOD3-3

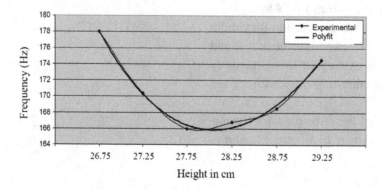

result. This methodology can be adopted for any reconfigurable Stewart platform to be developed for specific applications. An experimental test rig has been developed based on the design parameters obtained from the simulation results. An algorithm has been developed and implemented based on kinematic path control for conducting the experiments to find the tracking performance. The data set obtained from simulation for straight line and circular contours are used for experimentation.

Error analysis is used to establish the validity of the proposed approach which is developed based on variable geometry. The tracking performance of the platform for the trajectory with maximum stiffness is compared with the performance for second maximum stiffness. For this case, an increase in error percentage is observed. For circular contour, the maximum error is found to be 0.34 mm for maximum stiffness trajectory and 0.47 mm for the trajectory chosen with second maximum stiffness. Similar results are observed for straight line trajectory. Experiments are also performed to identify the natural frequency of the developed Stewart platform to examine the frequency used in simulation studies. Trajectories of other nature could be used for the analysis of tracking performance with inclusion of workspace as a modular parameter. A comprehensive methodology should be identified for conducting the experiments for vibration isolation application.

REFERENCES

Chen, I.-M. (2000). Realization of a rapidly reconfigurable robotic Workcell. *Journal of Japan Society of Precision Engineering, 66*(7), 1024–1030. doi:10.2493/jjspe.66.1024.

Chen, I.-M. (2001). Rapid response manufacturing through a rapidly reconfigurable robotic Workcell. *Robotics and Computer-integrated Manufacturing, 17*, 199–213. doi:10.1016/S0736-5845(00)00028-4.

Chen, I.-M., & Burdick, J. W. (1998). Enumerating non-isomorphic assembly configurations of modular robotic systems. *The International Journal of Robotics Research, 17*(7), 702–719. doi:10.1177/027836499801700702.

Chen, I.-M., & Yang, G. (1997). Kinematic calibration of modular reconfigurable robots using product-of-exponentials formula. *Journal of Robotic Systems, 14*(11), 807–821. doi:10.1002/(SICI)1097-4563(199711)14:11<807::AID-ROB4>3.0.CO;2-Y.

Chen, I.-M., & Yang, G. (1998). Automatic model generation for modular reconfigurable robot dynamics. *Transactions of the ASME, 120*, 346–352.

Chen, I.-M., Yang, G., & Kang, I.-G. (1999). Numerical inverse kinematics for modular reconfigurable robots. *Journal of Robotic Systems, 16*(4), 213–225. doi:10.1002/(SICI)1097-4563(199904)16:4<213::AID-ROB2>3.0.CO;2-Z.

Chen, I.-M., Yang, G., Tan, C. T., & Yeo, S. H. (2001). A local POE model for robot kinematic calibration. *Mechanism and Machine Theory, 36*, 1215–1239. doi:10.1016/S0094-114X(01)00048-9.

Dasgupta, B., & Mruthyunjaya, T. S. (2000). The Stewart platform manipulator - A review. *Mechanism and Machine Theory, 35*, 15–40. doi:10.1016/S0094-114X(99)00006-3.

Du Plessis, L. J., & Snyman, J. A. (2006). An optimally re-configurable Planar Gough-Stewart machining platform. *Mechanism and Machine Theory, 41*, 334–357. doi:10.1016/j.mechmachtheory.2005.05.007.

Fichter, E. F. (1986). A Stewart platform-based manipulator: General theory and practical construction. *The International Journal of Robotics Research, 5*(2), 157–182. doi:10.1177/027836498600500216.

Griffis, M., & Duffy, J. (1989). A forward displacement analysis of a class of Stewart platforms. *Journal of Robotic Systems*, *6*(6), 703–720. doi:10.1002/rob.4620060604.

Hanieh, A. A. (2003). *Active isolation and damping of vibrations via Stewart platform*. Unpublished doctoral dissertation, Active Structures Laboratory, Universit'e Libre de Bruxelles, Brussels, Belgium.

Huang, T., Wang, J., Gosselin, C. M., & Whitehouse, D. (1999). Determination of closed form solution to the 2-D-orientation workspace of Gough–Stewart parallel manipulators. *IEEE Transactions on Robotics and Automation*, *15*(6), 1121–1125. doi:10.1109/70.817675.

Leger, C. (1999). *Automated synthesis and optimization of robot configurations: An evolutionary approach*. Unpublished doctoral dissertation, The Robotics Institute, Carnegie Mellon University, Pittsburgh, PA.

Lin, C., Tang, X., Shi, J., & Duan, G. (2003). Workspace analysis of reconfigurable parallel machine tool based on setting-angle of spherical joint. In *Proceedings of the International Conference on Systems, Man and Cybernetics* (pp. 4945-4950).

McCloy, D., & Harris, D. M. J. (1986). Robotics: An introduction. New York, NY: A Halsted Press book, Open University Press.

Nanua, P., & Waldron, K. (1989). Direct kinematic solution of a Stewart platform. In *Proceedings of the IEEE International Conference on Robotics and Automation* (pp. 431-437).

Raghavan, M. (1993). The Stewart platform of general geometry has 40 configurations. *Journal of Mechanical Design*, *115*, 277–282. doi:10.1115/1.2919188.

Satheesh, G. K. (2009). *Characterization of reconfigurable Stewart platform*. Unpublished doctoral dissertation, Indian Institute of Technology Madras, Madras, India.

Satheesh, G. K., & Nagarajan, T. (2008). Experimental investigations on reconfigurable Stewart platform for contour generation. In *Proceedings of the 3rd International Conference on Sensing Technology* (pp. 292-296).

Satheesh, G. K., & Nagarajan, T. (2011). Reconfigurable Stewart platform for spiral contours. *Journal of Applied Sciences*, *11*, 1552–1558. doi:10.3923/jas.2011.1552.1558.

Satheesh, G. K., Nagarajan, T., & Srinivasa, Y. G. (2009). Characterization of reconfigurable Stewart platform for contour generation. *Robotics and Computer-integrated Manufacturing*, *25*, 721–731. doi:10.1016/j.rcim.2008.06.001.

Simaan, N., & Shoham, M. (2001). Singularity analysis of a class of composite serial in-parallel robots. *IEEE Transactions on Robotics and Automation*, *17*(3), 301–311. doi:10.1109/70.938387.

Simaan, N., & Shoham, M. (2003). Geometric interpretation of the derivatives of parallel robot's Jacobian matrix with application to stiffness control. *ASME Journal of Mechanical Design*, *125*, 33–42. doi:10.1115/1.1539514.

Stewart, D. (1965). A platform with six degrees of freedom. *Proceedings - Institution of Mechanical Engineers*, *180*(1), 371–386. doi:10.1243/PIME_PROC_1965_180_029_02.

Tsai, L. W. (1999). *Robotic analysis: The mechanics of serial and parallel manipulators*. New York, NY: John Wiley & Sons.

Wavering, A. J. (1998). Parallel kinematic machine research at NIST: Past, present, and future. In *Proceedings of the First European-American Forum on Parallel Kinematic Machines: Theoretical Aspects and Industrial Requirements*, Italy (pp. 1-13).

Xi, F. (2001). A comparison study on hexapods with fixed-length legs. *International Journal of Machine Tools & Manufacture, 41*, 1735–1748. doi:10.1016/S0890-6955(01)00038-4.

Ying, J. P., & Liang, C. G. (1994). The forward displacement analysis of a kind of special platform manipulator mechanisms. *Mechanism and Machine Theory, 29*(1), 1–9. doi:10.1016/0094-114X(94)90015-9.

Chapter 20
Parallel Architecture Manipulators for Use in Masticatory Studies

Madusudanan Sathia Narayanan
University at Buffalo, USA

Xiaobo Zhou
University at Buffalo, USA

Srikanth Kannan
University at Buffalo, USA

Frank Mendel
University at Buffalo, USA

Venkat Krovi
University at Buffalo, USA

ABSTRACT

There is considerable scientific and commercial interest in understanding the mechanics of mastication. In this paper, the authors develop quantitative engineering tools to enable this process by: (i) designing a general purpose mastication simulator test-bed based on parallel architecture manipulator, capable of producing the requisite motions and forces; and (ii) validating this simulator with a range of test-foods, undergoing various mastication cycles under controlled and monitored circumstances. Such an implementation provides a test bed to quantitatively characterize the mastication based on "chewability index". Due to the inherent advantages of locating actuators at the base (ground) in terms of actuator efforts and structural rigidity as well as benefits of using prismatic sliders compared to revolute actuators, the 6-P-U-S system was chosen. A detailed symbolic kinematic analysis was then conducted. For the practical implementation of the test-bed, the analytical Jacobian was examined for singularities and the design was adapted to ensure singularity free operation. A comprehensive parametric study was undertaken to obtain optimal design parameters for desired workspace and end effector forces. Experiments captured jaw motion trajectories using the high speed motion capture system which served as an input to the hardware-in-the-loop simulator platform.

DOI: 10.4018/978-1-4666-3634-7.ch020

INTRODUCTION

The goal of this research is to develop an experimental testbed for analyzing the masticatory jaw motions of animals (including humans) and establish the quantitative relationship between relevant geometric parameters (tooth geometries, numbers and types) as well as regimen parameters (joint forces, motions) during the mastication process for further detailed studies. Such an understanding would be of tremendous importance from different perspectives. From a biological science perspective, it is critical to understand the physiological variability in mastication across individuals in a species as well as across the entire population. It is also very useful to know how various animals preprocess food while chewing and biting so that the research scientists can hypothesize behavioral analyses of how certain breeds of animals kill their prey with an accurate estimate of biting forces (Signore, Krovi, & Mendel, 2005) and understand the resulting muscles functioning to draw interesting conclusions. From an engineering perspective, such a study on masticatory performance would enable to quantitatively assess the "chewability or performance index" due to variations in food

properties as well as facilitate the parametric design of dentitions/prosthetic additions.

To this end our work focused on developing quantitative engineering tools to enable this process (Figure 1). Specifically, we focused on (i) designing a general purpose mastication simulator test-bed based on parallel architecture manipulator, capable of producing the motions and forces encountered; and (ii) validating this simulator with a range of test-foods undergoing varying mastication cycles under controlled and carefully monitored circumstances. Such an implementation could provide a testbed to quantitatively characterize the mastication based on "chewability index" factor for a wide range of applications. Building a generic engineering test-bed that can be used to study mastication for these diverse goals offers significant challenges. The multiple facets of the overall project are shown in Figure 1 – in this paper, we will focus only on the highlighted sections. Specifically, various design variants of parallel manipulators were examined for jaw simulators before finalizing on 6-P-U-S keeping in mind the high force requirements as well as desirable workspace required for various masticatory motions. A detailed workspace analysis

Figure 1. Project flowchart

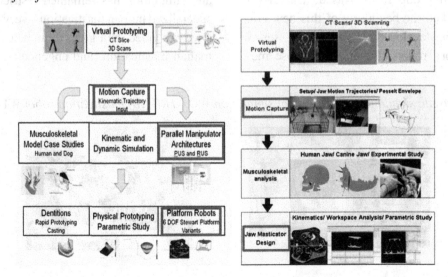

based on the Jacobian measures is considered to be significant to ensure desired characteristics for the manipulator that is presented in mathematical background section. We then drive the kinematic model with the jaw motion trajectories obtained from the motion capture (MoCap) analysis system. Finally, the optimal configuration and workspace analysis results of 6-P-U-S parallel manipulators will be presented in results section. This paper also discusses the validation performed on the physical platforms using the recorded MoCap data.

LITERATURE REVIEW

Robotic Mastication Test-Rig

Many research groups have concentrated on accurately simulating the jaw motion by building a physical robotic manipulator and conducting real time studies on biting and food texture properties. However, the first ever device built for tracking the jaw motion was in mid 1950s (Posselt, 1957). The apparatus (known as the Gnatho-thesiometer) permits measurements (at three points) in the three main planes on a freely movable cast of a lower jaw and he presented a simple comparative study on error obtained between mounting various casts manufactured by different techniques. The research group from Waseda-University actually marked the beginning of this new era. The group has developed a "Waseda Jaw" series of masticatory robots (Figure 2a), whose me-

chanical structures resemble those of the human masticatory system, especially muscle positions, mandibular movements and sensors in muscles and under the teeth. The same group had also succeeded in extending their application of WJ series of robots (Takanobu & Takanishi, 2003) to treating TMJ disorders with their new WY series of robots (Figure 2b).

Even though, their robots are predominantly 3-DOF, the authors mention in one of their works that human masticatory system is similar in configuration to 6-DOF parallel mechanical manipulator driven by linear or rotary actuators (Takanobu, Kuchiki, & Takanishi, 1995). We would like to make use of this insight in our research to model a general purpose robotic masticator. Another research group named BioMouth was actively involved in jaw modeling and dynamic analysis. They implemented a jaw model prototype based on parallel manipulator analogy as in Figure 3a and then mapped into a 6 DOF Stewart platform like dental simulator (Xu, Bronlund et al., 2008) as in Figure 3b. They also performed dynamic simulation studies using commercial software (SimMechanics/Cosmos Motion) on jaw kinematics, actuator forces and food textures more specific to human mastication (Xu, Torrence et al., 2008).

Most of the research work on jaw modeling and simulation has remained restricted to very specialized human mastication – very few studies have expanded their goals to include other mammalian mastication (and engendered significant

Figure 2. Waseda simulators: (a) 3D robot simulation WOJ-1RII, (b) mastication robot WY-5RII

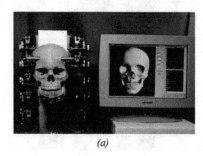

(a)

(b)

Figure 3. (a) Platform mechanism-type robotic model; (b) crank actuated-type robotic model

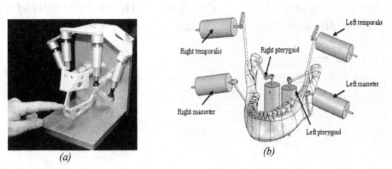

(a)

(b)

variability of mastication motions/forces). In this manuscript we would like to address this topic and arrive at a way to model and build a generic mastication simulator based of a 6-DOF parallel architecture manipulator.

6-DOF Parallel Architecture Manipulators

In the robotics literature, much of the earlier work related to kinematic- and workspace- analysis were primarily based on serial manipulators and simple parallel manipulators (Kong & Gosselin, 2007; Tsai, 1999). A detailed study of singularities and workspace analysis of parallel manipulator were discussed in Gosselin and Angeles (1990). Some of these were also based on studying the kinematics and workspace of parallel manipulators for a range of mechanisms (Joshi & Tsai, 2003). Numerical workspace analyzes for Stewart platform were carried out in Merlet and Gosselin (1999) and for other 6-DOF parallel platforms (Bonev & Gosselin, 2000; Chuang & Chien, 2007).

Numerical computation methods provided a starting point but tend to be limited by numerical errors. Moreover, always additional constraints need to be enforced in the algorithm to maintain the accuracy limits. Hence, developing closed-form analytical expressions for forward and inverse kinematics is really important. However, this proved to be a daunting task and still remains a much-researched problem for parallel architecture

devices. Symbolic code-generation of kinematic and dynamic modeling was later introduced that are capable of generating virtual prototypes based on standard approaches such as Lagrangian method (Durrbaum, Klier, & Hahn, 2002; Kecskemethy, Krupp, & Hiller, 1997; Merlet & Gosselin, 1999).

There are also several known examples of 6-P-U-S systems— Hexaglide Robot (Figure 4a) at ETH Zurich (Joshi & Tsai, 2003), the Hexa Slide Manipulator (Figure 4b) (Bonev & Ryu, 2001), the active wrist (Figure 4c) proposed in Bonev and Gosselin (2000). In Hopkins and Williams (2002), design of a 6-P-S-U platform (Figure 4d) with six legs having six vertical guides is discussed. Each leg comprises of a prismatic joint, spherical joint and universal joint that is used to connect the links to move the platform in six dimensional space. The six prismatic joints for all cases in this figure move with respect to the base and hence are not articulating. Closed form dynamics solutions were developed in Kim and Ryu (2002) and Rao, Saha, and Rao (2006) for a general 6-P-U-S manipulator.

MATH BACKGROUND

System Definition

The platform manipulator considered in this paper, a 6-P-U-S manipulator, is similar in architecture to the one considered in Narayanan, Chakravarty,

Figure 4. Examples of 6-P-U-S parallel manipulators: (a) hexaglide robot (Honegger, Codourey, & Burdet, 1997), (b) hexaslide machine (Bonev & Ryu, 2001), (c) active wrist (Merlet & Gosselin, 1991), and (d) 6-P-S-U platform (Hopkins & Williams, 2002)

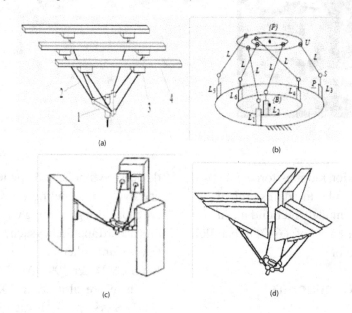

Shah, and Krovi (2010) and Zhao and Gao (2009). A schematic diagram of the generalized manipulator is shown in Figure 5 and a detailed schematic for each limb subsystem for our specialized configuration is shown in Figure 6. The point O is the global origin of this system and a right handed base reference frame XYZ with center

Figure 5. Schematic of ith limb of 6-P-U-S hexapod with reference frames attached

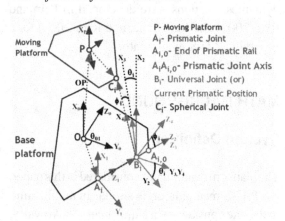

$\{O\}$ is attached to the base. All other frames are defined with respect to this frame. The top platform is defined by the position of its center of mass, $\{P\}$ and the 3D orientation of the plate in terms of Euler angles with respect to the global frame. The position of the platform is given by the cartesian coordinates, $(x, y, z)^T$ measured from the frame XYZ and the orientation angles (Euler angles about relative z-, y- and z- axes) are expressed as $(\alpha, \beta, \gamma)^T$. The matrix R_P^O that transforms rotation from platform, $\{P\}$, with reference to global origin, $\{O\}$, can be obtained as Equation 1 in Box 1.

The combined six dimensional vector representation of the position and orientation of the platform is defined as its pose:

$$\bar{X} = \begin{bmatrix} x & y & z & \alpha & \beta & \gamma \end{bmatrix}^T.$$

The system has six limbs and in all the subsequent discussions, the subscript i indexes each limb, $\forall i = 1 \cdots 6$. The center of the spherical joint

Figure 6. Front and side view of ith Limb of 6-P-U-S manipulator

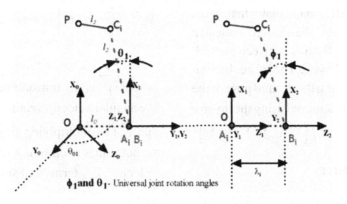

Box 1.

$$
R_P^O = \begin{bmatrix} \cos\alpha\cos\beta\cos\gamma - \sin\alpha\sin\gamma & -\cos\alpha\cos\beta\sin\gamma - \sin\alpha\cos\gamma & \cos\alpha\sin\beta \\ \sin\alpha\cos\beta\cos\gamma + \cos\alpha\sin\gamma & -\sin\alpha\cos\beta\sin\gamma + \cos\alpha\cos\gamma & \sin\alpha\sin\beta \\ -\sin\beta\cos\gamma & \sin\beta\sin\gamma & \cos\beta \end{bmatrix} \tag{1}
$$

Table 1. Geometry of the 6-P-U-S

Symbol	Values	Unit
Radius of prismatic origin L_o	0.1	m
Radius of base platform, L_b	0.125	m
Intermediate Link Length, L_2	0.120	m
Radius of top platform, L_p	0.080	m
Base attachment points (in frame O)		
$A_{1,0}$ & $A_{6,0}$	[0, +/-0.1133, 0.0528]	m
$A_{2,0}$ & $A_{3,0}$	[0, +/-0.1024, 0.0717]	m
$A_{4,0}$ & $A_{5,0}$	[0,+/-0.0109, -0.1245]	m
Platform attachment points (in frame P)		
C_1	*[0.08, 0.0693, -0.04]*	m
C_2	*[0.8, 0.0, -0.08]*	m
C_3	*[0.8, -0.0693, -0.04]*	m

in i^{th} limb will be denoted by a platform attachment point, C_i and the corresponding base attachment points are indicated as A_i. The start and end points of the rails of the ith prismatic joint are given by A_i and $A_{i,0}$ respectively. The point B_i indicates the current slider position which is also the center of universal joint of the ith limb lying along $A_i A_{i,0}$ (rail axis, i). The universal joint angles can be characterized in terms of θ_i and φ_i as in Figure 6.

The prismatic joints are considered as active joint coordinates (λ_i) of this manipulator that drives the top platform through the desired cartesian space trajectories. The distance between point A_i and B_i will be denoted by λ_i. Therefore, by manipulating the coordinates, the desired pose of the platform $\left(\overline{X}\right)$, P can be achieved using the inverse kinematics mapping $\overline{X} = f\left(\lambda_1, \lambda_2, \lambda_3, \lambda_4, \lambda_5, \lambda_6\right)$.

Position Kinematics

For the position kinematics, the typical loop closure method (Tsai, 1999) was used for an ith limb of the 6-P-U-S platform to obtain vector equations similar to that of the Stewart Platform kinematic equations in Tsai (2000).

$$\overrightarrow{OA_i} + \overrightarrow{A_iB_i} + \overrightarrow{B_iC_i} = \overrightarrow{OP} + \overrightarrow{PC_i} \qquad (2)$$

The analytical procedure to solve an inverse kinematic problem and determine the joint coordinates of each limb, is already outlined in Zhao and Gao (2009). The solution of the resulting quadratic equation can, hence, be determined as:

$$\lambda_i = \hat{a}_i^T \left(\overrightarrow{d_i}\right) - \sqrt{\hat{a}_i^T \left(\overrightarrow{d_i}\right) - \overrightarrow{d_i}^T \left(\overrightarrow{d_i}\right) + \left(l_2\right)^2} \qquad (3)$$

where, $\overrightarrow{d_i}$ is the leg vector as $\overrightarrow{B_iC_i}$, $\hat{a}_i$ be the unit vector along the rail axis slider $\overrightarrow{A_iB_i}$.

Velocity Kinematics

Equation (2) is differentiated w.r.t. time as follows to obtain the velocity kinematics equation as follows:

$$\dot{\lambda}_i \hat{a}_i + \overrightarrow{\omega}_{B_iC_i} \times \overrightarrow{B_iC_i} = {}^{O}\begin{bmatrix} \overrightarrow{\dot{p}} \end{bmatrix} + R_O^P \cdot {}^{P}\begin{bmatrix} \overrightarrow{\dot{c}} \end{bmatrix}$$

$$\overrightarrow{V_P} + \overrightarrow{\omega}_P \times {}^{O}\left[\overrightarrow{PC_i}\right] = \dot{\lambda}_i \hat{a}_i + \overrightarrow{\omega}_{B_iC_i} \times \overrightarrow{B_iC_i}, \forall i = 1\cdots 6$$

$$(4)$$

where, $\overrightarrow{V_P}$ is the translational velocity and $\overrightarrow{\omega}_P$ is the angular velocity vectors of the platform in 3D space. Dot-multiplying Equation (4) with $\overrightarrow{B_iC_i}$ on both sides to eliminate the passive leg angular velocity, $\overrightarrow{\omega}_{B_iC_i}$ terms and simplifying,

$$\left[\overrightarrow{B_iC_i} \quad {}^{O}\left(\overrightarrow{PC_i}\right) \times \overrightarrow{B_iC_i}\right]\begin{bmatrix} \overrightarrow{V_P} \\ \overrightarrow{\omega}_P \end{bmatrix} = \dot{\lambda}_i \left(\hat{a}_i \cdot \overrightarrow{B_iC_i}\right)$$

$$\forall i = 1\cdots 6$$

$$(5)$$

$$\Rightarrow \dot{\lambda}_i = \left(J_{link}\right)_i \begin{bmatrix} \overrightarrow{V_P} \\ \overrightarrow{\omega}_P \end{bmatrix}, \forall i = 1\cdots 6 \qquad (6)$$

where, the link Jacobian matrix $\left(J_{link}\right)_i$ can be obtained as

$$\left(J_{link}\right)_i = \left[\left(J_q\right)_i\right]^{-1} \left(J_X\right)_i \qquad (7)$$

where,

$$\left(J_X\right)_i = \left[\overrightarrow{B_iC_i} \quad {}^{O}\left(\overrightarrow{PC_i}\right) \times \overrightarrow{B_iC_i}\right] \text{ and } \left(J_q\right)_i = \hat{a}_i \cdot \overrightarrow{B_iC_i}$$

The link Jacobian matrix (Equation (7)) for the 6-P-U-S manipulator relates the prismatic joint velocity in the ith limb to the task space velocities of platform. Thus, by determining the Equation (6) for all the six limbs and cascading those into a single matrix will result in platform Jacobian matrix, J_P as:

$$J_P = \left(J_q\right)^{-1} J_X \qquad (8)$$

where,

$$J_q = \begin{bmatrix} \hat{a}_1 \cdot \overline{B_1C_1} & \hat{a}_2 \cdot \overline{B_2C_2} & \hat{a}_3 \cdot \overline{B_3C_3} & \hat{a}_4 \cdot \overline{B_4C_4} & \hat{a}_5 \cdot \overline{B_5C_5} & \hat{a}_6 \cdot \overline{B_6C_6} \end{bmatrix}_{6\times6}$$

and

$$J_X = \begin{bmatrix} \overline{B_1C_1} & {}^O\left(\overline{PC_1}\right) \times \overline{B_1C_1} \\ \overline{B_2C_2} & {}^O\left(\overline{PC_2}\right) \times \overline{B_2C_2} \\ \vdots & \vdots \\ \overline{B_6C_6} & {}^O\left(\overline{PC_6}\right) \times \overline{B_6C_6} \end{bmatrix}_{6\times6}$$

Specialization

The active prismatic joints of a generalized 6-P-U-S are located at any point and aligned along any direction in space. However, the location and orientation of the slider axis can be constrained to realize a design that would be structurally superior and mathematically easier to tackle compared to other designs discussed earlier. Specifically, it was aimed to modify the design that would lead to zero or minimal singular configurations. For this, a completely symbolic form of expressions for J_q and J_X was developed using Equation (8) to examine singularity conditions under which the Jacobian, J_p of a generalized 6-P-U-S manipulator loses its full rank. It is already shown (Gosselin & Angeles, 1990) that J_p can become singular only if J_q is singular (Type I singularity) or J_X is singular (Type II singularity) or both of them are singular (Type III singularity):

Type I Singularity

It gives rise to a set of singular configurations when the prismatic axes vector $\hat{a}_i$ becomes perpendicular to the leg vector $\overline{B_iC_i}$ for any one or more than one pair of a prismatic axis and the corresponding leg. Under this configuration, $\hat{a}_i \cdot \overline{B_iC_i} = 0$ should be true for one or more than one values of i. It can thus be ensured by limiting

the value of the universal joint within the range of $+/- \pi/2$ for both the joint axes' rotations that this manipulator is void of type I singularities within its workspace. It is also expected that enforcing such a constraint on the universal joint axes' rotations may only result in minimal reduction of the overall manipulator workspace.

Type II Singularity

This can occur when J_X degenerates from the full rank i.e., when $\overline{B_iC_i}$ and $\overline{PC_i}$ are collinear or any row of J_X is linearly dependent on the other. However, by design the former situation will not occur under the complete set of operating conditions.

In order to design the manipulator with a superior structure, different prismatic actuator alignments were considered and on analysis, the configuration as in Figure 7 was finalized. In order to enhance the structural stability and simplify the model further, it was assumed that a pair of legs attach to a common attachment point in the top platform. Thus, the specialized 6-P-U-S has three two-leg subsystems (TLS) as in Figure 8 connected together at three attachment points (C_j, $j=1, 2, 3$) in the top platform.

Suppose an arbitrary load (F') acts on top of the platform resulting in action of a force, F along each leg on one of the three subsystems. The link vector $\overline{B_iC_i}$ will serve as a force only member as it is attached to the top platform by spherical and universal joints and will not transmit any moments.

Considering the free-body diagram in Figure 9, it can be observed that in each two-leg subsystem a major portion of the forces is transmitted to the ground. The symmetry of the structure will also cause the component of the forces along the slider axes within each two-leg subsystem to be largely counterbalanced (and thereby significantly reducing the actuation requirements). Thus, by carefully choosing the geometric and design parameters, a superior architecture for 6-P-U-S type manipulator was achieved. Further discussion on 6-P-U-S will be pertaining to this specialized configuration only.

Figure 7. Top view of 6-P-U-S manipulator and orientation of prismatic joint axes

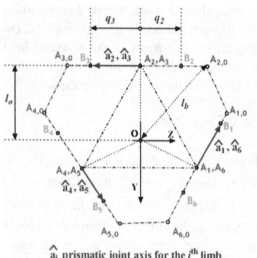

$\hat{a}_i$ prismatic joint axis for the ith limb

B_i- Universal Joint (or) $A_{i,0}$- End of Prismatic Rail
Current Prismatic Position $A_iA_{i,0}$- Prismatic Joint Axis

Figure 8. Two-leg subsystem (TLS) of the specialized 6-P-U-S manipulator

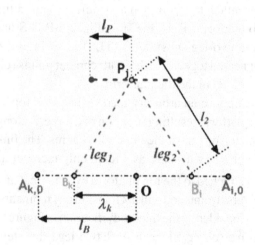

This architecture is also advantageous in terms of computations required to solve the inverse kinematics. Since a pair of sliders can be defined using the same joint axes, $\hat{a}_i$ for this specialized configuration, there will be only three unique active joint axes as in Figure 7 and will result in simplified symmetric solutions for inverse kine-

matics. This will be evident from the joint trajectory plots in results sections. To augment our understanding of the geometry and workspace of this configuration, a detailed CAD model (using SolidWorks) and a physical prototype was also developed as shown in Figure 10. The numerical values of the specialized 6-P-U-S are also available in the Table 1.

Posselt Envelope

The motions of the jaw, namely opening and closing, actually comprise of both translation and rotation. In the case of human jaws, the translational and rotational components are not combined equally through the whole motion. In the initial phase of the opening cycle, the movement is primarily rotational, but translation is prominent approximately after first 20 mm of jaw opening (distance between upper and lower incisor tooth tips).

Maximum opening occurs when this distance is about 50 mm, which is actually limited by temporomandible ligament and joint capsule. The closing action begins with a phase in which posterior translation predominates and the jaw closes with translation until about two-thirds of maximal opening is attained. At this time, the condyles and discs will return either to the height or the posterior slope of the articular eminence and immediately leads to closing of jaw as a smooth combination of translatory and rotary motion. The final occlusal position is then attained primarily by rotational motion.

In the biomechanics literature, for describing mandibular movements and different phases of a jaw motion cycle, the trajectory of a point located between the incisal edges of the lower (mandibular) central incisors, termed the incisal point, is usually used. The Posselt diagram or envelope (as in Figure 11) clearly outlines the extent of movements of this incisal point. The initial position of the incisal point is called centric occlusion, and at this position, the occlusal surfaces are in

Figure 9. Free body diagram for a two-leg subsystem of a specialized 6-P-U-S manipulator (indices 'i' and 'k' are used to indicate the positions of a pair of the sliders in each subsystem)

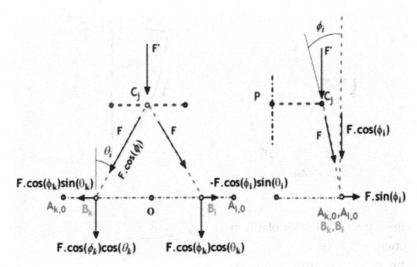

Figure 10. 3D CAD model and physical prototype of 6-DOF P-U-S hexapod system

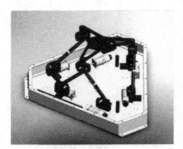 

maximum contact. Along the sagittal plane, the Posselt envelope can be divided into four segments.

In the first segment, called the maximal rear opening (MRO, 1-2) period, the jaw rotates approximately 10 degrees about an axis that intersects the center of the condyles. If the mandible opens farther, a protrusion starts, followed by the final rear opening (FRO, 2-3) period which can be considered as the combined movement of the rotation about the axis and a protrusion simultaneously until the maximal opening (MO- 3) is reached. The maximal frontal path (MFP- 3, 4) is then described as a rotation around the condyle axis accompanied by maximal protrusion. At the upper border, the mandible is only in maximum protrusion and can return to its initial position of

centric occlusion by a retrusive path primarily by translation (RP, 4-1).

Transformation Matrix between MoCap Coordinates and Platform Coordinates

The 3D coordinates of the three points of the mandible (left TMJ (L), right TMJ (R) and incisor tooth tip (P) as in Figure 12) obtained from MoCap system is defined with respect to the reference frame $\{D\}$ of the calibration grid used for MoCap experiments (not shown in the figure). These 3D trajectories must be transformed to the base reference frame of the platform carefully as

Figure 11. (a) Sagittal plane- posselt envelope, (b) frontal plane: posselt envelope (jaw motion cycles: 1-2: MRO, 2-3- FRO, 3- MO, 3-4: MFP, 4-1- RP)

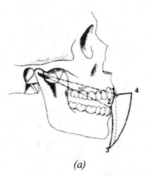

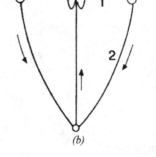

(a)   (b)

it will be used to drive the end-effector platform in the final simulation.

In our case, this is done in two steps— (i) transform the MoCap trajectories obtained over time from the calibration frame $\{D\}$ to a reference frame attached to the mid-point (A_0) of left and right TMJ points at time = 0 s (ii) transform the trajectories obtained from (i) to the base reference frame of the platform $\{O\}$ by a pure translation and if necessary, by angular offsets. The temporary reference frame $\{A_0\}$ required for (i) can be obtained as follows:

Consider the location of left TMJ $\left(\bar{X}_L^D\left(0\right)\right)$, right TMJ $\left(\bar{X}_R^D\left(0\right)\right)$ as well as incisor tooth tip $\left(\bar{X}_P^D\left(0\right)\right)$ defined w.r.t to frame $\{D\}$ at initial time i.e., when time is zero as shown in the Figure 13a. In reality, these points correspond to the position of jaw just before the start of MoCap. This can be extracted from the first row of points contained in the table of MoCap coordinates. Using these initial time coordinates, define $\hat{v}_1$ as a unit vector along the line joining $\bar{X}_L^D\left(0\right)$ and $\bar{X}_R^D\left(0\right)$ and $\hat{v}_2$ as a unit vector along the line joining $\bar{X}_L^D\left(0\right)$ to $\bar{X}_P^D\left(0\right)$. Then the orthonormal unit vectors $\left(\hat{e}_1^T, \hat{e}_2^T, \hat{e}_3^T\right)$ of the frame $\{A_0\}$ can be determined as:

$$\hat{e}_3 = \frac{\hat{v}_1 \times \hat{v}_2}{\left\|\hat{v}_1 \times \hat{v}_2\right\|} \tag{9}$$

$$\hat{e}_2 = \hat{e}_3 \times \hat{e}_1 \quad (or) \quad \hat{e}_2 = \frac{\hat{v}_2 - \left(\hat{v}_2 \cdot \hat{v}_1\right)\hat{v}_1}{\left|\hat{v}_2 - \left(\hat{v}_2 \cdot \hat{v}_1\right)\hat{v}_1\right|},$$

where

$$\hat{e}_1 = \hat{v}_1 \tag{10}$$

$$T_{A_0}^D = \begin{bmatrix} R & \bar{X}_M^D\left(0\right) \\ 0_{3\times1} & 1 \end{bmatrix}, \tag{11}$$

Figure 12. Plane of jaw motion measured using MoCap system

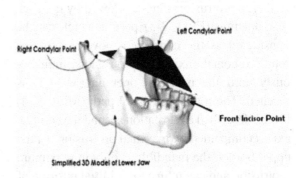

$$where, R = \begin{bmatrix} | & | & | \\ \hat{e}_1 & \hat{e}_2 & \hat{e}_3 \\ | & | & | \end{bmatrix}, and$$

$$\bar{X}_M^D(0) = \frac{\bar{X}_L^D(0) + \bar{X}_R^D(0)}{2} = \left(x_M(0), y_M(0), z_M(0)\right)^T \tag{12}$$

where, $T_{A_0}^D$ – transformation matrix of the jaw reference frame $\{A_0\}$ with respect to frame $\{D\}$. Using $\left(T_{A_0}^D\right)^{-1}$ it is then possible to transform the entire time-dependent coordinates from $\{D\}$ to $\{A_0\}$,

$$\bar{X}_j^{A_0}(t) = \left(T_{A_0}^D\right)^{-1} \bar{X}_j^D(t) \tag{13}$$

where, j: index for all motion capture points = {L, R, P}

$\bar{X}_j^{A_0}(t)$ is thus the computed time-dependent relative coordinates of points w.r.t. frame $\{A_0\}$. Now consider a frame $\{E_0\}$ defined at frame $\{P\}$ of the platform at $t = 0s$ (Figure 13b). Then, $\bar{X}_j^{A_0}(t)$ can be considered equivalent to expressing the relative coordinates with time in the frame $\{E_0\}$ i.e., $\bar{X}_j^{A_0}(t) = \bar{X}_j^{E_0}(t)$.

If the axes of $\{E_0\}$ are aligned with those of global reference frame, $\{O\}$, then the transformation, $T_{E_0}^O$ is only a pure translation and can be easily determined. However, under certain situations, the axes alignment of $\{E_0\}$ will have to be changed to ensure that the plane of jaw motion at the mean orientations about any axes lie close to the plane of zero orientation of the platform and hence, the transformation $T_{E_0}^O$ can be determined by computing the corresponding angles from $\bar{X}_j^{A_0}(t)$.

The next step is to calculate the platform position and orientation from the MoCap trajectories $\bar{X}_j^{A_0}(t)$ that correspond to TMJ positions and rotations. As mentioned earlier, the trajectories $\bar{X}_j^{A_0}(t)$, $j = \{L, R, P\}$ are used to compute the resulting TMJ translations and orientations. Since $T_{E_0}^O$ is already determined, we only have to compute the transformation of plane of jaw motion at time, t w.r.t. $\{E_0\}$, i.e., $T_{E_t}^{E_0}$. By taking average of $\bar{X}_L^{A_0}(t)$ and $\bar{X}_R^{A_0}(t)$, it is possible to locate the origin at time, t as $\bar{A}_t^{A_0}$ for frame $\{E_t\}$. Then the difference $\bar{A}_t^{A_0} - \bar{A}_0$ provides the 3-DOF translation component for the transformation matrix, $T_{E_t}^{E_0}$. Using the plane of jaw motion, the unit vector of the coordinate axes can be easily determined from $\bar{X}_j^{A_0}(t)$ and the rotation matrix $\left(R_t^{A_0}\right)$ can be formulated based on Equation (11).

Hence, $T_{E_t}^{E_0}$ is given as:

Figure 13. Transformation between MoCap and platform reference frames (a) MoCap transformation (b) platform transformation

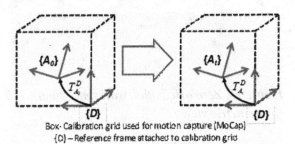

Box- Calibration grid used for motion capture (MoCap)
{D} – Reference frame attached to calibration grid

(a)

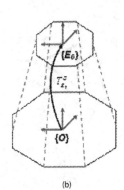

(b)

$$T_{E_t}^{E_0} = \begin{bmatrix} R_t^{A_0} & \bar{A}_t^{A_0} - \bar{A}_0 \\ 0_{3\times 1} & 1 \end{bmatrix} \qquad (14)$$

By composing $T_{E_0}^{O}$ and $T_{E_t}^{E_0}$, the combined transformation matrix from $\{O\}$ to $\{E_t\}$ can then be obtained, from which the final platform pose vector can be extracted. In our case, the parameterization of relative ZYZ axes rotation by α, β and γ respectively is followed. The resulting rota-

tion matrix can be given as in Equation (1). By equating the individual terms of R_t^O and R, the angle parameters $(\alpha, \beta, \gamma)^T$ can be easily found as:

$$T_{E_t}^{O} = T_{E_0}^{O} \cdot T_{E_t}^{E_0} = \begin{bmatrix} R_t^O & A_t^O \\ 0_{3\times 1} & 1 \end{bmatrix} \qquad (15)$$

$$\bar{X}_P^O = A_t^O$$

Figure 14. Zero-yaw-pitch workspace limits on Y-Z plane for pitch = -18, -9, 0, 9, 18 degrees (a) top view (b) isometric view

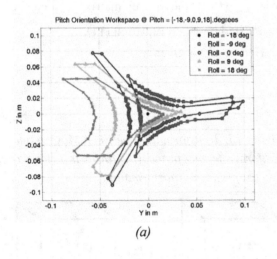

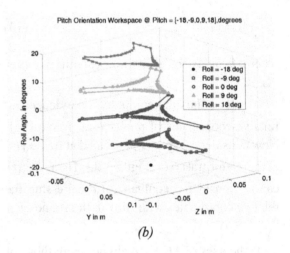

(a)

(b)

Figure 15. Zero orientation workspace limits on Y-Z plane for X = 0.08, 0.09, 0.10, 0.11 m (a) top view (b) isometric view

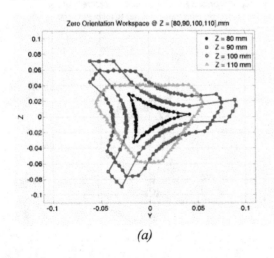

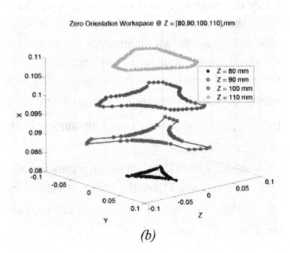

(a)

(b)

Figure 16. Constant orientation workspace measure: (a) surface and (b) contour plots, for x=0.070 α=0, β=0, γ=0; grid: y=z=-0.025:0.025

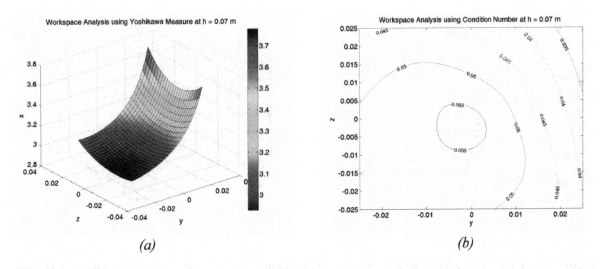

(a)   (b)

Figure 17. Constant orientation workspace measure: (a) surface and (b) contour plots, for x=0.090, α=0, β=0, γ=0; grid: y=z=-0.025:0.025

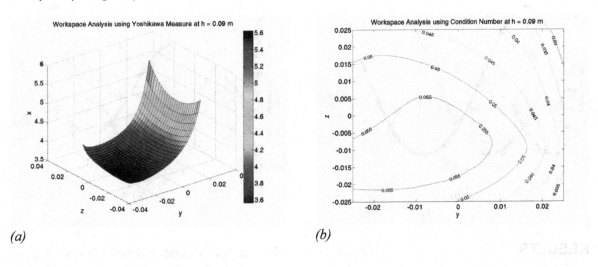

(a)   (b)

$$R = R^o_, = \begin{bmatrix} r_{11} & r_{12} & r_{13} \\ r_{21} & r_{22} & r_{23} \\ r_{31} & r_{32} & r_{33} \end{bmatrix} \Rightarrow \begin{cases} \alpha = atan2\left(r_{23}, r_{13}\right) \\ \beta = atan2\left(\sqrt{r_{23}^2 + r_{13}^2}, r_{33}\right) \\ \gamma = atan2\left(r_{32}, -r_{31}\right) \end{cases}$$

However, due to controlled nature of our experiments, the jaw movements are considered to comprise of 3-DOF translation and 2-DOF rotations (in sagittal as well as transverse planes about axes $\hat{e}_1$ and $\hat{e}_3$ respectively), which gives overall of 5-DOF. Though this might be an approximation, for almost all real-life mastications, the rotation about $\hat{e}_2$ is actually very minimal even for animals and hence, this can be considered as a safe assumption and simplifies the problem further with only two rotation parameters.

Figure 18. Actuator force estimates for quasi-static simulation for a straight line trajectory x = 0.008*sin(wt) m, y=z=0, zero orientation (a) actuator forces (b) actuator displacements

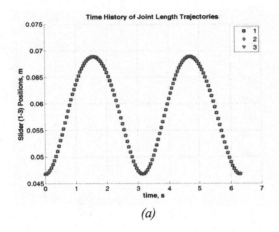
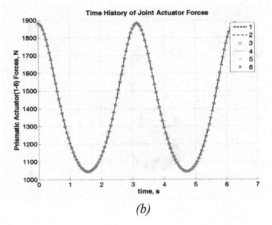

(a) *(b)*

Figure 19. Actuator force estimates for quasi-static simulation for a circular trajectory x = 0.08m, y=0.01*cos(t) m, z=0.01*sin(t) m (a) actuator forces (b) actuator displacements

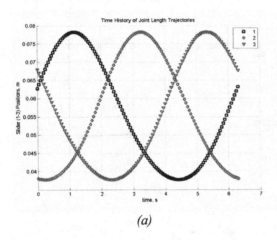
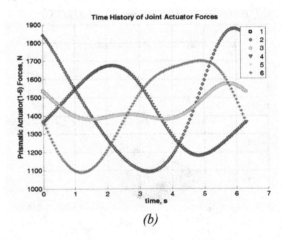

(a) *(b)*

RESULTS

Workspace Envelopes: Translation and Orientation

For the specialized design of the manipulator system, the workspace envelopes were computed using inverse kinematics routine and discretizing the polar workspace region to search for the boundary points (the point at which the inverse kinematics of Equation (3) fails to yield real solutions or fails to satisfy the actuator limits) along each radial directions from 0 to 2π radians. To obtain a more accurate estimate of the workspace envelopes, a finer discretized task space is desired or volume computation methods such as in Shah, Narayanan, Lee, and Krovi (2010) can be used.

We also plot 2D sliced workspaces for each value of the 3rd coordinate (the vertical coordinate or roll-pitch-yaw angles) for a variety of cases. However, we show only for the two cases:

1. Zero orientation Translation Workspace (Figure 14)
2. Zero Yaw-Pitch with Roll Orientation Workspace (Figure 15)

Figure 20. Actuator force estimates for quasi-static simulation for a yaw trajectory yaw = 10*sin(t) m, x=y=z=0, zero pitch and roll (a) actuator displacements (b) actuator forces

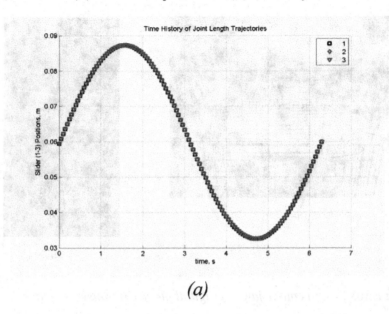

(a)

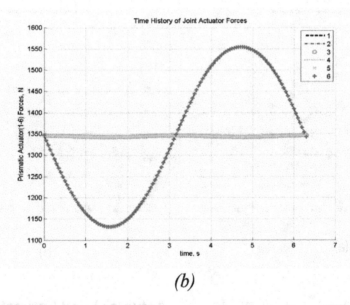

(b)

Workspace Analysis and Workspace Based Measures

While, by design this manipulator possesses minimal workspace singularities, the fluctuations in the quality of workspace using a range of Jacobian based measures (including Yoshikawa's measure (Yoshikawa, 1998), inverse condition number measure etc) were studied. The analytical

Jacobian derived earlier facilitates performance of this workspace analysis and yields more accurate results than a numerical implementation.

Initially, the 3D-cartesian space was divided into 2D slices and each slice is then discretized into a sequence of uniformly spaced grid points spanning each 2D planes. The inverse kinematics was solved and analytical Jacobian was then evaluated at each of the grid points and the corresponding

Figure 21. (a) Animal motion capture setup (b) recorded MoCap frames with partial and full marker tracking

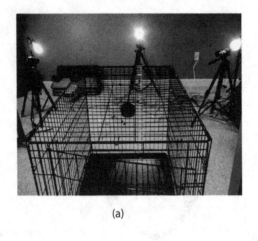

(a)

(b)

Figure 22. Posselt envelope of a canine jaw (a) sagittal plane (b) transverse plane

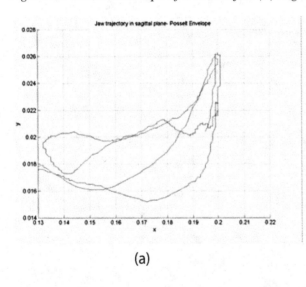

(a)

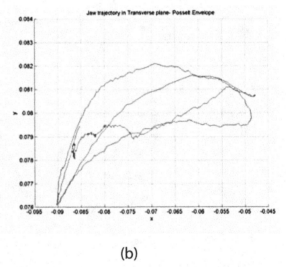

(b)

workspace measures were calculated. We show here only the Yoshikawa-measure (product of all the singular measures of the platform Jacobian matrix) to estimate the workspace quality of the specialized configuration (shown in Figure 16 and Figure 17). Since the visualization of 6-DOF workspace is not possible, only the constant orientation workspace of 6-P-U-S manipulator is presented here (for other types of workspace analysis refer (Narayanan, 2008).

Actuator Load Estimates by Static Analysis

In this section, the effect of Jacobian matrix on actuation requirements was studied by quasi-static analysis. The platform Jacobian matrix was used to compute the actuator forces for a known external load in task space assuming the platform is moving at near zero velocities. The quasi-static analysis was actually implemented by

Figure 23. Rapid prototypes and metal casts of canine dentitions (upper and lower mandible)

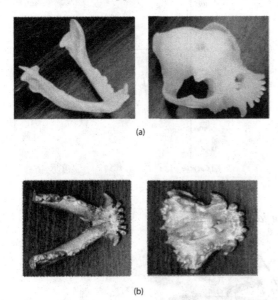

(a)

(b)

including the desired trajectory information into the formulation and computing actuator forces at each point in the trajectory.

A subset of sample case studies are presented in Figure 18 through 22 for a constant vertical load of 1000 N along negative X-axis and zero loads and moments on all other directions for different trajectory tracking problems. Figure 18 indicates the static actuator load estimates for a vertical line trajectory. Similar results for the circular trajectory in YZ plane with radius = 0.01 m and yaw motion trajectory with an amplitude of 10 degrees are shown in Figure 19 and Figure 20 respectively.

Motion Capture Case Study

As this masticatory simulator was built to be a generic system, it is necessary to validate the physical system against a set of MoCap trajectories of different animals. This simulator was also tested against human mastications but not reported due to space limitations. Hence, a set of case studies were performed to record the jaw motions of various breeds of canines and the following sections discuss the results based on such studies. In this case, a passive-optical-marker-based MoCap system was used to capture the masticatory motions of animals. The typical setup used for such a study is shown in Figure 21a. From the recorded videos, position of the markers located on the predefined anatomical landmark on animals (analogous to the humans as shown in Figure 12) were computed. Since in the case of animals, it is difficult to maintain the posture of the head, problems related to trajectory computation occur such as occlusions, swapping and loss of markers Figure 21b.

Hence, the recorded videos was thoroughly post-processed for such errors and finally, time-based cartesian trajectories of the three points – L, R and P were determined that served as kinematic inputs for our simulator. Based on the MoCap data for a series of complete chewing cycles, Posselt envelope of canines were obtained as in Figure 22. This envelope provides an estimate of jaw motion envelope that needs to be contained completely within the workspace of our mastica-

Figure 24. Masticatory simulator setup with force-transducer and casted dentition

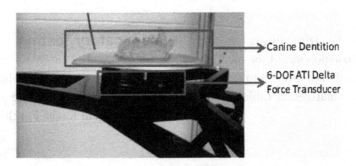

Figure 25. Raw mocap data (z-coordinate and velocity) of chewing cycle canine

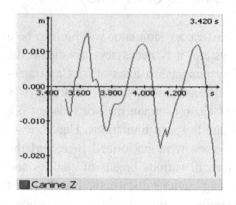 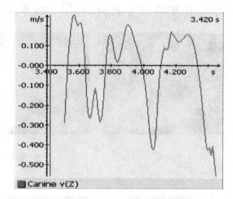

Figure 26. Masticatory simulator mounted with bulldog jaw while chewing

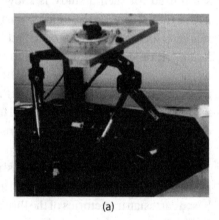 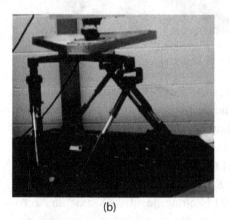

(a) (b)

Figure 27. MoCap frames at normalized time, t= 0, 0.25, 0.5, 0.75, 1 units for a chewing cycle

Representative Case Study

tory simulator. Thus, the sizing dimensions and geometry of various components of the parallel-platform were obtained taking this into account which was already summarized in Table 1.

Masticatory Testing

The robotic simulator is also required to be capable of estimating the bite forces as it is made to chew different foods. For this purpose, a 6-DOF force-torque transducer (ATI Delta) was mounted on top of the platform which can provide the corresponding average bite-force estimates during the chewing cycle. However, in order to accurately reproduce the mastication cycle (motion and food resistance) accurate dentitions of different animals were needed to be developed. For this purpose, CT scans and 3D laser scans were used to obtain highly detailed 3D models of skull and mandible of canine jaws which were then converted into rapid-prototypes and metal casts (Figure 23) to be used. These physical prototypes were then mounted on the platform with the force-transducer during mastication experiments as in Figure 24.

With the masticator setup established, the platform was driven using canine MoCap trajectories that is shown in Figure 25 for different types of food (from hard to soft) and resulting bite force estimates were also obtained during this hardware-in-the-loop implementation (Figure 26). The sequence of trajectory motions as well as bite-force measurements during testing of the simulator was recorded by our MoCap system and the resulting sequence of chewing cycles of the simulator as recorded by the MoCap cameras is shown Figure 27.

DISCUSSION

Thus, the parallel-architecture based mastication simulator was designed, developed and tested by hardware-in-the-loop jaw motion tracking experiments. Since the simulator had the actuators located at the base, high torque-range actuators were used. This provided the capability of exerting high end effector forces and moments to the simulator that are required for reproducing mastication cycles of certain breeds of animals. Further, by using the force-torque transducer the physical simulator could also measure bite-forces while performing chewing cycles. Hence, using the motion capture trajectories of canines

this simulator was then driven to chew various forms of food and the corresponding results were demonstrated. However, the concept of developing a "chewability index" requires a more comprehensive motion analysis across different forms of animal foods as well as various breeds. Our testbed now offers an automated means of performing both parametric sweep studies as well as careful Design-of-Experiments studies to help characterize these aspects and this work is currently underway.

REFERENCES

Bonev, I., & Gosselin, C. M. (2000). A geometrical algorithm for the computation of the constant orientation workspace of 6-R-U-S parallel manipulators. In *Proceedings of the ASME Design Engineering Technical Conference*, Baltimore, MD.

Bonev, I., & Ryu, J. (2001). A geometrical method for computing the constant-orientation workspace of 6-PRRS parallel manipulators. *Mechanism and Machine Theory*, 36(1), 1–13. doi:10.1016/S0094-114X(00)00031-8.

Chuang, H.-Y., & Chien, K.-H. (2007). Analysis of workspace and singularity of a slide equilateral triangle parallel manipulator. *Journal of System Design and Dynamics*, 1(4), 724–735. doi:10.1299/jsdd.1.724.

Durrbaum, A., Klier, W., & Hahn, H. (2002). Comparison of automatic and symbolic differentiation in mathematical modeling and computer simulation of rigid body systems. *Multibody System Dynamics*, 7(4), 331–355. doi:10.1023/A:1015523018029.

Gosselin, C., & Angeles, J. (1990). Singularity analysis of closed-loop kinematic chains. *IEEE Transactions on Robotics and Automation*, 6(3), 281–290. doi:10.1109/70.56660.

Honegger, M., Codourey, A., & Burdet, E. (1997). Adaptive control of the hexaglide, a 6 dof parallel manipulator. In *Proceedings of the IEEE International Conference on Robotics and Automation*, Albuquerque, NM (pp. 543-548).

Hopkins, B. R., & Williams, R. L. (2002). Kinematics, design and control of 6-P S U platform [Research Article]. *Industrial Robot: An International Journal*, 29(5), 443–451. doi:10.1108/01439910210440264.

Joshi, S. A., & Tsai, L.-W. (2003). The kinematics of a class of 3-DOF, 4-legged parallel manipulators. *Journal of Mechanical Design*, 125(1), 52–60. doi:10.1115/1.1540992.

Kecskemethy, A., Krupp, T., & Hiller, M. (1997). Symbolic processing of multiloop mechanism dynamics using closed-form kinematics solutions. *Multibody System Dynamics*, 1(1), 23–45. doi:10.1023/A:1009743909765.

Kim, J.-P., & Ryu, J. (2002). Inverse kinematics and dynamics of 6-DOF P-U-S type parallel manipulators. *KSME International Journal*, 16(1), 13–23.

Kong, X., & Gosselin, C. (2007). *Type synthesis of parallel mechanisms (Vol. 33)*. Berlin, Germany: Springer-Verlag.

Merlet, J.-P., & Gosselin, C. (1991). New architecture for a six-degree-of-freedom parallel manipulator. *Mechanism and Machine Theory*, 26(2), 77–90. doi:10.1016/0094-114X(91)90023-W.

Merlet, J.-P., & Gosselin, C. (1999). Determination of 6D workspaces of Gough-type parallel manipulator and comparison between different geometries. *The International Journal of Robotics Research*, 18(1), 902–916. doi:10.1177/02783649922066646.

Narayanan, M. S. (2008). *Analysis of parallel manipulator architectures for use in masticatory studies*. Unpublished master's thesis, University at Buffalo (SUNY), Buffalo, NY.

Narayanan, M. S., Chakravarty, S., Shah, H. L., & Krovi, V. N. (2010). Kinematic-, workspace- and static- analysis of a 6-P-U-S parallel manipulator. In *Proceedings of the ASME International Design Engineering Technical Conferences*, Montreal, QC, Canada.

Posselt, U. (1957). An analyzer for mandibular positions. *The Journal of Prosthetic Dentistry*, 7(3), 368–374. doi:10.1016/S0022-3913(57)80082-1.

Rao, A. B. K., Saha, S. K., & Rao, P. V. M. (2006). Dynamics modeling of hexaslides using the decoupled natural orthogonal complement matrices. *Multibody System Dynamics*, 15(2), 159–180. doi:10.1007/s11044-005-9003-1.

Shah, H. L., Narayanan, M. S., Lee, L., & Krovi, V. (2010). CAD-enhanced workspace optimization for parallel manipulators: A case study. In *Proceedings of the IEEE Conference on Automation Science and Engineering*, Toronto, ON, Canada (pp. 816-821).

Signore, M. J. D., Krovi, V. N., & Mendel, F. C. (2005). Virtual prototyping and hardware-in-the-loop testing for musculoskeletal system analysis. In *Proceedings of the IEEE International Conference on Mechatronics and Automation*, Niagara Falls, Canada (pp. 394-399).

Takanobu, H., Kuchiki, N., & Takanishi, A. (1995). Control of rapid closing motion of a robot jaw using nonlinear spring mechanism. In *Proceedings of the International Conference on Intelligent Robots and Systems*, Pittsburgh, PA (Vol. 1).

Takanobu, H., & Takanishi, A. (2003). Dental robotics and human model. In *Proceedings of the 1st International IEEE EMBS Conference on Neural Engineering*, Capri Island, Italy (pp. 671-674).

Tsai, L.-W. (1999). *Robot analysis: the mechanics of serial and parallel manipulators*. New York, NY: John Wiley & Sons.

Tsai, L.-W. (2000). Solving the inverse dynamics of Stewart-Gough manipulator by the principle of virtual work. *Journal of Mechanical Design*, *122*(1), 3–9. doi:10.1115/1.533540.

Xu, W. L., Bronlund, J. E., Potgieter, J., Foster, K. D., Röhrle, O., Pullan, A. J., & Kieser, J. A. (2008). Review of the human masticatory system and masticatory robotics. *Mechanism and Machine Theory*, *43*(11), 1353–1375. doi:10.1016/j.mechmachtheory.2008.06.003.

Xu, W. L., Torrence, J. D., Chen, B. Q., Potgieter, J., Bronlund, J. E., & Pap, J. S. (2008). Kinematics and experiments of a life sized masticatory robot for characterizing food texture. *IEEE Transactions on Industrial Electronics*, *55*(5), 12. doi:10.1109/TIE.2008.918641.

Yoshikawa, T. (1998). Manipulability of robotic mechanisms. *The International Journal of Robotics Research*, *4*(2), 3–9. doi:10.1177/027836498500400201.

Zhao, Y., & Gao, F. (2009). Inverse dynamics of the 6-DOF-out-parallel manipulator by means of the principle of virtual work. *Robotica*, *27*(2), 259–268. doi:10.1017/S0263574708004657.

This work was previously published in the International Journal of Intelligent Mechatronics and Robotics, Volume 1, Issue 4, edited by Bijan Shirinzadeh, pp. 100-122, copyright 2011 by IGI Publishing (an imprint of IGI Global).

Compilation of References

Adept Inc. (2009). *Industrial robots.* Retrieved from http://www.adept.com/products/robots

Aguerrevere, D., Choudhury, M., & Barreto, A. (2004). *Portable 3D sound/sonar navigation system for blind individuals.* Paper presented at the 2nd LACCEI International Latin American Caribbean Conference on Engineering Technology, Miami, FL.

Aguirre, G., Acevedo, M., Carbone, G., & Ottaviano, E. (2003). Kinematic and dynamic analysis of a 3 DOF parallel manipulator by symbolic formulations. In Ambrosio, J. A. C. (Ed.), *Multibody dynamics.* Lisbon, Portugal.

Akopoulos, D., Boddhu, S. K., & Bourbakis, N. (2007). A 2D vibration array as an assistive device for visually impaired. In *Proceedings of the 7th IEEE International Conference on Bioinformatics and Bioengineering,* Boston, MA (Vol. 1, pp. 930-937).

Albus, J. S., Bostelman, R. V., & Dagalakis, N. (1993). The NIST ROBOCRANE. *Journal of Robotic Systems, 10*(5), 709–724. doi:10.1002/rob.4620100509.

Alici, G., & Shirinzadeh, B. (2004). Optimum synthesis of a parallel manipulator based on kinematic isotropy and force balancing. *Robotica, 22*(1), 97–108. doi:10.1017/S0263574703005216.

Alp, A. B., & Agrawal, S. K. (2002). Cable suspended robots: Feedback controllers with positive inputs. In *Proceedings of the American Control Conference,* Anchorage, AK (pp. 815-820).

Amooshahi, Y., & Hadian, H. (2009). Dynamic modeling and control of a novel 4-DOF parallel manipulator. In *Proceedings of the 18th Annual International Conference on Mechanical Engineering,* Tehran, Iran.

Angeles, J. (2007). *Fundamentals of Robotic Mechanical Systems* (3rd ed.). New York: Springer.

Asif, U., & Iqbal, J. (2010, November 24-26). Modeling and Simulation of Biologically Inspired Hexapod Robot using SimMechanics. In *Proceedings of the IASTED International Conference on Robotics (Robo 2010),* Phuket, Thailand.

Audette, R., Balthazaar, J., Dunk, C., & Zelek, J. (2000). *A stereo-vision system for the visually impaired.* Guelph, ON, Canada: University of Guelph.

Bai-chao, C., Rong-ben, W., Lu, Y., Li-sheng, J., & Lie, G. (2007). Design and simulation research on a new type of suspension for lunar rover. In *Proceedings of the IEEE International Symposium on Computational Intelligence in Robotics and Automation,* Jacksonville, FL (pp. 173-177).

Bamberger, H., & Shoham, M. (2004). A new configuration of a six degrees-of-freedom parallel robot for MEMS fabrication. In *Proceedings of the IEEE International Conference on Robotics and Automation,* New Orleans, LA (pp. 4545-4550).

Bamberger, H., & Shoham, M. (2007). A novel six degrees-of-freedom parallel robot for MEMS fabrication. *IEEE Transactions on Robotics, 23*(2). doi:10.1109/TRO.2006.889493.

Barraquand, J., & Latombe, J.-C. (1991). Robot motion planning: A distributed representation approach. *The International Journal of Robotics Research, 10*(6), 628–649. doi:10.1177/027836499101000604.

Barvinok, A., & Samorodnitsky, A. (2007). Random weighting, asymptotic counting, and inverse is operimetry. *Israel Journal of Mathematics, 158*(1), 159–191. doi:10.1007/s11856-007-0008-8.

Bashash, S., & Jalili, N. (2008). Adaptive robust control strategy for coupled parallel-kinematics piezo-flexural micro and nano-positioning stages. *IEEE/ASME Transactions on Mechatronics*, *14*(1), 11–20. doi:10.1109/TMECH.2008.2006501.

Beccari, C. V., Farella, E., Liverani, A., Morigia, S., & Rucci, M. (2010). A fast interactive reverse-engineering system. *Computer Aided Design*, *42*(10), 860–873. doi:10.1016/j.cad.2010.06.001.

Behzadipour, S., & Khajepour, A. (2006). Cable-based robot manipulators with translational degrees of freedom. *Industrial Robotics: Theory, Modeling and Control*, 211-236.

Behzadipour, S., & Khajepour, A. (2004). Design of reduced DOF parallel cable-based robots. *Mechanism and Machine Theory*, *39*(10), 1051–1065. doi:10.1016/j.mechmachtheory.2004.05.003.

Behzadipour, S., & Khajepour, A. (2005). A new cable-based parallel robot with three degrees of freedom. *Journal of Multibody System Dynamics*, *13*(4), 371–383. doi:10.1007/s11044-005-3985-6.

Beji, L., Abichou, A., & Pascal, M. (1998, May). Tracking control of a parallel robot in the task space. In *Proceedings of the IEEE International Conference on Robotics & Automation*, Leuven, Belgium (pp. 2309-2304).

Belote, L. (2006). *Low vision education and training: Defining the boundaries of low vision patients. A personal guide to the VA Visual Impairment Services Program*. San Francisco, CA: Visual Impairment Service Team.

Benham, T., & Benjamin, J. (1963). Active energy radiating systems: An electronic travel aid. In *Proceedings of the International Congress on Technology and Blindness* (pp. 167-176).

Ben-Horin, R., & Shoham, M. (1996). Construction of a six degrees-of-freedom parallel manipulator with three planarly actuated links. In *Proceedings of the ASME Design Conference*.

Ben-Horin, R., Shoham, M., & Djerassi, S. (1998). Kinematics, dynamics and construction of a planarly actuated parallel robot. *Robotics and Computer-integrated Manufacturing*, *14*(2), 163–172. doi:10.1016/S0736-5845(97)00035-5.

Benjamen, J. M., Ali, N. A., & Schepis, A. F. (1973). A laser cane for the blind. *Proceedings of the San Diego Biomedical Symposium*, *12*, 53–57.

Bergander, A. (2003). *Control, wear testing & integration of stick-slip micropositioning*. Unpublished doctoral dissertation, Ecole Polytechnique Fédérale de Lausanne, Lausanne, Switzerland.

Bergander, A., Driesen, W., Lal, A., Varidel, T., Meizoso, M., Bleuler, H., et al. (2004). Position Feedback for Microrobots based on Scanning Probe Microscopy. In *Proceedings of the International Conference on Intelligent Robots and Systems* (Vol. 2, pp. 1734-1739).

Bergander, A., Driesen, W., Varidel, T., & Breguet, J.-M. (2003). Development of Miniature Manipulators for Applications in Biology and Nanotechnologies. In *Proceeding of Microrobotics for Biomanipulation Workshop, IEEE/RSJ International Conference on Intelligent Robots and Systems* (pp. 11-35).

Bergander, A., Driesen, W., Varidel, T., & Breguet, J.-M. (2003). Monolithic piezoelectric push-pull actuators for inertial drives. In *Proceedings of 2003 International Symposium on Micromechatronics and Human Science*. doi:10.1063/1.1139566

Bessonov, A. P., & Umnov, N. V. (1973). The analysis of gaits in six-legged vehicle according to their static stability. In *Proceedings of the Symposium on Theory and Practice of Robots and Manipulators*, Udine, Italy (pp. 1-9).

Beukeboom, J. J. A., van Dixhoorn, J. J., & Meerman, J. W. (1985). Simulation of mixed bond graphs and block diagrams on personal computer using TUTSIM. *Journal of the Franklin Institute*, *319*(1-2), 257–267. doi:10.1016/0016-0032(85)90079-1.

Beurle, R. L. (1951). Electronic guiding aids for blind people. *The British Journal of Psychology*, *42*(1-2), 164–171.

Bhattacharya, S., & Talapatra, S. (2005). Robot motion planning using neural networks: a modified theory. *International Journal of Lateral Computing*, *2*(1), 9–13.

Bieg, L. F. X. (1999). *U.S. Patent No. 5,901,936: Six degree-of-freedom multi-axes positioning apparatus*. Washington, DC: U. S. Patent and Trademark Office.

Blasch, B. B., Long, R. G., & Griffin-Shirely, N. (1999). National evaluation of electronic travel aids for blind and virtually impaired individuals: Implications for design. In *Proceedings of the 12ᵗʰ Annual Conference Rehabilitation Engineering Society of North America*, New Orleans, LA (pp. 133-134).

Blasch, B. B., Wiener, W. R., & Welsh, W. R. (1997). *Foundations of orientation and mobility* (2nd ed.). New York, NY: AFB Press.

Bonev, I. A. (2009). *Delta parallel robot - the story of success*. Retrieved from http://www.parallemic.org/Reviews/Review002.html

Bonev, I. A., & Ryu, J. (1999). Orientation workspace analysis of 6-DOF parallel manipulators. In *Proceedings of the ASME Design Engineering Technical Conference*.

Bonev, I. A., Zlatanov, D., & Gosselin, C. M. (2002). Advantages of the modified Euler angles in the design and control of PKMs. In *Proceedings of the 2002 Parallel Kinematic Machines International Conference*, Chemnitz, Germany (pp. 171-188).

Bonev, I., & Gosselin, C. M. (2000). A geometrical algorithm for the computation of the constant orientation workspace of 6-R-U-S parallel manipulators. In *Proceedings of the ASME Design Engineering Technical Conference*, Baltimore, MD.

Bonev, I. A., & Ryu, J. (2001). A new approach to orientation workspace analysis of 6-DOF parallel manipulators. *Mechanism and Machine Theory*, *36*(1), 15–28. doi:10.1016/S0094-114X(00)00032-X.

Bonev, I., & Ryu, J. (2001). A geometrical method for computing the constant-orientation workspace of 6-PRRS parallel manipulators. *Mechanism and Machine Theory*, *36*(1), 1–13. doi:10.1016/S0094-114X(00)00031-8.

Borenstein, J., & Ulrich, I. (1997). The GuideCane – A computerized travel aid for the active guidance of blind pedestrians. In *Proceedings of the IEEE International Conference on Robotics and Automation*, Albuquerque, NM (pp. 1283-1288).

Borenstein, J., & Koren, Y. (1985). Error eliminating rapid ultrasonic firing for mobile robot obstacle avoidance. *IEEE Transactions on Robotics and Automation*, *11*(1), 132–138. doi:10.1109/70.345945.

Borenstein, J., & Koren, Y. (1985). Obstacle avoidance with ultrasonic sensors. *IEEE Journal on Robotics and Automation*, *4*(2).

Borenstein, J., & Koren, Y. (1991). The vector field histogram – Fast obstacle avoidance for mobile robots. *IEEE Journal on Robotics and Automation*, *7*(3), 278–288. doi:10.1109/70.88137.

Bosscher, P., Riechel, A. T., & Ebert-Uphoff, I. (2006). Wrench-feasible workspace generation for cable-driven robots. *IEEE Transactions on Robotics*, *22*(5), 890–903. doi:10.1109/TRO.2006.878967.

Bouchard, S., & Gosselin, C. M. (2007). Workspace optimization of a very large cable-driven parallel mechanism for a radiotelescope application. In *Proceedings of the IDETC/CIE ASME International Design Engineering Technical Conferences & Computers and Information in Engineering Conference*, Las Vegas, NV (pp. 1-7).

Bourbakis, N. G., & Kavraki, D. (1996). Intelligent assistants for handicapped people's independence: Case study. In *Proceedings of the IEEE International Joint Symposium on Intelligent Systems* (pp. 337-344).

Bouzit, M., Chaibi, A., De Laurentis, K. J., & Mavroidis, C. (2004). Tactile feedback navigation handle for the visually impaired. In *Proceedings of the International Mechanical Engineering Congress and Exposition*, Anaheim, CA (pp. 13-19).

Brabin, J. A. (1982). New developments in mobility and orientation aids for the blind. *IEEE Transactions on Bio-Medical Engineering*, *29*, 285–290. doi:10.1109/TBME.1982.324945.

Brau, E., Gosselin, F., & Lallemand, J. P. (2005). Design of a singularity free architecture for cable driven haptic interfaces. In *Proceedings of the First Joint Eurohaptics Conference and Symposium on Haptic Interfaces for Virtual Environment and Teleoperator Systems* (pp. 208-212).

Breedveld, P. C. (1979). *Irreversible thermodynamics and bond graphs: A synthesis with some practical examples*. Unpublished master's thesis, University of Twente, The Netherlands.

Breedveld, P. C. (2004). *Port-based modeling of mechatronic systems*. Amsterdam, The Netherlands: Elsevier.

Breguet, J.-M. (1998). *Actionneurs "stick and slip" pour micro-manipulateurs.* Unpublished doctoral dissertation, Ecole Polytechnique Fédérale de Lausanne, Lausanne, Switzerland.

Breguet, J.-M., Driesen, W., Kaegi, F., & Cimprich, T. (2007). Applications of piezo-actuated micro-robots in micro-biology and material science. In *Proceedings of the IEEE International Conference on Mechatronics and Automation* (pp. 57-62).

Briones-Leon, A., Carbone, G., & Ceccarelli, M. (2009). Control de posición y fuerza del robot capaman 2 bis en tareas de barrenado. In *Proceedings of the 7th Congreso Internacional en Innovación y Desarrollo Tecnológico*, Cuernavaca, Mexico.

Brodsky, V., Shoham, M., & Glozman, D. (1998). Double circular-triangular six degrees-of-freedom parallel robot. In *Proceedings of the 6th International Symposium on Advances in Robotic Kinematics*, Salzburg, Austria (pp. 155-164).

Brooks, R. A. (1989). A robot that walks: Emergent behavior from a carefully evolved network. *Neural Computation, 1*, 253–262. doi:10.1162/neco.1989.1.2.253.

Brown, L. D. (1986). *Fundamentals of statistical exponential families with applications in statistical decision theory.* Hayward, CA: Institute of Mathematical Statistics.

Bruckmann, T., Mikelsons, L., Hiller, M., & Schramm, D. (2007). A new force calculation algorithm for tendon-based parallel manipulators. In *Proceedings of the IEEE/ASME International Conference on Advanced Intelligent Mechatronics*, Zurich, Switzerland (pp. 1-6).

Brufau, J., Puig-Vidal, M., Lapez-Sanchez, J., Samitier, J., Driesen, W., Breguet, J.-M., et al. (2005). MICRON: Small Autonomous Robot for Cell Manipulation Applications. In *Proceedings of the IEEE International Conference on Robotics and Automation* (pp. 844-849).

Bruzzone, L. E., & Molfino, R. M. (2006). A geometric definition of rotational stiffness and damping applied to impedance control of parallel robots. *International Journal of Robotics and Automation, 21*(3), 197–205. doi:10.2316/Journal.206.2006.3.206-2838.

Bruzzone, L., & Callegari, M. (2010). Application of the Rotation Matrix Natural Invariants to Impedance Control of Purely Rotational Parallel Robots. *Advances in Mechanical Engineering*, 284976.

Caccavale, F., Natale, C., Siciliano, B., & Villani, L. (1999). Six-DOF impedance control based on angle/axis representations. *IEEE Transactions on Robotics and Automation, 15*(2), 289–300. doi:10.1109/70.760350.

Cai, T. T. (2005). One-sided confidence intervals in discrete distributions. *Journal of Statistical Planning and Inference, 131*(1), 63–88. doi:10.1016/j.jspi.2004.01.005.

Callegari, M. (2008). Design and Prototyping of a SPM Based on 3-CPU Kinematics. In J.-H. Ryu (Ed.), Parallel Manipulators: New Developments (pp. 171-198). Vienna, Austria: I-Tech publications.

Callegari, M., & Suardi, A. (2004). Functionally-Oriented PKMs for Robot Cooperation. In [Amsterdam, The Netherlands: IOS Press.]. *Proceedings of the Intelligent Autonomous Systems Conference, IAS-8*, 263–270.

Callegari, M., & Tarantini, M. (2003). Kinematic analysis of a novel translational platform. *ASME Journal of Mechanical Design, 125*, 308–315. doi:10.1115/1.1563637.

Campbell, N. A., & Reece, J. B. (2007). *Biology.* Upper Saddle River, NJ: Pearson.

Canales, C., Kaegi, F., Groux, C., Breguet, J.-M., Meyer, C., Zbinden, U., et al. (2008). A nanomanipulation platform for semi automated manipulation of nano-sized objects using mobile microrobots inside a Scanning Electron Microscope. In *Proceedings of the 17th World Congress International Federation of Automatic Control* (pp. 13737-13742).

Carbone, G., & Ceccarelli, M. (2005). Numerical and experimental analysis of the stiffness performances of parallel manipulators. In *Proceedings of the 2nd International Colloquium Collaborative Research Centre*, Braunschweig, Germany.

Carbone, G., Ceccarelli, M., & Cimpoeru, I. (2005). Experimental tests with a macro-milli robotic system. In *Proceedings of the 14th International Workshop on Robotics in Alpe-Adria-Danube Region*, Bucharest, Romania.

Carbone, G., Ceccarelli, M., Ottaviano, E., Checcacci, D., Frisoli, A., Avizzano, C., & Bergamasco, M. (2003). A study of feasibility for a macro-milli serial-parallel robotic manipulator for surgery operated by a 3 DOFS haptic device. In *Proceedings of the 12th International Workshop on Robotics in Alpe-Adria-Danube Region*, Cassino, Italy.

Carbone, G., & Ceccarelli, M. (2005). A serial-parallel robotic architecture for surgical tasks. *Robotica, 23*, 345–354. doi:10.1017/S0263574704000967.

Cardin, S., Thalmann, D., & Vexo, F. (2007). A wearable system for mobility improvement of visually impaired people. *The Visual Computer, 23*(2), 109–118. doi:10.1007/s00371-006-0032-4.

Carvalho, J. C. M., & Ceccarelli, M. (2001). A closed-form formulation for the inverse dynamics of a Cassino parallel manipulator. *Multibody System Dynamics, 5*(2), 185–210. doi:10.1023/A:1009845926734.

Ceccarelli, M. (1998). A stiffness analysis for CaPaMan (Cassino parallel manipulator). In *Proceedings of the Conference on New Machine Concepts for Handling and Manufacturing Devices on the Basis of Parallel Structures*, Braunschweig, Germany (VDI 1427, pp. 67-80).

Ceccarelli, M., & Decio, P. (1999). A dynamic analysis of Cassino parallel manipulator in natural coordinates. In *Proceedings of the International Workshop on Parallel Machines*, Milano, Italy (pp. 87-92).

Ceccarelli, M. (1997). A new 3 DOF spatial parallel mechanism. *Mechanism and Machine Theory, 32*(8), 89–902. doi:10.1016/S0094-114X(97)00019-0.

Ceccarelli, M. (1997). Displacement analysis of a Turin platform parallel manipulator. *Advanced Robotics, 11*, 17–31. doi:10.1163/156855397X00029.

Ceccarelli, M., Decio, P., & Jimenez, J. (2002). Dynamic performance of CaPaMan by numerical simulations. *Mechanism and Machine Theory, 37*, 241–266. doi:10.1016/S0094-114X(01)00079-9.

Chakraborti, S., & Li, J. (2007). Confidence interval estimation of a normal percentile. *The American Statistician, 61*(4), 331–336. doi:10.1198/000313007X244457.

Cheng, H., Yiu, Y. K., & Li, Z. X. (2003). Dynamics and control of redundantly actuated parallel manipulators. *IEEE Transactions on Mechatronics, 8*(4), 483–491. doi:10.1109/TMECH.2003.820006.

Chen, I.-M. (2000). Realization of a rapidly reconfigurable robotic Workcell. *Journal of Japan Society of Precision Engineering, 66*(7), 1024–1030. doi:10.2493/jjspe.66.1024.

Chen, I.-M. (2001). Rapid response manufacturing through a rapidly reconfigurable robotic Workcell. *Robotics and Computer-integrated Manufacturing, 17*, 199–213. doi:10.1016/S0736-5845(00)00028-4.

Chen, I.-M., & Burdick, J. W. (1998). Enumerating non-isomorphic assembly configurations of modular robotic systems. *The International Journal of Robotics Research, 17*(7), 702–719. doi:10.1177/027836499801700702.

Chen, I.-M., & Yang, G. (1997). Kinematic calibration of modular reconfigurable robots using product-of-exponentials formula. *Journal of Robotic Systems, 14*(11), 807–821. doi:10.1002/(SICI)1097-4563(199711)14:11<807::AID-ROB4>3.0.CO;2-Y.

Chen, I.-M., & Yang, G. (1998). Automatic model generation for modular reconfigurable robot dynamics. *Transactions of the ASME, 120*, 346–352.

Chen, I.-M., Yang, G., & Kang, I.-G. (1999). Numerical inverse kinematics for modular reconfigurable robots. *Journal of Robotic Systems, 16*(4), 213–225. doi:10.1002/(SICI)1097-4563(199904)16:4<213::AID-ROB2>3.0.CO;2-Z.

Chen, I.-M., Yang, G., Tan, C. T., & Yeo, S. H. (2001). A local POE model for robot kinematic calibration. *Mechanism and Machine Theory, 36*, 1215–1239. doi:10.1016/S0094-114X(01)00048-9.

Chhabra, R., & Emami, M. R. (2010). *Holistic system modeling in mechatronics*. Amsterdam, The Netherlands: Elsevier.

Chin, I.-H., & Li, C.-F. (2005, August 28-31). Smooth sliding mode tracking control of the Stewart platform. In *Proceedings of the IEEE Conference on Control Applications*, Toronto, ON, Canada (pp. 43-48).

Choi, Y., Sreenivasan, S. V., & Choi, B. J. (2008). Kinematic design of large displacement precision XY positioning stage by using cross strip flexure joints and over-constrained mechanism. *Mechanism and Machine Theory*, *43*(6), 724–737. doi:10.1016/j.mechmachtheory.2007.05.009.

Chua, L. O., & Yang, L. (1988). Cellular neural network: Theory. *IEEE Transactions on Circuits and Systems*, *35*(10), 1257–1272. doi:10.1109/31.7600.

Chuang, H.-Y., & Chien, K.-H. (2007). Analysis of workspace and singularity of a slide equilateral triangle parallel manipulator. *Journal of System Design and Dynamics*, *1*(4), 724–735. doi:10.1299/jsdd.1.724.

Chung, G. J., Choi, K. B., & Kyung, J. H. (2006). Development of precision robot manipulator using flexure hinge mechanism. In *Proceedings of the IEEE Conference on Robotics, Automation and Mechatronics*, Bangkok, Thailand (pp. 1-6).

Chu, S. K., & Pang, G. K. (2002). Comparison between different models of hexapod robot gait. *IEEE Transactions on Systems, Man, and Cybernetics. Part A, Systems and Humans*, *32*(6), 752–756. doi:10.1109/TSMCA.2002.807066.

Clark, J. E., & Cutkosky, M. R. (2006). The effect of leg specialization in a biomimetic hexapedal running robot. *ASME. Journal of Dynamic Systems, Measurement, and Control*, *128*, 26–35. doi:10.1115/1.2168477.

Clavel, R. (1988). Delta, a fast robot with parallel geometry. In *Proceedings of the 18th International Symposium on Industrial Robots*, Lausanne, Switzerland (pp. 91-100).

Connolly, C. I. (1994). Harmonic functions and collision probabilities. In *Proceedings of the IEEE International Conference on Robotics and Automation*, San Diego (pp. 3015-3019).

Connolly, C. I., Burns, J. B., & Weiss, R. (1990). Path planning using Laplace's equation. In *Proceedings of the IEEE International Conference on Robotics and Automation*, Cincinnati, OH (pp. 2101-2106).

Connolly, C. I. (1993). The application of harmonic functions to robotics. *Journal of Robotic Systems*, *10*(7), 931–946. doi:10.1002/rob.4620100704.

Craig, J. J. (2005). *Introduction to Robotics: Mechanics & Control* (3rd ed.). Upper Saddle River, NJ: Pearson.

Dakopoulos, D., & Bourbakis, N. (2008). Preserving visual information in low resolution images during navigation for visually impaired. In *Proceedings of the 1st International Conference on Pervasive Technologies related to Assistive Environments*, Athens, Greece (pp. 5-19).

Daniel, H. R. M., Somber-Murky, P., & Angeles, J. (1993). The kinematics of 3-DOF planar and spherical double-triangular parallel manipulator. In Angeles, J., & Kovacs, P. (Eds.), *Computational kinematics* (pp. 153–164). Norwell, MA: Kluwer Academic.

Das, A., Zhang, W. P., Popa, D., & Stephanou, H. (2007). μ^3: Multiscale, Deterministic Micro-Nano Assembly System for Construction of On-Wafer Microrobots. In *IEEE International Conference on Robotics and Automation* (pp. 461-466). doi:10.1109/ROBOT.2007.363829

Dasgupta, B., & Mruthyunjaya, T. S. (1998). Closed-form dynamic equations of the Stewart platform through the Newton-Euler approach. *Mechanism and Machine Theory*, *33*(7), 993–1012. doi:10.1016/S0094-114X(97)00087-6.

Dasgupta, B., & Mruthyunthaya, T. S. (2000). The Stewart platform manipulator: a review. *Mechanism and Machine Theory*, *35*(1), 15–40. doi:10.1016/S0094-114X(99)00006-3.

Davies, T. C., Burns, C. M., & Pinder, S. D. (2007). Mobility interfaces for the visually impaired: What's missing? In *Proceedings of the 7th ACM SIGCHI New Zealand Chapter's International Conference on Computer-Human Interaction: Design Centered HCI*, Hamilton, New Zealand (Vol. 254, pp. 41-47).

Davliakos, I., & Papadopoulos, E. (2008). Model-based control of a 6 DOF electro hydraulic Stewart-Gough platform. *Journal of Mechanism and Machine Theory*, *43*(11), 1385–1400. doi:10.1016/j.mechmachtheory.2007.12.002.

De Floriani, L., Faicidieno, B., & Pienovi, C. (1985). Delaunay-based representation of surface defined over arbitrarily shaped domains. *Computer Vision Graphics and Image Processing*, *32*(1), 127–140. doi:10.1016/0734-189X(85)90005-2.

de Wit, C. C., Olsson, H., Aström, K., & Lischinsky, P. (1995). A new model for control of systems with friction. *IEEE Transactions on Automatic Control, 40*, 419–425. doi:10.1109/9.376053.

Dekker, R., Khajepour, A., & Behzadipour, S. (2006). Design and testing of an ultra high-speed cable robot. *International Journal of Robotics and Automation, 21*(1), 25–34. doi:10.2316/Journal.206.2006.1.206-2824.

Dereje, S., & Mitra, R. (2010). Fuzzy logic tuned PID controller for Stewart platform manipulator. In *Proceedings of the International Conference on Computer Applications in Electrical Engineering and Recent Advances*, Roorkee, India (pp. 272-257).

Dereje, S., & Mitra, R. (2010). Neuro-fuzzy sliding mode control: design and stability analysis. *International Journal of Computational Intelligence Studies, 1*(3), 242–255. doi:10.1504/IJCISTUDIES.2010.034888.

Diao, X., & Ma, O. (2007). A method of verifying force-closure condition for general cable manipulators with seven cables. *Mechanism and Machine Theory, 42*(12), 1563–1576. doi:10.1016/j.mechmachtheory.2007.06.008.

Dodds, A. G., Armstrong, J. D., & Shingledecker, C. A. (1981). The Nottingham obstacle detector: development and evaluation. In *Proceedings of the 7th ACM SIGCHI New Zealand Chapter's International Conference on Computer-Human Interaction: Design Centered HCI*, Hamilton, New Zealand (Vol. 75, pp. 203-209).

Dogan, G. (2007). Bootstrapping for confidence interval estimation and hypothesis testing for parameters of system dynamics models. *System Dynamics Review, 23*(4), 415–436. doi:10.1002/sdr.362.

Dong, W., Sun, L. N., & Du, Z. J. (2007). Design of a precision compliant parallel positioner driven by dual piezoelectric actuators. *Sensors and Actuators. A, Physical, 135*, 250–256. doi:10.1016/j.sna.2006.07.011.

Dong, W., Sun, L. N., & Du, Z. J. (2008). Stiffness research on a high-precision, large-workspace parallel mechanism with compliant joints. *Precision Engineering, 32*, 222–231. doi:10.1016/j.precisioneng.2007.08.002.

Dong, Y., Zhang, L., & Lu, D. (2005). Mechanism Modelling and Simulation of Cobot Based on Simulink. *Machine Design and Research, 21*(4), 33–36.

Dongya, Z., Shaoyuan, L., & Feng, G. (2008). Fully adaptive feedforward feedback synchronized tracking control for Gough-Stewart platform systems. *International Journal of Control. Automation and Systems, 6*(5), 689–701.

Dorf, R., & Bishop, R. (2005). *Sistemas de control moderno*. Upper Saddle River, NJ: Pearson/Prentice Hall.

Driesen, W. (2008). *Concept, modeling and experimental characterization of the modulated friction inertial drive (MFID) locomotion principle*. Unpublished doctoral dissertation, Ecole Polytechnique Fédérale de Lausanne, Lausanne, Switzerland.

Du Plessis, L. J., & Snyman, J. A. (2006). An optimally re-configurable Planar Gough-Stewart machining platform. *Mechanism and Machine Theory, 41*, 334–357. doi:10.1016/j.mechmachtheory.2005.05.007.

Du, C. (1996). An algorithm for automatic Delaunay triangulation of arbitrary planar domains. *Advances in Engineering Software, 27*(1-2), 21–26. doi:10.1016/0965-9978(96)00004-X.

Duncan, A. J. (1986). *Quality control and industrial statistics* (5th ed.). Homewood, IL: R. D. Irwin.

Dupont, P., Hayward, V., Armstrong, B., & Altpeter, F. (2002). Single state elastoplastic friction models. *IEEE Transactions on Automatic Control, 47*, 787–792. doi:10.1109/TAC.2002.1000274.

Durrbaum, A., Klier, W., & Hahn, H. (2002). Comparison of automatic and symbolic differentiation in mathematical modeling and computer simulation of rigid body systems. *Multibody System Dynamics, 7*(4), 331–355. doi:10.1023/A:1015523018029.

Dwivedy, S. K., & Eberhard, P. (2006). Dynamic analysis of flexible manipulators, a literature review. *Mechanism and Machine Theory, 41*(7), 749–777. doi:10.1016/j.mechmachtheory.2006.01.014.

Easton, R. D. (1992). Inherent problems of attempts to apply sonar and vibrotactile sensory aid technology to the perceptual needs of the blind. *Optometry and Vision Science, 69*(1), 3–14. doi:10.1097/00006324-199201000-00002.

Ebert-Uphoff, I., & Voglewede, P. A. (2004). On the connections between cable-driven manipulators, parallel manipulators and grasping. In *Proceedings of the IEEE International Conference on Robotics & Automation*, New Orleans, LA (pp. 4521-4526).

Ebrahimi, I., Carretero, J. A., & Boudreau, R. (2007). 3-PRRR redundant planar parallel manipulator: Inverse displacement, workspace and singularity analyses. *Mechanism and Machine Theory*, *42*(8), 1007–1016. doi:10.1016/j.mechmachtheory.2006.07.006.

Edeler, C. (2008). Simulation and Experimental Evaluation of Laser-Structured Actuators for a Mobile Microrobot. In *Proceedings of the IEEE International Conference on Robotics and Automation (ICRA 2008)* (pp. 3118-3123). doi:10.1109/ROBOT.2008.4543685

Edeler, C. (2010). Dynamic-mechanical Analysis of Piezoactuators for mobile Nanorobots. In *Proceedings of the International Conference on New Actuators (ACTUATOR2010)*, Bremen, Germany (pp. 1003-1006).

Edeler, C., Meyer, I., & Fatikow, S. (in press). Simulation and Measurements of Stick-Slip-Microdrives for Nanorobots. In *Proceedings of the European Conference on Mechanism Science (EUCOMES 2010)*, Cluj-Napoca. *Romania*.

Eichhorn, V., Carlson, K., Andersen, K. N., Fatikow, S., & Bøggild, P. (2007). Nanorobotic Manipulation Setup for Pick-and-Place Handling and Nondestructive Characterization of Carbon Nanotubes. In *Proceedings of the IEEE International Conference on Intelligent Robots and Systems* (pp. 291-296). doi:10.1109/IROS.2007.4398979

Eichhorn, V., Fatikow, S., Wortmann, T., Stolle, C., Edeler, C., Jasper, D., et al. (2009). NanoLab: A Nanorobotic System for Automated Pick-and-Place Handling and Characterization of CNTs. In *Proceedings of the IEEE International Conference on Robotics and Automation* (pp. 1826-1831). doi:10.1109/ROBOT.2009.5152440

Erden, M. S., & Leblebicioglu, K. (2007). Analysis of wave gaits for energy efficiency. *Autonomous Robots*, *23*(3), 213–230. doi:10.1007/s10514-007-9041-z.

Erlbacher, E. (2000). *Force control basics*. Dallas, TX: Push Corp.

Fan, L., Wu, M. C., Choquette, K. D., & Crawford, M. H. (1997). Self-assembled microactuated XYZ stages for optical scanning and alignment. In *Proceedings of the International Solid State Sensors Actuators Conference*, Chicago, IL (pp. 319-322).

Fang, T. P., & Piegl, L. A. (1993). Algorithm for Delaunay triangulation and convex-hull computation using a sparse matrix. *Computer Aided Design*, *24*(8), 425–536. doi:10.1016/0010-4485(92)90010-8.

Fang, Y., & Zhao, L. (2006). Approximation to the distribution of LAD estimators for censored regression by random weighting method. *Journal of Statistical Planning and Inference*, *136*(4), 1302–1316. doi:10.1016/j.jspi.2004.09.010.

Fatikow, S., Eichhorn, V., Krohs, F., Mircea, I., Stolle, C., & Hagemann, S. (2007). Development of automated microrobat-based nanohandling stations for nanocharacterization. *Microsystem Technologies*, *14*, 463–474. doi:10.1007/s00542-007-0471-5.

Fattah, A., & Ghasemi, H. A. M. (2002). Isotropic design of spatial parallel manipulators. *The International Journal of Robotics Research*, *21*(9), 811–824. doi:10.1177/0278364902021009842.

Fattah, A., & Kasaei, G. H. (2000). Kinematics and dynamics of a parallel manipulator with a new architecture. *Robotica*, *18*(5), 535–543. doi:10.1017/S026357470000271X.

Ferrell, C. (1993). *Robust agent control of an autonomous robot with many sensors and actuators*. Unpublished master's thesis, MIT, Cambridge, MA.

Fichter, E. F. (1986). A Stewart platform- based manipulator: General theory and practical construction. *The International Journal of Robotics Research*, *5*(2), 157–182. doi:10.1177/027836498600500216.

Fichter, E. F., Kerr, D. R., & Jones, J. R. (2009). The Gough–Stewart platform parallel manipulator: a retrospective appreciation. *Proceedings of the Institution of Mechanical Engineers. Part C, Journal of Mechanical Engineering Science*, *223*(1). doi:10.1243/09544062JMES1137.

Flavien, P., Nicolas, A., Philippe, M., & Khalil, W. (2006). Vision based computed torque controller for parallel robots. In *Proceedings of the IEEE 32 Annual Conference on Industrial Electronics* (pp. 3851-3856).

Frantsevicha, L. I., & Cruse, H. (2005). Leg coordination during turning on an extremely narrow substrate in a bug *Mesoceruse marginatus* (Heteroptera, Coreidae). *Journal of Insect Physiology*, *51*(10), 1092–1104. doi:10.1016/j.jinsphys.2005.05.008.

Full, R. J., & Weinstein, R. B. (1992). Integrating the physiology, mechanics and behavior of rapid running ghost crabs: Slow and steady doesn't always win the race. *Integrative and Comparative Biology*, *32*(3), 382–395. doi:10.1093/icb/32.3.382.

Fuyang, T., Hongtao, W., & Hongli, S. (2009). Efficient numerical integration method for dynamic of flexible space robots system. In *Proceedings of the International Asia Conference on Informatics in Control, Automation and Robotics* (pp. 102-106).

Gabrielli, A. (2009). *Mini- robotic applications for miniaturized assembly tasks.* Unpublished doctoral dissertation, Polytechnic University of Marche, Ancona, Italy.

Gallardo, J., Rico, J. M., & Frisoli, A. (2003). Dynamics of parallel manipulators by means of screw theory. *Mechanism and Machine Theory*, *38*(11), 1113–1131. doi:10.1016/S0094-114X(03)00054-5.

Gallina, P., & Rosati, G. (2002). Manipulability of a planar wire driven haptic device. *Mechanism and Machine Theory*, *37*(2), 215–228. doi:10.1016/S0094-114X(01)00076-3.

Gao, F., Zhang, J., Chen, Y., & Jin, Z. (2003). Development of a new type of 6-DOF parallel micro-manipulator and its control system. In *Proceedings of the IEEE International Conference on Robotics, Intelligent Systems and Signal Processing* (Vol. 2, pp. 715-720).

Gao, S., Feng, Z., Zhong, Y., & Shirinzadeh, B. (2008). Random weighting estimation of parameters in generalized Gaussian distribution. *Information Sciences*, *178*(9), 2275–2281. doi:10.1016/j.ins.2007.12.011.

Gao, S., Zhang, J., & Zhou, T. (2003). Large numbers law for sample mean of random weighting estimation. *Information Sciences*, *155*(1-2), 151–156. doi:10.1016/S0020-0255(03)00158-0.

Gao, S., Zhang, Z., & Yang, B. (2004). The random weighting estimation of quantile process. *Information Sciences*, *164*(1-4), 139–146. doi:10.1016/j.ins.2003.10.002.

Gao, S., & Zhong, Y. (2010). Random weighting estimation of kernel density. *Journal of Statistical Planning and Inference*, *140*(9), 2403–2407. doi:10.1016/j.jspi.2010.02.009.

Gawthrop, P. J. (1991). *Bond graphs: A representation for mechatronic systems.* Amsterdam, The Netherlands: Elsevier.

Geike, T., & McPhee, J. (2003). Inverse dynamic analysis of parallel manipulators with full mobility. *Mechanism and Machine Theory*, *38*(6), 549–562. doi:10.1016/S0094-114X(03)00008-9.

Ghobakhloo, A., Eghtesad, M., & Azadi, M. (2006). Position control of Gough-Stewart-Gough platform using inverse dynamics with full dynamics. In *Proceedings of the IEEE 9th International Workshop on Advanced Motion Control* (pp. 50-55).

Ghorbel, F. H., Chételat, O., Gunawardana, R., & Longchamp, R. (2000). Modeling and set point control of closed chain mechanisms, theory and experiment. *IEEE Transactions on Control Systems Technology*, *8*(5), 801–815. doi:10.1109/87.865853.

Glasius, R., Komada, A., & Gielen, S. (1994). Population coding in a neural net for trajectory formation. *Network (Bristol, England)*, *5*(4), 549–563. doi:10.1088/0954-898X/5/4/007.

Glasius, R., Komoda, A., & Gielen, S. (1996). A biologically inspired neural net for trajectory formation and obstacle avoidance. *Biological Cybernetics*, *74*(6), 511–520. doi:10.1007/BF00209422.

Glasius, R., Komoda, A., & Gielen, S. C. A. M. (1995). Neural network dynamics for path planning and obstacle avoidance. *Neural Networks*, *8*(1), 125–133. doi:10.1016/0893-6080(94)E0045-M.

González-Mora, J. L., Rodríguez-Hernández, A., Rodríguez-Ramos, L. F., Díaz-Saco, L., & Sosa, N. (2009). Development of a new space perception system for blind people, based on the creation of a virtual acoustic space. In J. Mira & J. V. Sánchez-Andrés (Eds.), *Proceedings of the International Work-Conference on Engineering Applications of Bio-Inspired Artificial Neural Networks* (LNCS 1607, pp. 321-330).

Gosselin, C. (1988). *Kinematics analysis optimization and programming of parallel robot manipulators.* Unpublished doctoral dissertation, McGill University, Montreal, QC, Canada.

Gosselin, C. M. (1996). Parallel computational algorithms for the kinematics and dynamics of planar and spatial parallel manipulators. *ASME Journal of Dynamic Systems. Measurement and Control, 118*(1), 22–28. doi:10.1115/1.2801147.

Gosselin, C., & Angeles, J. (1990). Singularity analysis of closed-loop kinematic chains. *IEEE Transactions on Robotics and Automation, 6*(3), 281–290. doi:10.1109/70.56660.

Gosselin, C., & Angeles, J. (1991). A global performance index for the kinematic optimization of robotic manipulators. *ASME Journal of Mechanical Design, 113*(3), 220–226. doi:10.1115/1.2912772.

Gouttefarde, M., Merlet, J.-P., & Daney, D. (2006). Determination of the wrench-closure workspace of 6-DOF parallel cable-driven mechanisms. *Advances in Robot Kinematics*, 315-322.

Gouttefarde, M., & Gosselin, C. M. (2006). Analysis of the wrench-closure workspace of planar parallel cable-driven mechanisms. *IEEE Transactions on Robotics, 22*(3), 434–445. doi:10.1109/TRO.2006.870638.

Griffis, M., & Duffy, J. (1989). A forward displacement analysis of a class of Stewart platforms. *Journal of Robotic Systems, 6*(6), 703–720. doi:10.1002/rob.4620060604.

Grossberg, S. (1988). Nonlinear neural networks: principles, mechanisms, and architectures. *Neural Networks, 1*(1), 17–61. doi:10.1016/0893-6080(88)90021-4.

Guo, H. B., & Li, H. R. (2006). Dynamics analysis and simulation of a Stewart platform manipulator. *Proceedings of the Institution of Mechanical Engineers. Part C, Journal of Mechanical Engineering Science, 220*(1), 61–72. doi:10.1243/095440605X32075.

Hacot, H., Dubowsky, S., & Bidaud, P. (1998). Analysis and simulation of a rocker-bogie exploration rover. In *Proceedings of the Twelfth Symposium on Theory and Practice of Robots and Manipulators*, Paris, France.

Hadian, H., & Fattah, A. (2008). Best kinematic performance analysis of a 6-6 Cable-suspended parallel robot. In *Proceedings of the IEEE/ASME International Conference on Mechatronic and Embedded Systems and Applications*, Beijing, China (pp. 510-515).

Hadian, H., & Fattah, A. (2009). On the study of dexterity measure for a novel 3-DOF cable-driven parallel manipulator. In *Proceedings of the 17th Annual International Conference on Mechanical Engineering*, Tehran, Iran.

Hagemann, S., Krohs, F., & Fatikow, S. (2007). *Automated Characterization and Manipulation of Biological Cells by a Nanohandling Robot Station.* Poster presented at Nanotech Northern Europe Conference and Exhibition, Helsinki, Finland.

Hag, S. K., Young, M. C., & Kyo, I. L. (2005). Robust nonlinear task space control for 6 DOF parallel manipulator. *Automatica, 41*, 1591–1600. doi:10.1016/j.automatica.2005.04.014.

Hanieh, A. A. (2003). *Active isolation and damping of vibrations via Stewart platform.* Unpublished doctoral dissertation, Active Structures Laboratory, Universit'e Libre de Bruxelles, Brussels, Belgium.

Harris, S. E., & De Ruffieu, F. L. (1993). *Horse gaits, balance and movement.* New York, NY: Howell Book House.

Haukoos, J. S., & Lewis, R. J. (2005). Advanced statistics: Bootstrapping confidence intervals for statistics with ''difficult'' distributions. *Academic Emergency Medicine, 12*(4), 360–365. doi:10.1111/j.1553-2712.2005.tb01958.x.

Hemami, H., & Weimer, F. C. (1981). Modeling of nonholonomic dynamic systems with applications. *ASME Journal of Applied Mechanics, 48*, 177–182. doi:10.1115/1.3157563.

Henrich, D. (1997). Fast motion planning by parallel processing-a review. *Journal of Intelligent & Robotic Systems, 20*(1), 45–69. doi:10.1023/A:1007948727999.

Hernández-Martínez, E., Ceccarelli, M., Carbone, G., & López-Cajún, C. (2008). Simulación de un manipulador paralelo espacial de 3 grados de libertad. In *Proceedings of the 5th Congreso Bolivariano de Ingeniería Mecánica, II Congreso Binacional de Ingeniería Mecánica*, Cúcuta, Colombia.

Hodgkin, A. L., & Huxley, A. F. (1952). A quantitative description of membrane current and its application to conduction and excitation in nerve. *The Journal of Physiology, 117*(4), 500–544.

Hogan, N. (1985). Impedance Control: An Approach to Manipulation: Part I, II, III. *Journal of Dynamic Systems, Measurement, and Control, 107*(1), 1–24. doi:10.1115/1.3140702.

Hollar, S., Bergbreiter, S., & Pister, K. S. J. (2003). Bidirectional inchworm motors and two-DOF robot leg operation. In *Proceedings of the 12th International Solid State Sensors Actuators Conference*, Boston, MA (pp. 262-267).

Homma, K., Fukuda, O., Sugawara, J., Nagata, Y., & Usuba, M. A. (2003). Wire-driven leg rehabilitation system: Development of a 4-dof experimental system. In *Proceedings of the IEEE/ASME International Conference on Advanced Intelligent Mechatronics*, Kobe, Japan (pp. 908-913).

Honegger, M. (1998). Nonlinear adaptive control of a 6 DOF parallel manipulator. In *Proceedings of the MOVIC Conference*, Zurich, Switzerland (Vol. 3, pp. 961-966).

Honegger, M., Codourey, A., & Burdet, E. (1997). Adaptive control of the hexaglide, a 6 dof parallel manipulator. In *Proceedings of the IEEE International Conference on Robotics and Automation*, Albuquerque, NM (pp. 543-548).

Hopkins, B. R., & Williams, R. L. (2002). Kinematics, design and control of 6-P S U platform[Research Article]. *Industrial Robot: An International Journal, 29*(5), 443–451. doi:10.1108/01439910210440264.

Huang, J., Hiller, M., & Fang, S.-Q. (2007). Simulation Modeling of the Motion Control of a Two Degree of Freedom, Tendon Base, Parallel Manipulator in Operational Space Using MATLAB. *Journal of China University of Mining and Technology, 17*(2), 179–183. doi:10.1016/S1006-1266(07)60067-4.

Huang, T., Wang, J., Gosselin, C. M., & Whitehouse, D. (1999). Determination of closed form solution to the 2-D-orientation workspace of Gough–Stewart parallel manipulators. *IEEE Transactions on Robotics and Automation, 15*(6), 1121–1125. doi:10.1109/70.817675.

Huang, Z., Chen, L. H., & Li, Y. W. (2003). The singularity principle and property of Stewart parallel manipulator. *Journal of Robotic Systems*, 163–176. doi:10.1002/rob.10078.

Hub, A., Diepstraten, J., & Ertl, T. (2004). Design and development of an indoor navigation and object identification system for the blind. In *Proceedings of the 6th International ACM SIGACCESS Conference on Computers and Accessibility* (pp. 147-152).

Hudgens, J., & Tesar, D. (1991). Analysis of a fully-parallel six degree-of freedom micromanipulator. In *Proceedings of the 5th International Conference on Advanced Robotics: Robots in Unstructured Environments* (Vol. 1, pp. 814-820).

Hung, J. Y., Gao, W., & Hung, J. C. (1993). Variable structure control: A survey. *IEEE Transactions on Industrial Electronics, 40*(1), 2–19. doi:10.1109/41.184817.

Hwang, Y. K., & Ahuja, N. (1992). Gross motion planning-a survey. *ACM Computing Surveys, 24*(3), 219–291. doi:10.1145/136035.136037.

Iannone, S., Carbone, G., & Ceccarelli, M. (2008). Regulation and control of LARM Hand III. In *Proceedings of the International Symposium on Multibody Systems and Mechanics*, San Juan, Puerto Rico.

Ifukube, T., Sasaki, T., & Peng, C. (1991). A blind mobility aid modeled after echolocation of bats. *IEEE Transactions on Bio-Medical Engineering, 38*(5), 461–465. doi:10.1109/10.81565.

Innocenti, C., & Parenti-Castelli, V. (1993). Echelon Form Solution of Direct Kinematics for the General Fully-Parallel Spherical Wrist. *Mechanism and Machine Theory, 28*, 553–561. doi:10.1016/0094-114X(93)90035-T.

Iqbal, S., & Bhatti, A. I. (2007). Robust sliding-mode controller design for a Stewart platform. In *Proceedings of the International Bhurban Conference on Applied Sciences & Technology*, Islamabad, Pakistan (pp. 155-160).

Iqbal, S., & Bhatti, A. I. (2008). Dynamic analysis and robust control design for Gough-Stewart platform with moving payloads. In *Proceedings of the 17th World Congress the International Federation of Automatic Control*, Seoul, Korea (pp. 5324-5329).

Ito, K., Okamoto, M., Akita, J., Ono, T., Gyobu, I., Tagaki, T., et al. (2005). CyARM: An alternative aid device for blind persons. In Proceedings of Extended Abstracts on Human Factors in Computing Systems, Portland, OR (pp. 1483-1488).

Jabatan Kebajikan Masyarakat (JKM). (2000). *Buletin Perangkaan Kebajikan (1999)*. Kuala Lumpur, Malaysia: Percetakan Nasional Malaysia Berhad.

Jasper, D., & Edeler, C. (2008). Characterization, Optimization and Control of a Mobile Platform. In *Proceedings of the International Workshop on Microfactories* (pp. 143-148).

Jeck, T., & Cruse, H. (2007). Walking in Aretaon asperrimus. *Journal of Insect Physiology, 53*(7), 724–733. doi:10.1016/j.jinsphys.2007.03.010.

Jiang, Z. H. (1992). Compensability of end-effector motion errors and evaluation of manipulability for flexible space robot arms. In *Proceedings of the IEEE International Conference on System, Man, and Cybernetics* (pp. 486-491).

Jiang, Z. H. (1992). Kinematics and dynamics of flexible space robot arms. In *Proceedings of the IEEE/RSJ International Conference on Intelligent Robots and Systems Raleigh* (pp. 1681-1688).

Jiang, D., & Wang, L. C. (2006). An algorithm of NURBS surface fitting for reverse engineering. *International Journal of Advanced Manufacturing Technology, 31*(1-2), 92–97. doi:10.1007/s00170-005-0161-3.

Joe, B., & Simpson, R. B. (1986). Triangulation meshes for regions of complicated shape. *International Journal for Numerical Methods in Engineering, 23*(5), 751–778. doi:10.1002/nme.1620230503.

Johnson, L. A., & Higgins, C. M. (2006). A navigation aid for the blind using tactile-visual sensory substitution. In *Proceedings of the IEEE Annual International Conference of the Engineering in Medicine and Biology Society* (pp. 6289-6292).

Jokiel, B., Benavides, B. L., Bieg, L. F., & Allen, J. J. (2001). Planar and spatial three-degree-of-freedom microstages in silicon MEMS. In *Proceedings of the Annual Meeting of the American Society of Precision Engineering*, Crystal City, VA (pp. 32-35).

Joshi, S. A., & Tsai, L.-W. (2003). The kinematics of a class of 3-DOF, 4-legged parallel manipulators. *Journal of Mechanical Design, 125*(1), 52–60. doi:10.1115/1.1540992.

Kallio, P., Lind, M., Zhou, Q., & Koivo, H. N. (1998). A 3-DOF piezohydraulic parallel micromanipulator. In *Proceedings of the 1998 IEEE International Conference on Robotics and Automation* (Vol. 2, pp. 1823-1828).

Kanaya, M., & Tanaka, M. (1998). Path planning method for multi-robots using a cellular neural network. *Electronics and Communications in Japan-Part 3, 81*(3), 335-345.

Kane, T. R. (1961). Dynamics of nonholonomic systems. *ASME Journal of Applied Mechanics*, 574-578.

Karamete, B. K., Tokdemir, T., & Ger, M. (1997). Unstructured grid generation and a simple triangulation algorithm for arbitrary 2D geometries using object oriented programming. *International Journal for Numerical Methods in Engineering, 40*(2), 251–268. doi:10.1002/(SICI)1097-0207(19970130)40:2<251::AID-NME62>3.0.CO;2-U.

Karlis, D., & Patilea, V. (2008). Bootstrap confidence intervals in mixtures of discrete distributions. *Journal of Statistical Planning and Inference, 138*(8), 2313–2329. doi:10.1016/j.jspi.2007.10.026.

Karniadakis, G. E., & Kirby, R. M. II. (2003). *Parallel scientific computing in C++ and MPI: a seamless approach to parallel algorithms and their implementation*. New York: Cambridge University Press.

Karnopp, D. C. (2006). *System dynamics: Modeling and simulation of mechatronics systems*. New York, NY: John Wiley & Sons.

Karnopp, D. C., Pomerantz, M. A., Rosenberg, R. C., & van Dixhoorn, J. J. (Eds.). (1979). Bond graph techniques for dynamic systems in engineering and biology. *Journal of the Franklin Institute, 308*(3).

Karnopp, D., Margolis, D., & Rosenberg, R. (2000). *System dynamics- modeling and simulation of mechatronics systems*. New York, NY: John Wiley & Sons.

Kawamura, S., Choe, W., Tanaka, S., & Pandian, S. (1995). Development of an ultrahigh speed robot falcon using wire drive system. In *Proceedings of the IEEE International Conference on Robotics and Automation* (pp. 215-220).

Kawamura, S., Kino, H., & Won, C. (2000). High-speed manipulation by using a parallel wire-driven robots. *Robotica, 18*, 13–21. doi:10.1017/S0263574799002477.

Kawasaki, T., & Tanaka, H. (2010). Formation of a crystal nucleus from liquid. *Proceedings of the National Academy of Sciences of the United States of America, 107*(32), 14036–14041. doi:10.1073/pnas.1001040107.

Kecskemethy, A., Krupp, T., & Hiller, M. (1997). Symbolic processing of multiloop mechanism dynamics using closed-form kinematics solutions. *Multibody System Dynamics, 1*(1), 23–45. doi:10.1023/A:1009743909765.

Khan, W. A., Krovi, V. A., Saha, S. K., & Angeles, J. (2005). Recursive kinematics and inverse dynamics for a planar 3R parallel manipulator. *ASME Journal of Dynamic Systems. Measurement and Control, 127*(4), 529–536. doi:10.1115/1.2098890.

Khatib, O. (1986). Real-time obstacle avoidance for manipulators and mobile robots. *The International Journal of Robotics Research, 5*(1), 90–98. doi:10.1177/027836498600500106.

Kim, H. S., & Cho, Y. M. (2009). Design and modeling of a novel 3-DOF precision micro-stage. *Mechatronics, 19*, 598–608. doi:10.1016/j.mechatronics.2009.01.004.

Kim, H., Son, H., Roska, T., & Chua, L. O. (2002). Optimal path finding with space- and time-variant metric weights via multi-layer CNN. *International Journal of Circuit Theory and Applications, 30*(2-3), 247–270. doi:10.1002/cta.199.

Kim, J.-P., & Ryu, J. (2002). Inverse kinematics and dynamics of 6-DOF P-U-S type parallel manipulators. *KSME International Journal, 16*(1), 13–23.

Kim, J., Park, F. C., & Ryu, S. J. (2001). Design and analysis of a redundantly actuated parallel mechanism for rapid machining. *IEEE Transactions on Robotics and Automation, 17*(4), 423–434. doi:10.1109/70.954755.

Kohonen, T. (1982). Self-organized formation of topologically correct feature maps. *Biological Cybernetics, 43*(1), 59–69. doi:10.1007/BF00337288.

Kong, X., & Gosselin, C. (2005). Type synthesis of 4-dof sp-equivalent parallel manipulators: A virtual-chain approach. In *Proceedings of the CK International Workshop on Computational Kinematics*.

Kong, X., & Gosselin, C. (2007). *Type synthesis of parallel mechanisms (Vol. 33)*. Berlin, Germany: Springer-Verlag.

Koseki, Y., Tanikawa, T., & Koyachi, N. (2000). Kinematic analysis of translational 3-DOF micro parallel mechanism using matrix method. In *Proceedings of the IROS*, Kagawa, Japan (pp. 786-792).

Kota, S. (1999). Design of compliant mechanisms: Applications to MEMS. In *Proceedings of the SPIE Conference on Smart Electronics*, Newport Beach, CA (LNCS 3673, pp. 45-54).

Kott, P. S., & Liu, Y. K. (2009). One-sided coverage intervals for a proportion estimated from a stratified simple random sample. *International Statistical Review, 77*(2), 251–265. doi:10.1111/j.1751-5823.2009.00081.x.

Kulyukin, V. (2005). *Robotic guide for the blind*. Logan Hill, UT: Utah State University.

Latombe, J. C. (1991). *Robot motio planning*. Boston: Kluwer.

LAUMAS. (2009). *Datasheet células de carga para plataformas 150 x 150 mm*. Retrieved from http://www.laumas.com/celletutte_es.htm

Lawson, C. L. (1977). Software for C¹ surface interpolation. In Rice, J. R. (Ed.), *Mathematical software III* (pp. 161–194). New York, NY: Academic Press.

Lebedev, D. V., Steil, J. J., & Ritter, H. J. (2005). The dynamic wave expansion neural network model for robot motion planning in time-varying environment. *Neural Networks, 18*(3), 267–285. doi:10.1016/j.neunet.2005.01.004.

Lee, H. H. (2004). A new trajectory control of a flexible-link robot based on a distributed-parameter dynamic model. *International Journal of Control*, 546–553. doi:10.1080/00207170410001695656.

Lee, J., Yoon, H., & Yoon, T. H. (2011). High-resolution parallel multipass laser interferometer with an interference fringe spacing of 15 nm. *Optics Communications, 284*(5), 1118–1122. doi:10.1016/j.optcom.2010.10.073.

Lee, K. M., & Arjunan, S. (1991). A three-degrees-of-freedom micromotion in-parallel actuated manipulator. *IEEE Transactions on Robotics and Automation, 7*(5), 634–641. doi:10.1109/70.97875.

Lee, K. M., & Shan, D. K. (1988). Dynamic analysis of a three-degrees-freedom in-parallel actuated manipulator. *IEEE Transactions on Robotics and Automation, 4*(3), 361–367. doi:10.1109/56.797.

Lee, T. T., Liao, C. M., & Chen, T. K. (1988). On the stability Properties of Hexapod Tripod Gait. *IEEE Transactions on Robotics and Automation, 4*(4).

Leger, C. (1999). *Automated synthesis and optimization of robot configurations: An evolutionary approach.* Unpublished doctoral dissertation, The Robotics Institute, Carnegie Mellon University, Pittsburgh, PA.

Levine, S., Koren, Y., & Borenstein, J. (1990). NavChair control system for automatic assistive wheelchair navigation. In *Proceedings of the 4th International Conference of the Computer for Handicapped Persons*, Vienna, Austria.

Li, D., & Salcudean, S. E. (1997, April). Modeling, simulation, and control of a hydraulic Stewart platform. In *Proceedings of the IEEE International Conference on Robotics and Automation*, Albuquerque, NM (pp. 3360-3366).

Li, S., Gao, H., & Deng, Z. (2008). Mobility performance evaluation of lunar rover and optimization of rocker-bogie suspension parameters. In *Proceedings of the International Symposium on Systems and Control in Aerospace and Astronautics*, Shenzhen, China.

Li, Y., & Xu, Q. (2006). GA-based multi-objective optimal design of a planar 3-DOF cable-driven parallel manipulator. In *Proceedings of the International IEEE Conference on Robotics and Biomimetics*, Kunming, China (pp. 1360-1365).

Liaw, H. C., & Shirinzadeh, B. (2008). Enhanced adaptive motion tracking control of piezo-actuated flexure-based four-bar mechanisms for micro/nano manipulation. *Sensors and Actuators. A, Physical, 147*(1), 254–262. doi:10.1016/j.sna.2008.03.020.

Liaw, H. C., & Shirinzadeh, B. (2009). Neural network motion tracking control of piezo-actuated flexure-based mechanisms for micro-/nanomanipulation. *IEEE/ASME Transactions on Mechatronics, 14*(5), 517–527. doi:10.1109/TMECH.2009.2005491.

Liaw, H. C., Shirinzadeh, B., & Smith, J. (2007). Enhanced sliding mode motion tracking control of piezoelectric actuators. *Sensors and Actuators. A, Physical, 138*(1), 194–202. doi:10.1016/j.sna.2007.04.062.

Liaw, H. C., Shirinzadeh, B., & Smith, J. (2008). Robust motion tracking control of piezo-driven flexure-based four-bar mechanism for micro/nano manipulation. *Mechatronics, 18*(2), 111–120. doi:10.1016/j.mechatronics.2007.09.002.

Liaw, H. C., Shirinzadeh, B., & Smith, J. (2008). Sliding-mode enhanced adaptive motion tracking control of piezoelectric actuation systems for micro/nano manipulation. *IEEE Transactions on Control Systems Technology, 16*(4), 826–833. doi:10.1109/TCST.2007.916301.

Licheng, W., Fuchun, S., Zengqi, S., & Wenjing, S. (2002). Dynamic modeling control and simulation of flexible dual arm space robot. In *Proceedings of the IEEE TENCON* (pp. 1282-1285).

Li, M., Huang, T., & Mei, J. P. (2005). Dynamic formulation and performance comparison of the 3-dof modules of two reconfigurable PKMs – the TriVariant and the Tricept. *ASME Journal of Mechanical Design, 127*(6), 1129–1136. doi:10.1115/1.1992511.

Lin, C., Tang, X., Shi, J., & Duan, G. (2003). Workspace analysis of reconfigurable parallel machine tool based on setting-angle of spherical joint. In *Proceedings of the International Conference on Systems, Man and Cybernetics* (pp. 4945-4950).

Lin, C.-Y., Liou, C.-S., & Lai, J.-Y. (1997). A surface-lofting approach for smooth-surface reconstruction from 3D measurement data. *Computers in Industry, 34*(1), 73–85. doi:10.1016/S0166-3615(96)00082-6.

Lindemann, R. A. (2005). Dynamic testing and simulation of the mars exploration rover. In *Proceedings of the ASME International Design Engineering Technical Conference and Computers and Information in Engineering Conference* (pp. 24-28).

Lindemann, R. A., Bickler, D. B., Harrington, B. D., Ortiz, G. M., & Voorhees, C. J. (2006). Mars exploration rover mobility development. *IEEE Robotics & Automation Magazine*, 19–26. doi:10.1109/MRA.2006.1638012.

Liu, D., Xu, Y., & Fei, R. (2003). Study of an intelligent micromanipulator. *Journal of Materials Processing Technology, 139*, 77–80. doi:10.1016/S0924-0136(03)00185-7.

Liu, M. J., Li, C. X., & Li, C. N. (2000). Dynamics analysis of the Gough-Stewart platform manipulator. *IEEE Transactions on Robotics and Automation, 16*(1), 94–98. doi:10.1109/70.833196.

Liu, X.-J., Wang, J., Gao, F., & Wang, L.-P. (2002). Mechanism design of a simplified 6-DOF 6-RUS parallel manipulator. *Robotica*, 81–89.

Liu, Z., & DeVoe, D. L. (2001). Micromechanism fabrication using silicon fusion bonding. *Robotics and Computer-integrated Manufacturing, 17*(1), 131–137. doi:10.1016/S0736-5845(00)00046-6.

Li, Y., & Xu, Q. (2010). A totally decoupled piezo-driven XYZ flexure parallel micropositioning stage for micro/nanomanipulation. *IEEE Transactions on Automation Science and Engineering, 8*(2), 265–279. doi:10.1109/TASE.2010.2077675.

Li, Z. X., & Bui, T. D. (1998). Robot path planning using fluid model. *Journal of Intelligent & Robotic Systems, 21*(1), 29–50. doi:10.1023/A:1007963408438.

Lobontin, N., & Garcia, E. (2003). Analytical model of displacement amplification and stiffness optimization for a class of flexure-based compliant mechanisms. *Computers & Structures, 81*(32), 2797–2810. doi:10.1016/j.compstruc.2003.07.003.

Lofving, S. (1998). Extending the cane range using laser technique. In *Proceedings of the 9th International Orientation and Mobility Conference*.

Lozano-Perez, T., & Wesley, M. A. (1997). An algorithm for planning collision-free paths among polyhedral obstacles. *Communications of the ACM, 22*(10), 560–570. doi:10.1145/359156.359164.

Ludeman, L. (1986). *Fundamentals of digital signal processing*. New York, NY: John Wiley & Sons.

Luo, Z. H., & Guo, B. Z. (1997). Shear force feedback control of a single-link flexible robot with a revolute joint. *IEEE Transactions on Automatic Control, 42*(1).

Maeda, K., Tadokoro, S., Takamori, T., Hiller, M., & Verhoeven, R. (1999). On design of a redundant wire-driven parallel robot warp manipulator. In *Proceedings of the International Conference on Robotics and Automation* (pp. 895-900).

Majda, C., Borutzky, W., & Damic, V. (2009). Comparison of different formulations of 2D beam elements based on bond graph technique. *Simulation Modelling Practice and Theory, 17*(1), 107–124. doi:10.1016/j.simpat.2008.02.014.

Malaysian Association for Blind (MAB). (n. d.). *Tun Sambanthan*. Kuala Lumpur, Malaysia. *MAB*.

Malvern, B. J., Jr. (1973). The new C-5 laser cane for the blind. In *Proceedings of the Carnahan Conference on Electronic Prosthetics*, Lexington, KY (pp. 77-82).

Mandel, M., & Betensky, R. A. (2008). Simultaneous confidence intervals based on the percentile bootstrap approach. *Computational Statistics & Data Analysis, 52*(4), 2158–2165. doi:10.1016/j.csda.2007.07.005.

Mann, M. P., & Shiller, Z. (2005). Dynamic stability of a rocker bogie vehicle: Longitudinal motion. In *Proceedings of the IEEE International Conference on Robotics and Automation*, Barcelona, Spain (pp. 861-866).

Mano, M. M. (2006). *Digital design* (4th ed.). Upper Saddle River, NJ: Prentice Hall.

Mantalosa, P., & Zografos, K. (2008). Interval estimation for a binomial proportion: A bootstrap approach. *Journal of Statistical Computation and Simulation, 78*(2), 1251–1265. doi:10.1080/00949650701749356.

Mariotto, G., D'Angelo, M., & Shvets, I. V. (1999). Dynamic behavior of a piezowalker, inertial and frictional configurations. *The Review of Scientific Instruments, 70*, 3651–3655. doi:10.1063/1.1149972.

Martel, S., Sherwood, M., & Helm, C. de, W. G., Fofonoff, T., Dyer, R., et al. (2001). Three-legged wireless miniature robots for mass-scale operations at the sub-atomic scale. In *Proceedings of the IEEE International Conference on Robotics and Automation* (pp. 3423-3428).

Ma, W., & He, P. (1998). B-spline surface local updating with unorganized points. *Computer Aided Design, 30*(11), 853–862. doi:10.1016/S0010-4485(98)00042-6.

Ma, W., & Kruth, J. P. (1998). NURBS curve and surface fitting for reverse engineering. *International Journal of Advanced Manufacturing Technology, 14*(12), 918–927. doi:10.1007/BF01179082.

McCloy, D., & Harris, D. M. J. (1986). Robotics: An introduction. New York, NY: A Halsted Press book, Open University Press.

McGhee, R. B., & Frank, A. A. (1968). On the stability properties of quadruped creeping gaits. *Mathematical Biosciences*, *3*, 331–351. doi:10.1016/0025-5564(68)90090-4.

McGhee, R. B., & Iswandhi, G. I. (1979). Adaptive locomotion of a multi-legged robot over rough terrain. *IEEE Transactions on Systems, Man, and Cybernetics*, *9*(4), 176–182. doi:10.1109/TSMC.1979.4310180.

McPhee, J., Shi, P., & Piedboeuf, J. C. (2002). Dynamics of multibody systems using virtual work and symbolic programming. *Mathematical and Computer Modelling of Dynamical Systems*, *8*(3), 137–155. doi:10.1076/mcmd.8.2.137.8591.

Mehregany, M., Gabriel, K. J., & Trimmer, W. S. N. (1988). Integrated fabrication of polysilicon mechanisms. *IEEE Transactions on Electron Devices*, *35*(6), 719–723. doi:10.1109/16.2522.

Meijer, P. B. L. (1992). An experimental system for auditory image representations. *IEEE Transactions on Bio-Medical Engineering*, *39*(2), 112–121. doi:10.1109/10.121642.

Mendes, J., & Ceccarelli, M. (2001). A closed-form formulation for the inverse dynamics of Cassino parallel manipulator. *Multibody System Dynamics*, *5*, 185–210. doi:10.1023/A:1009845926734.

Meng, C., & Chen, F. L. (1996). Curve and surface approximation from CMM measurement data. *Computers & Industrial Engineering*, *30*(2), 211–225. doi:10.1016/0360-8352(95)00165-4.

Merlet, J. P. (2009). *Personal website*. Retrieved from http://www-sop.inria.fr/members/Jean-Pierre.Merlet/merlet_eng.html

Merlet, J. (1988). Force-feedback control of parallel manipulators. In. *Proceedings of the IEEE International Conference on Robotics and Automation*, *3*, 1484–1489.

Merlet, J. P. (1996). Redundant parallel manipulators. *Laboratory Robotics and Automation*, *8*(1), 17–24. doi:10.1002/(SICI)1098-2728(1996)8:1<17::AID-LRA3>3.0.CO;2-#.

Merlet, J. P. (2000). *Parallel Robots*. London, UK: Kluwer.

Merlet, J. P. (2006). *Parallel robots*. Dordrecht, The Netherlands: Springer-Verlag.

Merlet, J.-P., & Gosselin, C. (1991). New architecture for a six-degree-of-freedom parallel manipulator. *Mechanism and Machine Theory*, *26*(2), 77–90. doi:10.1016/0094-114X(91)90023-W.

Merlet, J.-P., & Gosselin, C. (1999). Determination of 6D workspaces of Gough-type parallel manipulator and comparison between different geometries. *The International Journal of Robotics Research*, *18*(1), 902–916. doi:10.1177/02783649922066646.

Meyer, C., Sqalli, O., Lorenz, H., & Karrai, K. (2005). Slip-stick step-scanner for scanning probe microscopy. *The Review of Scientific Instruments*, *76*. doi:10.1063/1.1927105.

Miller, K., & Clavel, R. (1992). The Lagrange-based model of Delta-4 robot dynamics. *Robotersysteme*, *8*(4), 49–54.

Ming, A., & Higuchi, T. (1994). Study on multiple degree-of-freedom positioning mechanism using wires (part 1) - concept, design and control. *International Journal of the Japanese Society for Precision Engineering*, *28*, 131–138.

Minoshima, K. (2010). High-precision absolute length metrology using fiber-based optical frequency combs. In *Proceedings of the International Conference on Electromagnetics in Advanced Applications*, Sydney, NSW, Australia (pp. 800-802).

Mohamed, M. G., & Gosselin, C. M. (2005). Design and analysis of kinematically redundant parallel manipulators with configurable platforms. *IEEE Transactions on Robotics*, *21*(3), 277–287. doi:10.1109/TRO.2004.837234.

Mohd Zubir, M. N., & Shirinzadeh, B. (2009). Development of a high precision flexure-based microgripper. *Precision Engineering*, *33*(4), 362–370. doi:10.1016/j.precisioneng.2008.10.003.

Montgomery, D. C. (2001). *Introduction to statistical quality control* (4th ed.). New York, NY: John Wiley & Sons.

Morris, C. N. (1982). Natural exponential families with quadratic variance functions. *Annals of Statistics*, *10*(1), 65–80. doi:10.1214/aos/1176345690.

Motamedi, M., Vossoughi, G., Ahmadian, M. T., Rezaei, S. M., Zareinejad, M., & Saadat, M. (2010). Robust adaptive control of a micro telemanipulation system using sliding mode-based force estimation. In *Proceedings of the American Control Conference* (pp. 2811-2816).

Mukherjee, A. (2006). *Users manual of SYMBOLS Shakti.* Retrieved from http://www.htcinfo.com/

Mukherjee, A., Karmarkar, R., & Samantaray, A. K. (2006). *Bondgraph in modeling simulation and fault identification.* New Delhi, India: I. K. International Publishing House Pvt.

Müller, A. (2005). Internal preload control of redundantly actuated parallel manipulators-its application to backlash avoiding control. *IEEE Transactions on Robotics, 21*(4), 668–677. doi:10.1109/TRO.2004.842341.

Muller, A., & Maißer, P. (2003). A Lie-group formulation of kinematics and dynamics of constrained MBS and its application to analytical mechanics. *Multibody System Dynamics, 9*(4), 311–352. doi:10.1023/A:1023321630764.

Munassypov, R., Grossmann, B., Magnussen, B., & Fatikow, S. (1996). Development and Control of piezoelectric actuators for a mobile micromanipulation system. In *Proceedings of the 5th International Conference on New Actuators (ACTUATOR1996)* (pp. 213-216).

Murotsu, Y., Tsujio, S., Senda K., & Hayashi, M. (1992). Trajectory control of flexible manipulators on a free flying space robot. *IEEE Control Systems*, 51-57.

Murthy, R., & Popa, D. O. (2009). A four degree of freedom microrobot with large work volume. In *Proceedings of the IEEE International Conference on Robotics and Automation* (pp. 1028-1033). doi:10.1109/ROBOT.2009.5152812

Murthy, R., Das, A., & Popa, D. (2008). ARRIpede: A stick-slip micro crawler/conveyor robot constructed via 2 ½D MEMS assembly. In *Proceedings of the IEEE/RSJ International Conference on Intelligent Robots and Systems* (pp. 34-40). doi:10.1109/IROS.2008.4651181

Mustafa, S. K., Yang, G., Yeo, S. H., Lin, W., & Chen, I. M. (2008). Self-calibration of a biologically inspired 7 DOF cable-driven robotic arm. *IEEE/ASME Transactions on Mechatronics, 13*, 66–75. doi:10.1109/TMECH.2007.915024.

Nag, I.-K., & Chong, W.-L. (1998). High speed tracking control of Gough-Stewart platform manipulator via enhanced sliding model controller. In *Proceedings of the International Conference on Robotics and Automation*, Leuven, Belgium (pp. 2716-2721).

Nakamura, Y., & Mukherjee, R. (1991). Non-holonomic path planning of space robots via a bidirectional approach. *IEEE Transactions on Robotics and Automation, 7*(4), 500–514. doi:10.1109/70.86080.

Nanua, P., & Waldron, K. (1989). Direct kinematic solution of a Stewart platform. In *Proceedings of the IEEE International Conference on Robotics and Automation* (pp. 431-437).

Narayanan, M. S. (2008). *Analysis of parallel manipulator architectures for use in masticatory studies.* Unpublished master's thesis, University at Buffalo (SUNY), Buffalo, NY.

Narayanan, M. S., Chakravarty, S., Shah, H. L., & Krovi, V. N. (2010). Kinematic-, workspace- and static- analysis of a 6-P-U-S parallel manipulator. In *Proceedings of the ASME International Design Engineering Technical Conferences*, Montreal, QC, Canada.

National Instruments. (2000). *LabVIEW user manual.* Austin, TX: National Instruments Corporation.

National Instruments. (2000). *LabVIEW measurement manual.* Austin, TX: National Instruments Corporation.

Nava, N., Carbone, G., & Ceccarelli, M. (2006). CaPaMan2bis as trunk module in CALUMA (Cassino low-cost hUMAnoid robot). In *Proceedings of the 2nd IEEE International Conference on Robotics, Automation and Mechatronics*, Bangkok, Thailand.

Negash, D. S., & Mitra, R. (2010). Integral sliding mode control for trajectory tracking control of Stewart platform manipulator. In *Proceedings of the IEEE-ICIIS International Conference on Industrial and Information Systems*, Surathkal, India (pp. 650-654).

Nenchev, D. N. (1993). A controller for a redundant fee-flying space robot with spacecraft attitude/manipulator motion coordination. In *Proceedings of the IEEE/RSJ International Conference on Intelligent Robots and Systems*, Yokohama, Japan (pp. 2108-2114).

Ng, C. C., Ong, S. K., & Nee, A. Y. C. (2006). Design and development of 3-DOF modular micro parallel kinematic manipulator. *International Journal of Advanced Manufacturing Technology*, *31*(1-2), 188–200. doi:10.1007/s00170-005-0166-y.

Niaritsiry, T. F., Fazenda, N., & Clavel, R. (2004). Study of the sources of inaccuracy of a 3DOF flexure hinge-based parallel manipulator. In *Proceedings of the IEEE International Conference on Robotics and Automation* (pp. 4091-4096).

Niguyen, C. C., Zhen, L. Z., & Sami, S. A. (1997). Efficient computation of forward kinematics and Jacobian matrix of a Gough-Stewart platform based manipulator. In. *Proceedings of the IEEE International Conference on Robotics and Automation*, *1*, 869–874.

Nokleby, S. B., Fisher, R., & Podhorodeski, R. P. (2005). Force capabilities of redundantly-actuated parallel manipulators. *Mechanism and Machine Theory*, *40*(5), 578–599. doi:10.1016/j.mechmachtheory.2004.10.005.

Noyong, M., Blech, K., Rosenberger, A., Klocke, V., & Simon, U. (2007). In situ nanomanipulation system for electrical measurements in SEM. *Measurement Science & Technology*, *18*(84).

Oh, S., & Agrawal, S. (2005). Cable suspended planar robots with redundant cables: Controllers with positive inputs. *IEEE Transactions on Robotics*, *21*(3), 457–465. doi:10.1109/TRO.2004.838029.

Ong, C. J., & Gilbert, E. G. (1998). Robot path planning with penetration growth distance. *Journal of Robotic Systems*, *15*(2), 57–74. doi:10.1002/(SICI)1097-4563(199802)15:2<57::AID-ROB1>3.0.CO;2-R.

Oriolo, G., Ulivi, G., & Vendittelli, M. (1998). Real-time map building and navigation for autonomous robots in unknown environments. *IEEE Transactions on Systems, Man, and Cybernetics. Part B, Cybernetics*, *28*(3), 316–333. doi:10.1109/3477.678626.

Ottaviano, E., & Ceccarelli, M. (2002). Optimal design of CaPaMan (Cassino parallel manipulator) with a specified orientation workspace. *Robotica*, *20*, 159–166. doi:10.1017/S026357470100385X.

Ottaviano, E., & Ceccarelli, M. (2006). An application of a 3-DOF parallel manipulator for earthquake simulations. *IEEE Transactions on Mechatronics*, *11*(2), 240–246. doi:10.1109/TMECH.2006.871103.

Ouyang, P. R., Tjiptoprodjo, R. C., Zhang, W. J., & Yang, G. S. (2008). Micromotion devices technology: The state of arts review. *International Journal of Advanced Manufacturing Technology*, *38*(5-6), 463–478. doi:10.1007/s00170-007-1109-6.

Ozguner, F., Tsai, S. J., & McGhee, R. B. (1984). An approach to the use of terrain-preview information in rough terrain locomotion by a hexapod walking machine. *The International Journal of Robotics Research*, *3*(2), 134–146. doi:10.1177/027836498400300211.

Palmer, L. R., Diller, E. D., & Quinn, R. D. (2009). Design of a wall-climbing hexapod for advanced maneuvers. In *Proceedings of the International Conference on Intelligent Robots and Systems*, St. Louis, MO (pp. 625-630).

Pal, P. (2008). A reconstruction method using geometric subdivision and NURBS interpolation. *International Journal of Advanced Manufacturing Technology*, *38*(3-4), 296–308. doi:10.1007/s00170-007-1102-0.

Pal, P., & Ballav, R. (2007). Object shape reconstruction through NURBS surface interpolation. *International Journal of Production Research*, *45*(2), 287–307. doi:10.1080/00207540600688481.

Panda and Edie. (1999). *The guide horse program*. Retrieved from http://guidehorse.org/

Park, H., & Lee, J. M. (2004). Adaptive impedance control of a haptic interface. *Mechatronics*, *14*(3), 237–253. doi:10.1016/S0957-4158(03)00040-0.

Paros, J. M., & Weisbord, L. (1965). How to design flexure hinges. *Machine Design*, *37*, 151–157.

Pashkevich, A., Kazheunikau, M., & Ruano, A. E. (2006). Neural network approach to collision free path-planning for robotic manipulators. *International Journal of Systems Science*, *37*(8), 555–564. doi:10.1080/00207720600783884.

Passarello, C. E., & Huston, R. L. (1973). Another look at nonholonomic systems. *ASME Journal of Applied Mechanics*, 101-104.

Pathak, P. M., Kumar, P., Mukherjee, A., & Dasgupta, A. (2008). A scheme for robust trajectory control of space robots. *Simulation Modelling Practice and Theory*, *16*, 1337–1349. doi:10.1016/j.simpat.2008.06.011.

Paynter, H. M. (1961). *Analysis and design of engineering systems*. Cambridge, MA: MIT Press.

Paynter, H. M. (1992). An epistemic prehistory of bond graphs. In Breedveld, P. C., & Dauphin-Tanguy, G. (Eds.), *Bond graphs for engineers* (pp. 3–17). Amsterdam, The Netherlands: Elsevier.

Peng, L. M., Chen, Q., Liang, X. L., Gao, S., Wang, J. Y., & Kleindiek, S. et al. (2004). Performing probe experiments in the SEM. *Micron (Oxford, England)*, *35*, 495–502. doi:10.1016/j.micron.2003.12.005.

Perreault, S., & Gosselin, C. (2007). Cable-driven parallel mechanisms: application to a locomotion interface. *Journal of Mechanical Design*, *130*(10), 102301–102309. doi:10.1115/1.2965607.

Pham, H. H., & Chen, I. M. (2005). Stiffness modeling of flexure parallel mechanism. *Precision Engineering*, *29*, 467–478. doi:10.1016/j.precisioneng.2004.12.006.

PHYSIK. (2010). *INSTRUMENTE*. Retrieved from http://www.physikinstrumente.com/

Pister, K. S. J., Judy, M. W., Burgett, S. R., & Fearing, R. S. (1992). Microfabricated hinges. *Sensors and Actuators. A, Physical*, *33*(3), 249–256. doi:10.1016/0924-4247(92)80172-Y.

Piyawattanametha, W., Fan, L., Lee, S. S., Su, J. G. D., & Wu, M. C. (1998). MEMS technology for optical crosslink for micro/nano satellites. In *Proceedings of the International Conference on Integrated Nano/Microtechnology for Space and Biomedical Applications*, Houston, TX.

Pohl, D. W. (1987). Dynamic piezoelectric translation devices. *The Review of Scientific Instruments*, *58*, 54–57. doi:10.1063/1.1139566.

Popov, V. L. (2009). Kontaktmechanik und Reibung. Berlin: Springer. doi: doi:10.1007/978-3-540-88837-6.

Posselt, U. (1957). An analyzer for mandibular positions. *The Journal of Prosthetic Dentistry*, *7*(3), 368–374. doi:10.1016/S0022-3913(57)80082-1.

Pusey, J., Fattah, A., Agrawal, S., & Messina, E. (2004). Design and workspace analysis of a 6-6 cable-suspended parallel robot. *Mechanism and Machine Theory*, *39*(7), 761–778. doi:10.1016/j.mechmachtheory.2004.02.010.

Qi, Y. Y., Zhao, M. R., & Lin, Y. C. (2007). Study on a positioning and measuring system with nanometer accuracy. In *Proceedings of the First International Conference on Integration and Communication of Micro and Nanosystems* (pp. 1573-1577).

Qi, Z., McInroy, J. E., & Jafari, F. (2007). Trajectory tracking with parallel robots using low chattering, fuzzy sliding mode controller. *Journal of Intelligent & Robotic Systems*, *48*(3), 333–356. doi:10.1007/s10846-006-9084-y.

Quoy, M., Moga, S., & Gaussier, P. (2003). Dynamical neural networks for planning and low-level robot control. *IEEE Transactions on Systems, Man, and Cybernetics. Part A, Systems and Humans*, *33*(4), 523–532. doi:10.1109/TSMCA.2003.809224.

Rabenorosoa, Clévy, Lutz, Bargiel, & Gorecki, (2009). A micro-assembly station used for 3D reconfigurable hybrid MOEMS assembly. In *Proceedings of the IEEE International Symposium on Assembly and Manufacturing* (pp. 95-100).

Raghavan, M. (1993). The Stewart platform of general geometry has 40 configurations. *Journal of Mechanical Design*, *115*, 277–282. doi:10.1115/1.2919188.

Rakotondrabe, M., Haddab, Y., & Lutz, P. (2009). Development, modeling, and control of a micro-/nanopositioning 2-DOF stickSlip device. *IEEE/ASME Transactions on Mechatronics*, *14*(6), 733–745. doi:10.1109/TMECH.2009.2011134.

Rao, A. B. K., Saha, S. K., & Rao, P. V. M. (2006). Dynamics modeling of hexaslides using the decoupled natural orthogonal complement matrices. *Multibody System Dynamics*, *15*(2), 159–180. doi:10.1007/s11044-005-9003-1.

Reljin, B., Krstic, I., Kostic, P., Reljin, I., & Kandic, D. (2004). CNN applications in modelling and solving non-electrical problems. In Slavova, A., & Mladenov, V. (Eds.), *Cellular neural networks: Theory and Applications* (pp. 135–172). New York: Nova Science Publishers.

Roennau, A., Kerscher, T., Ziegenmeyer, M., Zöllner, J. M., & Dillmann, R. (2009). Adaptation of a six-legged walking robot to its local environment. In Kozlowski, K. R. (Ed.), *Robot motion and control* (pp. 155–164). Berlin, Germany: Springer-Verlag. doi:10.1007/978-1-84882-985-5_15.

Ronkanen, P., Kallio, P., & Koivo, H. N. (2007). Simultaneous Actuation and Force Estimation Using Piezoelectric Actuators. In *Proceedings of the 2007 IEEE International Conference on Mechatronics and Automation* (pp. 3261-3265).

Rosati, G., Gallina, P., Masiero, S., & Rossi, A. (2005). Design of a new 5 DOF wire-based robot for rehabilitation. In *Proceedings of the IEEE 9th International Conference on Rehabilitation Robotics*, Chicago, IL (pp.430-433).

Rosenberg, R. C. (1965). *Computer-aided teaching of dynamic system behavior.* Unpublished doctoral dissertation, MIT, Cambridge, MA.

Rosenberg, R. C. (1974). *A user's guide to ENPORT-4.* New York, NY: John Wiley & Sons.

Rosenberg, R. C., & Karnopp, D. C. (1983). *Introduction to physical system dynamics.* New York, NY: McGraw-Hill.

Roska, T., Chua, L. O., Wolf, D., Kozek, T., Tetzlaff, R., & Puffer, F. (1995). Simulating nonlinear waves and partial differential equations via CNN-Part I: basic techniques. *IEEE Transactions on Circuits and Systems*, *42*(10), 807–815. doi:10.1109/81.473590.

Ryu, J. W., & Gweon, D. G. (1997). Error analysis of a flexure hinge mechanism induced by machining imperfection. *Precision Engineering*, *21*(2), 83–89. doi:10.1016/S0141-6359(97)00059-7.

Ryu, J. W., & Gweon, D. G. (1997). Optimal design of a flexure hinge based $XY\theta$ wafer stage. *Precision Engineering*, *21*(1), 18–28. doi:10.1016/S0141-6359(97)00064-0.

Ryu, K. (1997). A criterion on inclusion of stress stiffening effects in flexible multibody dynamic system simulation. *Computers & Structures*, *62*(6), 1035–1048. doi:10.1016/S0045-7949(96)00285-4.

Sadati, S. H., Alipour, K., & Behroozi, M. (2008). A combination of neural network and ritz method for robust motion planning of mobile robots along calculated modular paths. *International Journal of Robotics and Automation*, *23*(3), 187–198.

Saeidpourazar, R., & Jalili, N. (2006). Modeling and observer-based robust tracking control of a nano/micromanipulator for nanofiber grasping applications. In *Proceedings of the ASME International Mechanical Engineering Congress and Exposition* (pp. 969-977).

Sainarayanan, G., Nagarajan, R., & Yaacob, S. (2001). Interfacing vision sensor for real-time application in MATLAB. In Proceedings of Scored, Malaysia.

Samanta, B., & Devasia, S. (1988). Modeling and control of flexible manipulators using distributed actuators: A bond graph approach. In *Proceeding of the IEEE International Workshop on Intelligent Robots Systems* (pp. 99-104).

Sapidis, N., & Perucchio, R. (1991). Delaunay triangulation of arbitrarily shaped planar domains. *Computer Aided Geometric Design*, *8*(6), 421–437. doi:10.1016/0167-8396(91)90028-A.

Satheesh, G. K. (2009). *Characterization of reconfigurable Stewart platform.* Unpublished doctoral dissertation, Indian Institute of Technology Madras, Madras, India.

Satheesh, G. K., & Nagarajan, T. (2008). Experimental investigations on reconfigurable Stewart platform for contour generation. In *Proceedings of the 3rd International Conference on Sensing Technology* (pp. 292-296).

Satheesh, G. K., & Nagarajan, T. (2011). Reconfigurable Stewart platform for spiral contours. *Journal of Applied Sciences*, *11*, 1552–1558. doi:10.3923/jas.2011.1552.1558.

Satheesh, G. K., Nagarajan, T., & Srinivasa, Y. G. (2009). Characterization of reconfigurable Stewart platform for contour generation. *Robotics and Computer-integrated Manufacturing*, *25*, 721–731. doi:10.1016/j.rcim.2008.06.001.

Schott, W. (2010). Developments in homodyne interferometry. *Key Engineering Materials*, *437*, 84–88. doi:10.4028/www.scientific.net/KEM.437.84.

Schreiner, J. N. (2004). *Adaptations by the locomotor systems of terrestrial and amphibious crabs walking freely on land and underwater.* Unpublished master's thesis, Louisiana State University, Eunice, LA.

Schuldt, T., Gohlke, M., Weise, D., Peters, A., Johann, U., & Braxmaier, C. (2010). High-resolution dimensional metrology for industrial applications. In *Proceedings of the 9th International Symposium on Measurement Technology and Intelligent Instruments*, Saint Petersburg, Russia (Vol. 437, pp. 113-117).

Se-Han, L., Jae-Bok, S., Woo-Chun, C., & Daehie, H. (2003). Position control of a Stewart platform using inverse dynamics control with approximate dynamics. *Mechatronics, 13*, 605–619. doi:10.1016/S0957-4158(02)00033-8.

Seiler, A., Balendran, V., Sivayoganathan, K., & Sackfield, A. (1996). Reverse engineering from uni-directional CMM scan data. *International Journal of Advanced Manufacturing Technology, 11*(4), 276–284. doi:10.1007/BF01351285.

Selig, J. M. (2005). *Geometric fundamentals of robotics* (2nd ed.). New York: Springer.

Selig, J. M., & McAree, P. R. (1999). Constrained robot dynamics II: Parallel machines. *Journal of Robotic Systems, 16*(9), 487–498. doi:10.1002/(SICI)1097-4563(199909)16:9<487::AID-ROB2>3.0.CO;2-R.

Senda, K., & Murotsu, Y. (2000). Methodology for control of a space robot with flexible links. *IEEE Proceedings on Control Theory Applications, 147*(6), 562–568. doi:10.1049/ip-cta:20000870.

Seshadri, C., & Ghosh, A. (1993). Optimum path planning for robot manipulators amid static and dynamic obstacles. *IEEE Transactions on Systems, Man, and Cybernetics, 23*(2), 576–584. doi:10.1109/21.229471.

Shah, H. L., Narayanan, M. S., Lee, L., & Krovi, V. (2010). CAD-enhanced workspace optimization for parallel manipulators: A case study. In *Proceedings of the IEEE Conference on Automation Science and Engineering*, Toronto, ON, Canada (pp. 816-821).

Shaowen, F., & Yu, Y. (2006). Non-linear robust control with partial inverse dynamic compensation for a Stewart platform manipulator. *International Journal of Modelling. Identification and Control, 1*(1), 44–51. doi:10.1504/IJMIC.2006.008647.

Shelley, T. (2007). Extreme measurements become routine. *Eureka, 27*(2), 29–30.

Shiang, W., Cannon, D., & Gorman, J. (1999). Dynamic analysis of the cable array robotic crane. In *Proceedings of the IEEE International Conference on Robotics and Automation*.

Shiang, W., Cannon, D., & Gorman, J. (2000). Optimal force distribution applied to a robotic crane with flexible cables. In *Proceedings of the IEEE International Conference on Robotics and Automation*.

Shiou, F. J., Chen, C. J., Chiang, C. J., Liou, K. J., Liao, S. C., & Liou, H. C. (2010). Development of a real-time closed-loop micro-/nano-positioning system embedded with a capacitive sensor. *Measurement Science & Technology, 21*(5). doi:10.1088/0957-0233/21/5/054007.

Shirinzadeh, B., Teoh, P. L., Tian, Y., Dalvand, M. M., Zhong, Y., & Liaw, H. C. (2010). Laser interferometry-based guidance methodology for high precision positioning of mechanisms and robots. *Robotics and Computer-integrated Manufacturing, 26*(1), 74–82. doi:10.1016/j.rcim.2009.04.002.

Shoval, S., Borenstein, J., & Koren, Y. (1993). The NavBelt – A computerized travel aid for the blind. In *Proceedings of the Rehabilitation Engineering and Assistive Technology Society of North America Conference*, Las Vegas, NV (pp. 240-242).

Shoval, S., Borenstein, J., & Koren, Y. (1994). Mobile robot obstacle avoidance in a computerized travel aid for the blind. In *Proceedings of the IEEE Conference on Robotics and Automation*, San Diego, CA (pp. 2023-2029).

Shoval, S., Borenstrin, J., & Koren, Y. (1998). The NavBelt – A computerized travel aid for the blind based on mobile robotics technology. *IEEE Transactions on Bio-Medical Engineering, 45*(11), 1376–1386. doi:10.1109/10.725334.

Siciliano, B., & Villani, L. (2000). *Robot force control.* Boston: Kluwer.

Signore, M. J. D., Krovi, V. N., & Mendel, F. C. (2005). Virtual prototyping and hardware-in-the-loop testing for musculoskeletal system analysis. In *Proceedings of the IEEE International Conference on Mechatronics and Automation*, Niagara Falls, Canada (pp. 394-399).

Simaan, N., & Shoham, M. (2001). Singularity analysis of a class of composite serial in-parallel robots. *IEEE Transactions on Robotics and Automation, 17*(3), 301–311. doi:10.1109/70.938387.

Simaan, N., & Shoham, M. (2003). Geometric interpretation of the derivatives of parallel robot's Jacobian matrix with application to stiffness control. *ASME Journal of Mechanical Design, 125*, 33–42. doi:10.1115/1.1539514.

Sirouspour, M. R., & Salcudean, S. E. (2001). Nonlinear control of hydraulic robots. *IEEE Transactions on Robotics and Automation, 17*(2), 173–192. doi:10.1109/70.928562.

Sloan, S. W. (1987). A fast algorithm for constructing Delaunay triangulations in the plane. *Advances in Engineering Software, 9*(1), 34–55. doi:10.1016/0141-1195(87)90043-X.

Smith, S. T. (2000). *Flexures: Elements of Elastic Mechanisms.* Boca Raton, FL: CRC Press.

Smith, S. T., Badami, V. G., & Dale, J. S. (1997). Elliptical flexure hinges. *The Review of Scientific Instruments, 68*(3), 1474–1483. doi:10.1063/1.1147635.

Sokolov, A., & Xirouchakis, P. (2007). Dynamics analysis of a 3-dof parallel manipulator with R-P-S joint structure. *Mechanism and Machine Theory, 42*(5), 541–557. doi:10.1016/j.mechmachtheory.2006.05.004.

Sommargren, G. E. (1987). A new laser measurement system for precision metrology. *Precision Engineering, 9*(4), 179–184. doi:10.1016/0141-6359(87)90075-4.

Song, S. M., & Waldron, K. J. (1987). An analytical approach for gait study and its applications on wave gaits. *The International Journal of Robotics Research, 6*(2), 60–71. doi:10.1177/027836498700600205.

Spanner, K., & Vorndran, S. (2003, July 20-24). Advances in piezo-nanopositioning technology. In *Proceedings of the IEEE/ASME International Conference on Advanced Intelligent Mechatronics* (pp. 1338-1343).

Speich, J. E., & Goldfarb, M. (1998). A three degree-of-freedom flexure-based manipulator for high resolution spatial micromanipulation. In *Proceedings of the SPIE Conference on Microbiotics and Micromanipulation.*

Spenko, M. J., Haynes, G. C., Saunders, J. A., Cutkosky, M. R., Rizzi, A. A., Full, R. J., & Koditschek, D. E. (2008). Biologically inspired climbing with a hexapedal robot. *Journal of Field Robotics, 25*(4-5), 223–242. doi:10.1002/rob.20238.

Srinivasan, M., & Ruina, A. (2006). Computer optimization of a minimal biped model discovers walking and running. *Nature, 439*, 72–75. doi:10.1038/nature04113.

Staicu, S. (2009). Power requirement comparison in the 3-RPR planar parallel robot dynamics. *Mechanism and Machine Theory, 44*(5), 1045–1057. doi:10.1016/j.mechmachtheory.2008.05.009.

Starke, S. D., Robilliard, J. J., Weller, R., Wilson, A. M., & Pfau, T. (2009). Walk-run classification of symmetrical gaits in the horse: a multidimensional approach. *Journal of the Royal Society of London, 6*(33), 335–342.

Stewart, D. (1965). A platform with six degrees of freedom. *Proceedings - Institution of Mechanical Engineers, 180*(1), 371–386. doi:10.1243/PIME_PROC_1965_180_029_02.

Stieber, M. E., Vukovich, G., & Petriu, E. (1997). Stability aspects of vision based control for space robots. In *Proceedings of the IEEE International Conference on Robotics and Automation* (pp. 2771-2776).

Subramanian, G., Raveendra, V. V. S., & Kamath, M. G. (1994). Robust boundary triangulation and Delaunay triangulation of arbitrary planar domains. *International Journal for Numerical Methods in Engineering, 37*(10), 1779–1789. doi:10.1002/nme.1620371009.

Sueur, C., & Dauphin-Tanguy, G. (1992). Bond-graph modeling of flexible robots: The residual flexibility. *Journal of the Franklin Institute, 329*(6), 1109–1128. doi:10.1016/0016-0032(92)90006-3.

Sugimoto, K. (1987). Kinematics and dynamic analysis of parallel manipulator by means of motor algebra. *ASME Journal of Mechanisms, Transmissions, and Automation in Design, 109*(1), 3–7. doi:10.1115/1.3258783.

Surdilovic, D., & Bernhardt, R. (2004). STRING-MAN: a new wire robot for gait rehabilitation. In *Proceedings of the IEEE International Conference on Robotics and Automation*, Berlin, Germany (pp. 2031-2036).

Su, Y. X., & Duan, B. Y. (2000). The application of the Stewart platform in large spherical radio telescopes. *Journal of Robotic Systems, 17*(7), 375–383. doi:10.1002/1097-4563(200007)17:7<375::AID-ROB3>3.0.CO;2-7.

Su, Y. X., Duan, B. Y., Zheng, C. H., Zhang, Y. F., Chen, G. D., & Mi, J. W. (2004). Disturbance-rejection high-precision motion control of a Stewart platform. *IEEE Transactions on Control Systems Technology, 12*(3). doi:10.1109/TCST.2004.824315.

Svestka, P., & Overmars, M. H. (1997). Motion planning for carlike robots using a probabilistic approach. *The International Journal of Robotics Research, 16*(2), 119–145. doi:10.1177/027836499701600201.

Tabib-Azar, M. (1998). *Microactuators: Electrical, magnetic, optical, mechanical, chemical & smart structures.* Norwell, MA: Kluwer Academic.

Takahashi, Y., & Tsubouchi, O. (2000). Tension control of wire suspended mechanism and application to bathroom cleaning robot. In *Proceedings of the 39th SICE Annual Conference*, Iizuka, Japan (pp. 143-147).

Takanobu, H., & Takanishi, A. (2003). Dental robotics and human model. In *Proceedings of the 1st International IEEE EMBS Conference on Neural Engineering*, Capri Island, Italy (pp. 671-674).

Takanobu, H., Kuchiki, N., & Takanishi, A. (1995). Control of rapid closing motion of a robot jaw using nonlinear spring mechanism. In *Proceedings of the International Conference on Intelligent Robots and Systems*, Pittsburgh, PA (Vol. 1).

Takemura, F., Enomoto, M., Tanaka, T., Denou, K., Kobayashi, Y., & Tadokoro, S. (2005). Development of the balloon-cable driven robot for information collection from sky and proposal of the search strategy at a major disaster. In *Proceedings of the IEEE/ASME International Conference on Advanced Intelligent Mechatronics*, Monterey, CA (pp. 658-663).

Tandirci, M., Angeles, J., & Darcovich, J. (1992). The Role of Rotation Representations in Computational Robot Kinematics. In *Proceedings of the IEEE Intl. Conf. Robotics and Automation*, Nice, France (pp.344-349). Washington, DC: IEEE Press.

Tang, X., Pham, H. H., Li, Q., & Chen, I.-M. (2004). Dynamic analysis of a 3-DOF flexure parallel micromanipulator. In *Proceedings of the 2004 IEEE Conference on Robotics, Automation and Mechatronics* (Vol. 1, pp. 95-100).

Tang, P. B., Huber, D., Akinci, B., Lipman, R., & Lytle, A. (2010). Automatic reconstruction of as-built building information models from laser-scanned point clouds: A review of related techniques. *Automation in Construction, 19*(7), 829–843. doi:10.1016/j.autcon.2010.06.007.

Tanikawa, T., Arai, T., & Koyachi, N. (1999). Development of small-sized 3 DOF finger module in micro hand for micro manipulation. In *Proceedings of the IEEE/RSJ International Conference on Intelligent Robots and Systems* (pp. 876-881).

Thoma, J. U. (1975). *Introduction to bond graphs and their applications.* Oxford, UK: Pergamon Press.

Thueer, T., & Siegwart, R. (2007). Characterization and comparison of rover locomotion performance based on kinematic aspects. In *Proceedings of the International Conference on Field and Service Robotics*, Chamonix, France (pp. 189-198).

Thueer, T., Siegwart, R., & Backes, P. G. (2008). Planetary vehicle suspension options. In *Proceedings of the IEEE Aerospace Conference* (pp. 1-13).

Thueer, T., Krebs, A., Siegwart, R., & Lamon, P. (2007). Performance comparison of rough-terrain robots- simulation and hardware. *Journal of Field Robotics*, 251–271. doi:10.1002/rob.20185.

Tian, Y., Shirinzadeh, B., & Zhang, D. (2008). Stiffness estimation of the flexure-based five-bar micro-manipulator. In *Proceedings of the 10th Intl. Conf. on Control, Automation, Robotics and Vision*, Hanoi, Vietnam (pp. 599-604).

Tian, Y., Shirinzadeh, B., & Zhang, D. (2008). Development and dynamic modeling of a flexure-based Scott-Russell mechanism for nano-manipulation. *Mechanical Systems and Signal Processing, 23*(3), 957–978. doi:10.1016/j.ymssp.2008.06.007.

Tian, Y., Shirinzadeh, B., & Zhang, D. (2009). A flexure-based mechanism and control methodology for ultra-precision turning operation. *Precision Engineering, 33*(2), 160–166. doi:10.1016/j.precisioneng.2008.05.001.

Tian, Y., Shirinzadeh, B., & Zhang, D. (2009). Design and forward kinematics of the compliant micro-manipulator with lever mechanisms. *Precision Engineering, 33*(4), 466–475. doi:10.1016/j.precisioneng.2009.01.003.

Tian, Y., Shirinzadeh, B., & Zhang, D. (2010). Design and dynamics of a 3-DOF flexure-based parallel mechanism for micro/nano manipulation. *Microelectronic Engineering, 87*(2), 230–241. doi:10.1016/j.mee.2009.08.001.

Tian, Y., Shirinzadeh, B., & Zhang, D. (2010). Three flexure hinges for compliant mechanism designs based on dimensionless graph analysis. *Precision Engineering, 34*(1), 92–100. doi:10.1016/j.precisioneng.2009.03.004.

Trüper, T., Kortschack, A., Jähnisch, M., Hülsen, H., & Fatikow, S. (2004). Transporting Cells with Mobile Microrobots. *IEEE Proceedings of Nanobiotechnology, 151*, 145–150. doi:10.1049/ip-nbt:20040839.

Tsai, L., & Stamper, R. (1996). A parallel manipulator with only translational degrees of freedom. In *Proceedings of the 1996 ASME Design Engineering Technical Conference (MECH)* (p. 1152).

Tsai, L.-W. (1999). *Robot analysis: the mechanics of serial and parallel manipulators*. New York, NY: John Wiley & Sons.

Tsai, L.-W. (2000). Solving the inverse Dynamics of a Stewart-Gough Manipulator by the Principle of Virtual Work. *ASME J. Mech. Design, 122*(1), 3–9. doi:10.1115/1.533540.

Tsai, L.-W., & Tahmasebi, F. (1993). Synthesis and analysis of a new class of six-degree-of-freedom parallel minimanipulator. *Journal of Robotic Systems, 10*(5), 561–580. doi:10.1002/rob.4620100503.

Tuohy, S. T., Maekawa, T., Shen, G., & Patrikalakis, N. M. (1997). Approximation of measured data with interval B-splines. *Computer Aided Design, 29*(11), 791–799. doi:10.1016/S0010-4485(97)00025-0.

Ueno, H., Xu, Y., & Yoshida, T. (1991). Modeling and control strategy of a 3-D flexible space robot. In *Proceedings of the IEEE/RSJ International Workshop on Intelligent Robots and Systems* (pp. 978-983).

Ulrich, I., & Borenstein, J. (2001). The guidecane – applying mobile robot technologies to assist the visually impaired people. *IEEE Transactions on Systems, Man, and Cybernetics. Part A, Systems and Humans, 31*(2), 131–136. doi:10.1109/3468.911370.

Umetani, Y., & Yoshida, K. (1989). Resolved motion rate control of space manipulators with generalized jacobian matrix. *IEEE Transactions on Robotics and Automation, 5*(3), 303–314. doi:10.1109/70.34766.

van Amerongen, J., & Breedveld, P. C. (2003). *Modelling of physical systems for the design and control of mechatronic systems*. Amsterdam, The Netherlands: Elsevier.

Van den Doel, K., Smilek, D., Bodnar, A., Chita, C., Corbett, R., Nekrasovski, D., & McGrenere, J. (2004). Geometric shape detection with sound view. In *Proceedings of the Tenth Meeting of the International Conference on Auditory Display*, Sydney, Australia (pp. 1-8).

van Dixhoorn, J. J. (1972). Network graphs and bond graphs in engineering modeling. *Annals of System Research, 2*, 22–38.

van Dixhoorn, J. J., & Evans, F. J. (Eds.). (1974). *Physical structure in system theory: Network approaches to engineering and economics* (p. 305). London, UK: Academic Press.

Varady, T., Martin, R. R., & Cox, J. (1997). An integrated reverse engineering approach to reconstructing free-form surfaces. *Computer Integrated Manufacturing Systems, 10*(1), 49–60. doi:10.1016/S0951-5240(96)00019-5.

Varady, T., Martin, R. R., & Cox, J. (1997). Reverse engineering of geometric models-an introduction. *Computer Aided Design, 29*(4), 255–268. doi:10.1016/S0010-4485(96)00054-1.

Vaz, A., & Samanta, B. (1991). Digital controller implementation for flexible manipulators using the concept of intelligent structures. In *Proceedings of the IEEE/RSJ International Workshop on Intelligent Robots and Systems*, Osaka, Japan (pp. 952-958).

Verhoeven, R., & Hiller, M. (2002). Estimating the controllable workspace of tendon-based Stewart platforms. In *Proceedings of the International Symposium on Advances in Robot Kinematics*, Portoroz, Slovenia (pp. 277-284).

Verhoeven, R., Hiller, M., & Tadokoro, S. (1998). Workspace of tendon-driven Stewart platforms: Basics, classification, details on the planar 2-dof class. In *Proceedings of the International Conference on Motion and Vibration Control* (pp. 871-876).

Versteeg, H. K., & Malalasekera, W. (1995). *An introduction to computational fluid dynamics: the finite volume method*. Harlow, UK: Longman Scientific & Technical.

Villegas, E. (2009). *Development of a force control for CaPaMan2bis as a terminal tool for drilling*. Cassino, Italy: University of Cassino, Laboratory of Robotics and Mechatronics.

Vogel, M., Stein, B., Pettersson, H., & Karrai, K. (2001). Low-temperature scanning probe microscopy of surface and subsurface charges. *Applied Physics Letters*, *78*, 2592–2594. doi:10.1063/1.1360780.

Voglewede, A. P., & Ebert-Uphoff, I. (2005). Application of the antipodal grasp theorem to cable-driven robots. *IEEE Transactions on Robotics*, *21*(4), 713–718. doi:10.1109/TRO.2005.844679.

Wald, A., & Wolfowitz, J. (1939). Confidence limits for continuous distribution functions. *Annals of Mathematical Statistics*, *10*(2), 105–118. doi:10.1214/aoms/1177732209.

Wang, D., Meng, M., & Liu, Y. (1999). Influence of shear, rotary inertia on the dynamic characteristics of flexible manipulators. In *Proceedings of the IEEE Pacific Rim Conference on Communications, Computers and Signal Processing* (pp. 615-618).

Wang, G., & Wang, C. (1997). Reconstruction of sculptured surface on reverse engineering. In *Proceedings of the ASME Design Engineering Technical Conference*, Sacramento, CA (pp14-17).

Wang, G., Zhang, L., Chen, D., Liu, D., & Chen, X. (2007). Modeling and Simulation of multi-legged walking machine prototype. In *Proceedings of the International Conference on Measuring Technology and Mechatronics Automation*.

Wang, H. P. (2010). *Mathematic model and trajectory of multiped gaits*. Unpublished master's thesis, National Sun Yat-sen University, Kaohsiung, Taiwan.

Wang, L., Deng, Z., Zhang, L., & Meng, Q. (2006). Analysis of Assistant Robotic Leg on MATLAB. In *Proceedings of the 2006 IEEE International Conference on Mechatronics and Automation* (pp.1092-1096).

Wang, H., & Xie, Y. (2009). Adaptive jacobian position/force tracking control of free-flying manipulators. *Robotics and Autonomous Systems*, *57*, 173–181. doi:10.1016/j.robot.2008.05.003.

Wang, J., & Gosselin, C. M. (1998). A new approach for the dynamic analysis of parallel manipulators. *Multibody System Dynamics*, *2*(3), 317–334. doi:10.1023/A:1009740326195.

Wang, J., & Gosselin, C. M. (2004). Kinematic analysis and design of kinematically redundant parallel mechanisms. *ASME Journal of Mechanical Design*, *126*(1), 109–118. doi:10.1115/1.1641189.

Wang, S. C., Hikita, H., Kubo, H., Zhao, Y.-S., Huang, Z., & Ifukube, T. (2003). Kinematics and dynamics of a 6 degree-of-freedom fully parallel manipulator with elastic joints. *Mechanism and Machine Theory*, *38*, 439–461. doi:10.1016/S0094-114X(02)00132-5.

Watanabe, Y., & Nakamura, Y. (1998). A space robot of the center-of-mass invariant structure. In *Proceedings of the IEEE/RSJ International Conference on Intelligent Robots and Systems*, Victoria, BC, Canada (pp. 1370-1375).

Wavering, A. J. (1998). Parallel kinematic machine research at NIST: Past, present, and future. In *Proceedings of the First European-American Forum on Parallel Kinematic Machines: Theoretical Aspects and Industrial Requirements*, Italy (pp. 1-13).

Wenger, P., & Chablat, D. (2000). Kinematic analysis of a new parallel machine tool: The orthoglide. In *Proceedings of the 7th International Symposium on Advances in Robot Kinematics*, Portoroz, Slovenia (pp. 305-314).

Wenger, P., & Chablat, D. (2002). Design of a three-axis isotropic parallel manipulator for machining applications: The Orthoglide. In *Proceedings of the Workshop on Fundamental Issues and Future Research Directions for Parallel Mechanisms and Manipulators*, Montreal, QC, Canada.

Wiens, G. J., Shamblin, S. A., & Oh, Y. H. (2002). Characterization of PKM dynamics in terms of system identification. *Journal of Multi-body Dynamics. Part K, 216*(1), 59–72.

Williams, R. L., Gallina, P., & Rossi, A. (2001). Planar cable-direct-driven robots, part ii: Dynamics and control. In *Proceedings of the ASME Design Technical Conference*.

Williams, R. L. II, & Gallina, P. (2002). Planar cable-direct-driven robots: design for wrench exertion. *Journal of Intelligent & Robotic Systems, 35*, 203–219. doi:10.1023/A:1021158804664.

Willms, A. R., & Yang, S. X. (2006). An efficient dynamic system for real-time robot-path planning. *IEEE Transactions on Systems, Man, and Cybernetics. Part B, Cybernetics, 36*(4), 755–766. doi:10.1109/TSMCB.2005.862724.

Wilson, D. M. (1966). Insect walking. *Annual Review of Entomology, 11*, 103–122. doi:10.1146/annurev.en.11.010166.000535.

Wilson, E. B. (1927). Probable inference, the law of succession, and statistical inference. *Journal of the American Statistical Association, 22*(158), 209–212. doi:10.2307/2276774.

Wohlhart, K. (2003). Mobile 6-SPS parallel manipulators. *Journal of Robotic Systems, 20*(8), 509–516. doi:10.1002/rob.10101.

Wolf, A., Ottaviano, E., Shoham, M., & Ceccarelli, M. (2004). Application of line geometry and linear complex approximation to singularity analysis of the 3-DOF CaPaMan parallel manipulator. *Mechanism and Machine Theory, 39*, 75–95. doi:10.1016/S0094-114X(03)00105-8.

Wong, F., Nagarajan, R., Yaacob, S., Chekima, A., & Belkhamza, N.-E. (2001). An image segmentation method using fuzzy – based threshold. In *Proceedings of the 6th International Symposium on Signal Processing and its Applications*, Kuala Lumpur, Malaysia (pp. 144–147).

World Health Organization (WHO). (1997). *Blindness and visual disability, part i of vii: General information*. Geneva, Switzerland: WHO.

Wu, H., Sun, F., Sun, Z., & Wu, L. (2004). Optimal trajectory planning of a flexible dual-arm space robot with vibration reduction. *Journal of Intelligent & Robotic Systems*, 147–163. doi:10.1023/B:JINT.0000038946.21921.c7.

Wu, T.-L., Chen, J.-H., & Chang, S.-H. (2008). A six-DOF prismatic-spherical-spherical parallel compliant nano-positioner. *IEEE Transactions on Ultrasonics, Ferroelectrics, and Frequency Control, 55*(12), 2544–2551. doi:10.1109/TUFFC.2008.970.

Xi, F. (2001). A comparison study on hexapods with fixed-length legs. *International Journal of Machine Tools & Manufacture, 41*, 1735–1748. doi:10.1016/S0890-6955(01)00038-4.

Xu, Q. S., & Li, Y. M. (2006). Stiffness modeling for an orthogonal 3-PUU compliant parallel micromanipulator. In *Proceeding of the 2006 IEEE International Conference on Mechatronics and Automation*, Luoyang, China (pp. 124–129).

Xu, Q., & Li, Y. (2008). Structure improvement of an XY flexure micromanipulator for micro/nano scale manipulation. In *Proceedings of the 17th IFAC World Conference* Seoul, Korea (pp. 12733-12738).

Xu, Q., & Li, Y. (2009). Global sliding mode-based tracking control of a piezo-driven XY micropositioning stage with unmodeled hysteresis. In *Proceedings of the IEEE International Conference on Intelligent Robots and Systems* (pp. 755-760).

Xu, Q., & Li, Y. (2010). Precise tracking control of a piezoactuated micropositioning stage based on modified Prandtl-Ishlinskii hysteresis model. In *Proceedings of the IEEE International Conference on Automation Science and Engineering* (pp. 692-697).

Xue, L., & Zhu, L. (2005). L1-norm estimation and random weighting method in a semiparametric model. *Acta Mathematicae Applicatae Sinica, 21*(2), 295–30. doi:10.1007/s10255-005-0237-8.

Xu, W. L., Bronlund, J. E., Potgieter, J., Foster, K. D., Röhrle, O., Pullan, A. J., & Kieser, J. A. (2008). Review of the human masticatory system and masticatory robotics. *Mechanism and Machine Theory, 43*(11), 1353–1375. doi:10.1016/j.mechmachtheory.2008.06.003.

Xu, W. L., Torrence, J. D., Chen, B. Q., Potgieter, J., Bronlund, J. E., & Pap, J. S. (2008). Kinematics and experiments of a life sized masticatory robot for characterizing food texture. *IEEE Transactions on Industrial Electronics, 55*(5), 12. doi:10.1109/TIE.2008.918641.

Yamada, K., & Tsuchiya, K. (1987). Formulation of rigid multibody system in space. *JSME International Journal*, *30*(268), 1667–1674.

Yamamoto, M., Yanai, N., & Mohri, A. (2004). Trajectory control of incompletely restrained parallel-wire-suspended mechanism based on inverse dynamics. *IEEE Transactions on Robotics*, *20*(5), 840–850. doi:10.1109/TRO.2004.829501.

Yang, C., Huang, Q., Jiang, H. O., Ogbobe, P., & Han, J. (2010). PD control with gravity compensation for a hydraulic 6-DOF parallel manipulator. *Mechanism and Machine Theory*, *45*(4), 666–677. doi:10.1016/j.mechmachtheory.2009.12.001.

Yang, J. M. (2009). Fault-tolerant gait planning for a hexapod robot walking over rough terrain. *International Journal of Intelligent & Robotic System*, *54*(4), 613–627. doi:10.1007/s10846-008-9282-x.

Yang, J. M., & Kim, J. H. (2000). A fault tolerant gait for a hexapod robot over uneven terrain. *IEEE Transactions on Systems, Man, and Cybernetics. Part B, Cybernetics*, *30*(1), 172–180. doi:10.1109/3477.826957.

Yang, Q. Z., Yin, X. Q., & Ma, L. Z. (2005). Establishing stiffness of prismatic pair in fully compliant parallel micro-robot using energy method. *Journal of Jiangsu University*, *26*(1), 12–15.

Yang, S. X., & Meng, M. (2003). Real-time collision-free motion planning of a mobile robot using a neural dynamics-based approach. *IEEE Transactions on Neural Networks*, *14*(6), 1541–1552. doi:10.1109/TNN.2003.820618.

Yau, H. (1997). Reverse engineering of engine intake ports by digitization and surface approximation. *International Journal of Machine Tools & Manufacture*, *37*(6), 871–875. doi:10.1016/S0890-6955(95)00100-X.

Yau, H., & Chen, J. (1997). Reverse engineering of complex geometry using rational B-splines. *International Journal of Advanced Manufacturing Technology*, *13*(8), 548–555. doi:10.1007/BF01176298.

Yeh, H. C., Ni, W. T., & Pan, S. S. (2005). Digital closed-loop nanopositioning using rectilinear flexure stage and laser interferometry. *Control Engineering Practice*, *13*(5), 559–566. doi:10.1016/j.conengprac.2004.04.019.

Yi, B. J., Na, H. Y., & Chung, G. B. (2002). Design and experiment of a 3 DOF parallel micro-mechanism utilizing flexure hinges. In *Proceedings of the IEEE International Conference on Robotics and Automation 2002* (Vol. 2, pp. 1167-1172).

Yi, B. J., Chung, G. B., Na, H.-Y., Kim, W. K., & Suh, I. H. (2003). Design and experiment of a 3-DOF parallel micromechanism utilizing flexure hinges. *IEEE Transactions on Robotics and Automation*, *19*(4), 604–612. doi:10.1109/TRA.2003.814511.

Ying, J. P., & Liang, C. G. (1994). The forward displacement analysis of a kind of special platform manipulator mechanisms. *Mechanism and Machine Theory*, *29*(1), 1–9. doi:10.1016/0094-114X(94)90015-9.

Yin, Z., Zhang, Y., & Jiang, S. (2003). A methodology of sculptured surface fitting from CMM measurement data. *International Journal of Production Research*, *41*(14), 3375–3384. doi:10.1080/00207540310000112815.

Yin, Z., Zhang, Y., & Jiang, S. (2003). Methodology of NURBS surface fitting based on off-line software compensation for errors of a CMM. *Precision Engineering*, *27*(3), 299–303. doi:10.1016/S0141-6359(03)00033-3.

Yiu, K. (2002). *Geometry, dynamics and control of parallel manipulators*. Unpublished doctoral dissertation, Hong Kong University of Science and Technology, Hong Kong.

Yokokohji, Y., Toyoshima, T., & Yoshikawa, T. (1993). Efficient computational algorithms for trajectory control of free-flying space robots with multiple arms. *IEEE Transactions on Robotics and Automation*, *9*(5), 571–579. doi:10.1109/70.258050.

Yongming, W., Xiaoliu, Y., & Wencheng, T. (2009). Analysis of obstacle-climbing capability of planetary exploration rover with rocker-bogie structure. In *Proceedings of the International Conference on Information Technology and Computer Science*, Kiev, Ukraine (pp. 329-332).

Yoshida, K. (2003). Engineering test satellite VII flight experiments for space robot dynamics and control: Theories on laboratory test beds ten years ago, now in orbit. *The International Journal of Robotics Research*, *22*(5), 321–335. doi:10.1177/0278364903022005003.

Yoshikawa, T. (1998). Manipulability of robotic mechanisms. *The International Journal of Robotics Research, 4*(2), 3–9. doi:10.1177/027836498500400201.

Yu, J. J., Bi, S. S., & Zong, G. H. (2002). Analysis for the static stiffness of a 3-DOF parallel compliant micromanipulator. *Chinese journal of mechanical engineering, 38*(4), 7-10.

Yuan, D., & Manduchi, R. (2004). A tool for range sensing and environment discovery for the blind. In. *Proceedings of the Conference on Computer Vision and Pattern Recognition, 3,* 39.

Yuan, X., & Yang, S. X. (2003). Virtual assembly with biologically inspired intelligence. *IEEE Transactions on Systems, Man and Cybernetics. Part C, Applications and Reviews, 33*(2), 159–167. doi:10.1109/TSMCC.2003.813148.

Yung, T., Yu-Shin, C., & Shih-Ming, W. (1999). Task space control algorithm for Gough-Stewart platform. In *Proceedings of the 38th Conference on Decision and Control*, Phoenix, AZ (pp. 3857-3862).

Zabalsa, I., & Ros, J. (2007). Aplicaciones actuales de los robots paralelos. In *Proceedings of the 8º Congreso Iberoamericano de Ingeniería Mecánica*, Navarra, Spain.

Zalama, E., Gaudiano, P., & Coronado, J. L. (1995). A real-time, unsupervised neural network for the low-level control of a mobile robot in a nonstationary environment. *Neural Networks, 8*(1), 103–123. doi:10.1016/0893-6080(94)00063-R.

Zalama, E., Gomez, J., Paul, M., & Peran, J. R. (2002). Adaptive behavior navigation of a mobile robot. *IEEE Transactions on Systems, Man, and Cybernetics. Part A, Systems and Humans, 32*(1), 160–169. doi:10.1109/3468.995537.

Zelinsky, A. (1994). Using path transforms to guide the search for findpath in 2D. *The International Journal of Robotics Research, 13*(4), 315–325. doi:10.1177/027836499401300403.

Zeng, Y., & He, G. (2009). Study on interferometric measurements of a high accuracy laser-diode interferometer with real-time displacement measurement and it's calibration. In *Proceedings of the 4th International Symposium on Advanced Optical Manufacturing and Testing Technologies: Optical Test and Measurement Technology and Equipment.*

Zesch, W., Buechi, R., Codourey, A., & Siegwart, R. Y. (1995). Inertial drives for micro- and nanorobots: two novel mechanisms. *SPIE, 2593,* 80–88. doi:10.1117/12.228638.

Zhang, C., & Yu, Y. Q. (2004). Dynamic analysis of planar cooperative manipulators with link flexibility. *Transactions of ASME, 126,* 442–448. doi:10.1115/1.1701875.

Zhang, Z., & Menq, C. H. (2007). Laser interferometric system for six-axis motion measurement. *The Review of Scientific Instruments, 78*(8). doi:10.1063/1.2776011.

Zhao, Y. J., & Gao, F. (2009). Dynamic performance comparison of the 8PSS redundant parallel manipulator and its non-redundant counterpart—the 6PSS parallel manipulator. *Mechanism and Machine Theory, 44*(5), 991–1008. doi:10.1016/j.mechmachtheory.2008.05.015.

Zhao, Y., & Gao, F. (2009). Inverse dynamics of the 6-DOF-out-parallel manipulator by means of the principle of virtual work. *Robotica, 27*(2), 259–268. doi:10.1017/S0263574708004657.

Zheng, Z. (1987). Random weighting method. *Acta Mathematicae Applicatae Sinica, 10*(2), 247–253.

Zhou, X. Y., Zhang, T., & Cheng, Y. H. (2009). Study of high precision and high measurement speed dual-longitudinal thermal frequency stabilization laser interferometer. *Dianzi Keji Daxue Xuebao/Journal of the University of Electronic Science and Technology of China, 38*(3), 451-454.

Zhu, Z. Q., Li, J. S., Gan, Z. X., & Zhang, H. (2005). Kinematic and dynamic modeling for real-time control of Tau parallel robot. *Mechanism and Machine Theory, 40*(9), 1051–1067. doi:10.1016/j.mechmachtheory.2004.12.024.

Zygo. (2010). *Corporation - Zygo error sources.* Retrieved from http://www.zygo.com/library/papers/ZMI_Error_Sources.pdf

About the Contributors

Shahin Sirouspour received the B.Sc. and M.Sc. degrees in electrical engineering from Sharif University of Technology, Tehran, Iran, in 1995 and 1997, respectively, and the Ph.D. degree in electrical engineering from the University of British Columbia, Vancouver, BC, Canada, in 2003. He then joined McMaster University, Hamilton, ON, Canada, where he is currently an Associate Professor in the Department of Electrical and Computer Engineering. His research interests include teleoperation control, haptics, robot-assisted medical intervention, advanced robot controls, machine vision, and medical image processing. He was on research leave at MDA Space Missions (Brampton, Ontario, Canada) during the period July 2010-June 2011. Dr. Sirouspour received the McMaster President's Award of Excellence in Graduate Supervision in 2008.

* * *

Ahad Ali is an Assistant Professor in Mechanical Engineering at the Lawrence Technological University. He earned B.S. in Mechanical Engineering from Bangladesh Institute of Technology, Khulna, and Masters in Systems and Engineering Management from Nanyang Technological University, Singapore and PhD in Industrial Engineering from University of Wisconsin-Milwaukee. Dr. Ali was Assistant Professor in Industrial Engineering at the University of Puerto Rico - Mayaguez, Visiting Assistant Professor in Mechanical, Industrial and Manufacturing Engineering at the University of Toledo, Lecturer in Mechanical Engineering at the Bangladesh Institute of Technology, Khulna. Dr. Ali has published journal and conference papers. He is member of IEEE, IIE, INFORMS, and SME.

Yasser Amooshahi received the MS degree with distinction in mechanical engineering from the Isfahan University of Technology, Iran, in 2008 and the BS degree in mechanical engineering from Yazd University, Iran, in 2006, and the. His research interests include reactionless Mechanisms, parallel robots, cable robots, biped robots, flight simulators. His current project is the design and fabrication of flight simulator based on novel control strategy.

Umar Asif received his Bachelor's degree in Mechanical Engineering from National University of Sciences & Technology, Rawalpindi, Pakistan in 2005. He completed his Master's degree in Mechatronics Engineering from the Department of Mechanical & Manufacturing Engineering, University of New South Wales, Sydney, Australia in 2007. He is now working in National Engineering and Scientific Commission as an engineering manager and working in affiliation with the Mechatronics research centre of National University of Sciences & Technology. His areas of interest are control systems, robotics and design & simulation of mechatronic systems.

Umesh Bhagat received his Bachelors of Engineering from University of Pune, India in 2003 and Masters of Computer science and IT from University of New south Wales, Sydney, Australia in 2008. Umesh posses 10 years experience in industrial automation and control system design. He has worked for major automotive giants such as TATA Motors and General Motors Corporation as technical lead for the area of Robotics and Plat floor systems. He is currently a PhD candidate at Department of Mechanical and Aerospace Engineering, Monash University, Australia. His research interest includes Micro/Nano manipulation, laser-based measurements and sensory-based controls, Robotics, and automated manufacturing.

Luca Bruzzone received the Laurea degree magna cum laude in mechanical engineering in 1997, with a specialisation in Industrial Automation and Robotics and a degree thesis on the design of a parallel robot co-operating in deburring tasks. In 1998 he joined the Material Handling division of Techint Italimpianti, working as mechanical, structural and automation designer of continuous ship unloaders and other machines. In 1999 he became researcher at the Department of Machinery Mechanics and Design of the University of Genova. His research interests include mechanics and control of robots and automation devices, and in particular of parallel kinematics machines, miniaturized assembly, mobile robotics, fractional-order control of mechatronic devices.

Massimo Callegari has worked in the R&D Departments of firms making business in the field of factory automation since 1986. In 1990 he joined the Department of Mechanics of the University of Genova as researcher; from 1998 to 2005 he has been Associate Professor at the Faculty of Engineering of the Polytechnic University of Marche in Ancona, where he actually serves as Professor of "Machine Mechanics" and "Robot Mechanics". His main research fields are the analysis and design of dynamic mechanical systems where mechanical structure and control architecture are considered with an integrated mechatronic approach; in particular he has worked on research topics related to vehicle dynamics, factory automation and robotics, with special interest for parallel kinematics machines.

Luca Carbonari took a laurea degree in Mechanical Engineering in December 2008 at Polytechnic University of Marche in Ancona, Italy, where he is currently attending the Ph.D. School in Engineering Sciences. His research topics cover vehicle dynamics and robotics, with a special interest for control and dynamic analysis of parallel kinematics machines.

Giuseppe Carbone has got the master degree with laude as mechanical engineer on 2000 and the PhD degree on 2004 both at University of Cassino (Italy). Since March 2005 Giuseppe Carbone is Assistant professor at the Dept. DiMSAT, School of Engineering, University of Cassino, Italy. He has been visiting scholar at Technical University of Braunschweig, Germany for six months in the year 1999 and at Humanoid Robotics Institute, Waseda University, Tokyo, Japan for about one year during the years 2002-2005. His research interests cover aspects of Mechanics of Manipulation and Grasp, Mechanics of Robots, Mechanics of Machinery. He is author and/or co-author of more than 170 papers that have been published in Proceedings of well-known peer-reviewed International Journals or Conferences.

Marco Ceccarelli is Full Professor of Mechanics of Machinery and Director of LARM, Laboratory of Robotics and Mechatronics at University of Cassino. He is member of Robotics Commission of IFToMM, the International Federation for the Promotion of Machine and Mechanism Science. He has written the books 'Fundamentals of Mechanics of Robotic Manipulation' in 2004 and 'Mecanismos' in 2008. He is current President of IFToMM. His research interests are in Mechanics of Mechanisms and Robots. He is author/co-author of 500 papers, presented at Conferences or published in journals, and he has edited 14 books as for conference proceedings and specific topics.

Christoph Edeler studied mechanical engineering at the Rheinisch-Westfälisch-Technische-Hochschule, RWTH-Aachen, Germany, with the major field mechanical design. He completed his degree with an external diploma thesis at the company CLAAS harvesting machines, Harsewinkel, Germany, finishing in 2006. Numerous internships accompanied his activities during his studies. Furthermore he worked as student assistant at the Institute for mechanical components, "Institut für Maschinenelemente," RWTH-Aachen, from 2003 to 2005. His main interests lay in measure and control engineering of several test benches. Since June 2006 he is with the Department of Microrobotics and Control Engineering, "Abteilung für Mikrorobotik und Regelungstechnik, AMiR" at the CvO University of Oldenburg as scientific researcher and PhD candidate. His research interests are piezo-driven micro-actuators for mobile nanohandling robots and modeling of stick-slip actuators including friction phenomena.

Sergej Fatikow studied computer science and electrical engineering at the Ufa Aviation Technical University in Russia, where he received his doctoral degree in 1988 with work on fuzzy control of complex non-linear systems. In 1990 he moved to the Institute for Process Control and Robotics at the University of Karlsruhe in Germany, where he worked as researcher and since 1994 as Head of the research group "Microrobotics and Micromechatronics". In 2001 he was invited to establish a new Division for Microrobotics and Control Engineering at the University of Oldenburg as a full professor. His research interests include micro- and nanorobotics, automated robot-based nanohandling in SEM, micro- and nanoassembly, AFM-based nanohandling, sensor feedback on the nanoscale, and neuro-fuzzy robot control.

Abbas Fattah has completed his PhD degree in mechanical engineering from McGill University in 1995. He has been in Department of Mechanical Engineering at Isfahan University of Technology as a faculty and with Department of Mechanical Engineering at University of Delaware as a visiting professor and research scholar. His research activities are now mainly on dynamics, design and control of special mechanical systems such as rehabilitation robots, bipeds, cable-suspended robots, and parallel mechanisms. He has published several patents and more than 70 articles in international journals and conferences. The results of his research have novelty in the field of robotics and mechanical systems.

Hamoon Hadian received the MS degree with first rank in mechanical engineering from Isfahan University of Technology, Iran, in 2007, and the BS degree with distinction in mechanical engineering from Kashan University, Iran, in 2005. He is currently in the Subsea R&D Institute at Isfahan University of Technology as a research scholar. His research interests include kinematics, parallel robots, cable-driven robots, dynamics and control of underwater robots. His current research focuses on the inspection robots for oil and gas piping, high voltage power transmission line, and boiler tubes that operate autonomously. Now, he is project manager of two important project at Isfahan University of Technology, i.e., design and implementation of a new robot for high voltage power transmission line; control strategy and implementation of an Autonomous underwater vehicle.

Innchyn Her was born in Taipei, Taiwan, in 1957. He obtained his B.S., M.S., and Ph.D. degrees from National Taiwan University (1979), the Pennsylvania State University (1983), and Purdue University (1987). He has been a professor at National Sun Yat-sen University since 1995. He has published in areas like mechanical design, robotics, automation, image processing, and crystallography.

Jonathan M. Her was born in Kaohsiung, Taiwan, in 1991. He is now working for his B.S. degree in Life Science in National Taiwan University. He is interested in physical and chemical biology, and entomology.

Eklas Hossain is a PhD student in Electrical Engineering at University of Wisconsin Milwaukee, USA. He earned his MS in Mechatronics Engineering from International Islamic University of Malaysia, Malaysia. He earned his BS in Electrical Engineering from Khulna University of Engineering and Technology, Bangladesh. His research interests renewable energy, power system, electrical machinery, Instrumentation, robotics, and linear and nonlinear control.

Javaid Iqbal did his PhD in Mobile Robots from University of New South Wales, Australia. He is a member of Association of Professional Engineers, Scientists and Managers, Australia. He is serving in the College of Electrical & Mechanical Engineering as a head of department of the Mechatronics Department of the university and as a professor. His areas of expertise are Linear Systems, Control Theory, Microprocessor based Design, Algorithm and computing, Advance Artificial Intelligence, Digital Image Processing, Machine Learning and Intelligent Machines.

X. Jia received B.Eng. degree and MSc in Mechanical Engineering from Hebei University of Technology, China, in 2000 and 2003 respectively. From 2003 to 2006, she worked as a teacher in Modern college of business and technology, China. From 2006 to now, she is an doctoral candidate in Mechanical Engineering at Tianjin University, her tasks include Micro/nano manipulation, Mechanical Kinematics and Dynamics, Finite Element Method (FEM).

Srikanth Kannan received the B.Eng. degree in mechanical engineering from the Vishvesharaiya National Institute of Technology, Nagpur (VNIT), India in 2006. He received the MS degree in mechanical engineering in 2008 from the State University of New York at Buffalo, Buffalo, NY, USA. He is currently working as Interactive Product Engineer at Simulia, Providence, Rhode Island, USA.

Raisuddin Khan received his BS Degree in Mechanical Engineering from Bangladesh Institute of Technology, Rajshahi, Bangladesh, and his MS and PhD Degree in Mechanical Engineering from BUET, Bangladesh. He is a Professor of Mechatronics Engineering at the International Islamic University of Malaysia, Malaysia. Dr. Khan is president of AA researcher group in Malaysia and involving research on robotics, control system, design and development various Mechatronics products and so on.

Venkat N. Krovi received his MS and PhD degrees in mechanical engineering from the University of Pennsylvania in 1995 and 1998, respectively. In 2001, he joined the Department of Mechanical and Aerospace Engineering at the State University of New York at Buffalo. His research interests include design, analysis and prototyping of novel articulated mechanical systems. Dr. Krovi is the recipient of the 2004 National Science Foundation CAREER Award.

Amit Kumar is an Assistant Professor in the Department of Mathematics, S. M. P. Government Girls PG College Meerut, Uttar Pradesh, India. He obtained his M.Sc (2002) and M.Phil (2005) Degree in Mathematics from Chaudhary Charan Singh University, Campus Meerut, India. He completed his PhD in Flexible Space Robot from IIT, Roorkee in 2010. His areas of interest are Bond Graph Modeling and Flexible Space Robots.

G. Satheesh Kumar received his PhD degree in IIT Madras, India in the field of Parallel manipulators. His research interest includes Robotics, mechatronics, active vibration isolation and machine tools. He has published over 20 journals and conference papers. He was an organizing member for the Indian National Robo Olympiad conducted for school children. He was actively involved in developing and testing prototypes of mobile robots developed in the laboratory, currently working on developmental work of Virtual Remote Robotics Laboratory at IIT Madras.

J. Liu received B.Eng. degree and MSc in Mechanical Engineering from Hebei University of Technology, China, in 2000 and 2005 respectively, and PhD degree in Lab. of Precision Measurement Technology at Tianjin University, China, in 2008. Now, he worked as a Senior Lecturer in Mechanical Engineering at Hebei University of Technology, China. His research interests include optical, mechanical and electronic integration, precision measuring technology.

Frank C. Mendel is an associate professor in the Department of Anatomical Sciences at the State University of New York at Buffalo, NY, USA. He received his MA and PhD degrees in anthropology from the University of California, Davis in 1971 and 1975, respectively. His research interests include forensic pathology of extant and extinct big cats and cat-like carnivores (sabertoothed forms), developing virtual clinical training platforms and determining best practices for managing sprains and other soft tissue injuries.

R. Mitra (BS, MS, M.Tech., PhD.) received his M.Tech. from IIT Kharagpur and his PhD in the area of Control Systems in 1976 from University of Roorkee. Besides post doctoral work at DFVLR laboratory in West Germany, he had been in the faculty of Department of Systems and Control, University of Technology, Baghdad and Thapar Engineering College, Patiala. For the past 40 years, Dr. Mitra is with the Department of Electronics and Computer Engineering, Indian Institute of Technology, Roorkee, teaching various courses in the area of control systems, electronic instrumentation and microprocessor application. His research interests are system theory, control.

Kazi Mostafa was born in Chittagong, Bangladesh on July 1, 1981. He obtained Bachelor of Engineering in Mechanical Engineering from Chittagong University of Engineering & Technology. He obtained his master of Engineering in Manufacturing Management from University of South Australia, Adelaide. Since September 2009 he is with the Department of Mechanical & Electro-mechanical Engineering, "Non Traditional Design Lab" at the National Sun Yat-sen University, Taiwan as PhD candidate. His research interests include computer vision, image processing and robotics.

Riza Muhida is a Lecturer and Researcher at Surya Research and Education Center, Banten, Indonesia. He was an Assistant Professor in the Department of Mechatronics Engineering at the International Islamic University Malaysia. He got his Bachelor in Engineering Physics from Institute of Technology Bandung, Indonesia. He earned diploma in Japanese from Kokusai Kotoba Gakuin, Japanese School, Japan. He received MS in Electrical Engineering from Osaka University, Japan. He earned his Dr. Eng in Semiconductor and Optical Electronics from Osaka University, Japan. His research areas are Thin Film of Solar Cells Design and Characterizations, Microelectronics, and Micro sensors (MEMS), Power Systems and Power Electronics, Building Integrating Photovoltaic (BIPV) and Renewable Energy and Energy Conversion.

T. Nagarajan received his M.Tech. degree from IIT Madras, India in the year 1969 and his PhD also from IIT Madras in the year 1980. Presently he is a Professor at IIT Madras. In his academic career spanning over 40 years he has completed many research projects and over 15 research scholars received their degrees under his guidance. He has published over 100 journal and conference papers in the fields of Robotics, Parallel manipulators, Precision gears, Pneumatics, Medical instrumentation, Multi-body dynamics and Mechatronics. His current research includes MEMS, Parallel manipulators and smart material actuators.

Madusudanan Sathia Narayanan received the B.Eng. degree in mechanical engineering from the College of Engineering Guindy (Anna University- Main Campus), India in 2006. He received the MS degree in mechanical engineering in 2008 from the State University of New York at Buffalo, Buffalo, NY, USA. He is currently pursuing his Ph.D. degree in mechanical engineering since 2008 from the same institution. His research interests include kinematics and control of parallel manipulators and application of haptics in surgical robotics.

Pushparaj Mani Pathak is an Assistant Professor at Mechanical & Industrial Engineering Department Indian Institute of Technology, Roorkee, since 2006. He obtained B. Tech from NIT Calicut, M. Tech in Solid Mechanics and Design from IIT Kanpur and PhD in area of Space Robotics from IIT Kharagpur. His research interests are space robotics, walking robots, in vivo robots dynamics and control.

Dereje Shiferaw (BS, M.Tech) received his B. Sc. Addis Ababa University, Ethiopia and M.Tech. from IIT Roorkee, Roorkee India in control and guidance. Since August 2007, he has been a PhD student in the department of Electronics and Computer Engineering, Indian Institute of Technology Roorkee, India and presently he has submitted his thesis to the department and is working as an assistant professor in Graphic Era University, Dehradun, India. His research interests are neural network, fuzzy logic systems and genetic algorithm, application of AI in nonlinear control and robotics.

Bijan Shirinzadeh received engineering qualifications: BE (Mechanical), BE (Aerospace), MSE (Mechanical), and MSE (Aerospace) from the University of Michigan, and PhD in Mechanical Engineering from University of Western Australia (UWA). He has held various positions in academia and industry including research fellow at the University of Western Australia, and senior research scientist at Commonwealth Scientific Industrial Research Organization (CSIRO), Australia. Dr Bijan Shirinzadeh is currently an associate professor and the Director of Robotics & Mechatronics Research Laboratory (RMRL) which he established in 1994, and the Director of Mechatronics Degree program in the Department of Mechanical and Aerospace Engineering at Monash University, Australia. His current research interests include modelling and planning/simulation in virtual reality, haptics, intelligent systems, medical robotics, laser-based measurements and sensory-based control, micro-nano manipulation systems, systems kinematics and dynamics, and automated fabrication and manufacturing.

N. Sukavanam is an Associate Professor in the Department of Mathematics, IIT Roorkee. He did his MSc from Government Arts College, Salem, University of Madras, India in 1979. He completed his PhD in Integral Operators from Indian Institute of Science, Bangalore in 1985. He worked as Scientist B at NSTL, DRDO, Vizag from 1984 to 1986. He was a Research Scientist in Department of Mathematics, IIT, Bombay from 1987 to 1990. He worked as lecturer at BITS, Pilani, Rajasthan from 1990 to 1996. He joined the Department of Mathematics as an Assistant Professor at IIT, Roorkee in 1996 and since then he has published over 40 research papers in the Journals and Conferences of repute. His area of research includes Nonlinear analysis (control theory), Robotics and Control.

Yanling Tian received B.Eng. degree in Mechanical Engineering from Northwest Institute of Light Industry, China, in 1997, and MSc and PhD degree in Mechanical Engineering from Tianjin University, China, in 2002 and 2005 respectively. From 2005 to 2006, he worked as a postdoctoral research fellow and then holds an Associate Professor position in the School of Mechanical Engineering at Tianjin University. From 2007 to 2009, He was employed as postdoctoral research fellow at Robotics and Mechatronics Research Laboratory, Department of Mechanical and Aerospace Engineering, Monash University, Australia. He also worked as short-term visiting scholar at Hongkong University of Science and Technology, China, and University of Warwick, UK, in 2001 and 2006, respectively. He worked as short-term visiting professor at Tohoku University, Japan in 2010. He has obtained the prestigious Alexander von Humboldt Fellowship for experienced researcher in 2010. Dr. Tian is the members of IEEE and Euspen. He serves as the associate editor of International Journal of Intelligent Mechatronics and Robotics, and reviewer for IEEE Transactions on Mechatronics, Sensors and Actuators A: Physical, Precision Engineering etc. He is also an assessor for the National Natural Science Foundation of China, Natural Science Foundation of Tianjin Municipality, and Natural Science Foundation of Heilongjiang Province. His research interests include Micro/nano manipulation, Mechanical Dynamics, Finite Element Method (FEM), Surface Metrology and Characterisation. He has accomplished couple of government and industry based projects, and published more than 50 peer reviewed papers.

Enrique Villegas was born in Merida, Venezuela in 1986. He obtained the bachelor degree of "Automation and Control Systems Engineer" at Universidad de Los Andes in 2009. He spent a period of 4 months at LARM-University of Cassino in 2009. He is currently working at Endress+Hauser Argentina as product manager of Tank Gauging and Process Solutions.

Xiaobu Yuan is an Associate Professor with the School of Computer Science, University of Windsor, ON, Canada. He received the B.Sc. degree in computer science from the University of Science and Technology of China, Hefei, in 1982, the M.S. degree in computer science from the Institute of Computing Technology, Academia Sinica, Beijing, China in 1984, and the Ph.D. degree also in computer science from University of Alberta, Edmonton, AB, Canada, in 1993. His research interests include advanced human/computer interaction, robotics, intelligent systems, and software engineering and testing.

D. Zhang received B. Eng. Degree in Mechanical Engineering from Shenyang University of Industry, China, in 1984, and MSc and PhD degree in Mechanical Engineering at Tianjin University, China, in 1990 and 1995 respectively. He worked as visiting scholar at Hong Kong University of Science and Technology, China, in 1999, University of Warwick, UK, in 2003, Tokyo Institute of Technology, Japan, in 2005, and Monash University, Australia, in 2009, respectively. Professor Zhang's research interests include Mechanical Dynamics, Surface Engineering, Modern Manufacturing Equipment and System.

Yongmin Zhong holds a university fellowship in the Department of Mechanical Engineering at Curtin University of Technology, Australia. Prior to joining Curtin University of Technology, he worked as a research fellow at Department of Mechanical and Aerospace Engineering, Monash University, Australia. He also worked as a research fellow at School of Computer Engineering, Nanyang Technological University, Singapore and a lecturer at Department of Aeronautical Manufacturing, Northwestern Polytechnical University, China. His research interest includes virtual reality and haptics, computational modelling, soft tissue modelling and surgery simulation, robotics, mechatronics, optimum estimation and control, and integrated navigation system.

Xiaobo Zhou received the B.Eng. degree in Department of Automation Engineering from the University of Science and Technology of China, Heifei, China in 2008. He is currently pursuing the PhDdegree in mechanical engineering since 2008 from the State University of New York at Buffalo, Buffalo, NY, USA. His research interests include modeling, analysis and hardware-in-the-loop testing of parallel cable manipulators and wheeled mobile robots.

Index